FOR ESTER BOSERUP

who broke new ground with her sense of
how smallholders intensify their cultivation
and with her vision of the
economics of agrarian change

ROBERT McC. NETTING

Smallholders, Householders

Farm Families and the Ecology of Intensive, Sustainable Agriculture

STANFORD UNIVERSITY PRESS
STANFORD, CALIFORNIA

Stanford University Press
Stanford, California

© 1993 by the Board of Trustees of the
Leland Stanford Junior University

Printed in the United States of America

CIP data appear at the end of the book

Original printing 1993
Last figure below indicates year of this printing:
04 03 02 01 00 99 98 97 96 95

Stanford University Press publications are distributed exclusively by Stanford University Press
within the United States, Canada, Mexico, and
Central America; they are distributed exclusively
by Cambridge University Press throughout the
rest of the world.

Preface: Book Building-Blocks

In MY PREVIOUS attempts at writing books, I cannot remember devoting much thought to how the text should be arranged. Ethnographic monographs with an ecological bent (Netting 1968, 1981) started naturally with chapters on physical environment and land use and finished with discussions of economic and social change. A textbook comparing different types of subsistence adaptation (Netting 1977, 1986) went from hunting and gathering to intensive cultivation (while all the time denying an evolutionary approach). The chapters in this book on smallholder farming and households have been written over a longer period of time, in fits and starts, mainly during the summers of 1989 and 1990, and then rearranged. Following the conventional, cut-and-dried scholarly saw "Theory comes first," I dutifully crammed definitions, generalizations, and critical appreciations of Ester Boserup, Malthus, and Marx into an early chapter and then went through some of the same procedures for the Chayanov model in a chapter following my own discussion of the farm-family household. Some informed advice and further reflection persuaded me that my desired audience of students, environmentalists, international developers, and agricultural policy advisers, as well as academic geographers, historians, economists, and anthropologists, might well prefer to begin with the on-the-ground realities of actual farming systems, labor, and property. This approach reproduced in part my own inductive experience of learning about specific cases of intensive agricultural practices and their functional links to household organization and land tenure.

Chapter 1 on the technology and knowledge of intensive farm practices and Chapter 2 on the farm-family household identify and describe the smallholder system, contrasting it with extensive types of land use and such agrarian social arrangements as the estate, the plantation, and the commune. Chapters 3, 4, and 5 broaden the comparative framework to consider data drawn largely from the economics literature on labor time, energy exchange and sustainability, and farm size and productivity. I have tried to explain those awkward, counterintuitive findings that small farms regularly produce more per unit area than bigger farms in the same area, and that they do so with greater energy efficiency and less environmental degradation.

Chapter 6 enters the confused and thorny thicket of property and land tenure, in support of the claim that households of intensive cultivators necessarily hold private, heritable rights in some resources, while exercising well-defined, corporate common property rights in others. Such tenure institutions and the differential returns to labor and management in different household enterprises account in part for obvious inequalities among smallholders, the subject of Chapter 7. Both upward and downward economic mobility over the household developmental cycle and between generations reflect the persistent internal dynamism of smallholder societies. Some of the same factors suggest reasons for the failure of the market economy in general and capitalism in particular to polarize smallholders into rich landlords and a landless proletariat of workers. All of the themes of smallholder population, intensive agriculture, and household organization are reprised in a modest approach to the richly textured history of Chinese agriculture in Chapter 8. The dramatic resurgence of the smallholder pattern in China after an era of socialist collectivization gives testimony to the essential social and economic soundness of the institution.

It may appear to be some sort of mock-populist arrogance or intellectual discrimination that causes me to shuffle the chapters on substantive theory to the back of the book. It is with no disrespect toward the great thinkers of the past and the present that I postponed my commentary on their ideas until after the evidence had been laid out. Rather it was owing to my own inability to come to grips with their more general and abstract formulations until I could measure them against some ethnographic and historic benchmarks. Those who feel more comfortable with an introduction presenting a theoretical context might want to skip immediately to Chapters 9 and 10. The fact that my own mean approximations have been informed and refined throughout by my understanding of Boserup and Julian Steward, Marx and A. V. Chayanov, Clifford Geertz and Marshall Sahlins is too obvious to require comment.

I cannot honestly say that I embarked on field research to test well-defined hypotheses or to develop the propositions of a theoretical paradigm. Although operating under the loose appellation of *cultural ecologist*, I neither adopted the biologists' procedures of measuring energetic inputs and outputs in caloric terms (Rappaport 1971; Hardesty 1975; Lee 1969) nor applied ecosystemic models of equilibrium and cybernetic feedback loops (Vayda and Rappaport 1968; Flannery 1968; Rappaport 1968) with scientific rigor. Problems of both method and theory in translating biological models into human ecology have been recognized by John Bennett (1976), Roy Ellen (1982), P. J. Richerson (1977), Emilio Moran (1984, 1990), and Benjamin Orlove (1980). My own orientation has been that of a rather pedestrian neofunctionalism, broadening the range of variables to include relevant factors of physical environment, population, land-use practices, and technology as they are interrelated with the social organization of agricultural production, and to suggest regularities of change through time in this system. Ecological functionalism need not be the analysis of a closed, static equilibrium system—it should instead be comparative and historical.

Theory became important to me inasmuch as it named concepts and provided generalizations of significant processes—for example, Steward's cultural ecology, Boserup's agricultural intensification and population pressure, Malthus's preventive checks, B. L. Turner and S. B. Brush's market demand, and Hans Binswanger's transaction costs in household and wage labor. Perhaps equally important to the effort of ethnological, cross-cultural comparison was my growing sense that certain theoretical positions seemed to run contrary to the accumulating evidence of the smallholder pattern. Marx appeared to neglect the role of demographic change and to substitute evolutionary conjectures about primitive communism for an understanding of the conditions that generate more individualized rights in property. The polarization of rural society into landowners and wage workers predicted by Lenin seemed not to take place among smallholders in a market economy (Hayami and Kikuchi 1982; White and Wiradi 1989; Eder 1990; Attwood 1992). Chayanov's emphasis on consumer/worker ratios and a subsistence economy isolated from the market and wage labor provided an ideal peasant-household type with few referents in the real world. Geertz's (1963, 1984) influential concept of involution restricted the range of adaptation that could take place in smallholder communities and attacked "economistic" explanation. Sahlins's (1972) domestic mode of production gave priority to kinship and political considerations and explicitly denied the role of "practical reason" in making sense of household farm production.

Perhaps most striking of all to me were the contradictions in the concept of global agricultural progress toward the unquestioned goal of industrialized, large-scale, scientific, energy-intensive, mechanized, specialized commodity-producing farms that dominated most development thinking in the half century after World War II. I touch on some of these issues in the concluding Epilogue. The theories of capitalist modernization and socialist revolution, often so different, appeared to be in implicit agreement that the smallholder had to go. The emerging problems of the farm-as-factory model in maintaining food production, supporting environmental sustainability, reducing economic costs and risk, mobilizing efficient labor, and providing incentives now threaten both agribusiness industrial farms and rural collectives. In this book I frame a smallholder alternative.

R. McC. N.

Acknowledgments

I GREW UP thinking that the word *synthetic* referred mainly to things produced artificially, like the man-made synthetic rubber that was one of the scientific triumphs of the World War II era. Anything that was synthetic was by definition not "natural," and there was some suggestion that it was ersatz, a less desirable substitute. For cultural anthropologists, the medium that seemed most direct and natural was the ethnographic report and analysis based on fieldwork (though postmodern sensibility has now exposed the artifice in such accounts). Conversely, synthesis, in the sense of "the combining of often diverse conceptions into a coherent whole," seemed necessarily derivative, speculative, and somehow ostentatious (one recalls Kent Flannery's ironic "Great Synthesizer"). Having at long last embarked, perhaps imprudently, on a synthetic and possibly quixotic venture, I am very much aware of how dependent I am on the work of others and on a professional lifetime of support and encouragement.

The field research that germinated most of my own ideas in this book was generously underwritten by grants from the Ford Foundation, the Social Science Research Council, the Guggenheim Foundation, the National Institute of Child Health and Human Development, and the National Science Foundation (most recently by Grant No. BNS–8318569). The opportunity to read widely and attempt to find common threads linking an interdisciplinary literature and my ethnographic findings in a more generalizing framework was provided by a year at the Center for Advanced Study in the Behavioral Sciences at Stanford. The intellectual freedom and stimulation of that period, in the company of colleagues like Dave Weber, Chris Hastorf, Richard Saller, Woody Powell, and Bill Bright, was invigorating and inspiring. A paper written at the Center synthesizing materials on population, permanent agriculture, and polities for a School of American Research Advanced Seminar organized by Stead Upham (Netting 1990) allowed an initial foray into some of the issues this book considers. Giving the Scott-Hawkins Lecture for the departments of History and Anthropology at Southern Methodist Univer-

sity in the fall of 1988 gave me the chance to outline the smallholder /
householder argument (Netting 1989), and I appreciate the efforts of David Weber and Ben Wallace in making this possible.

The actual writing of the manuscript began during 1989, when I was a
research scholar at the Workshop in Political Theory and Policy Analysis
and, simultaneously, visiting professor in the Department of Anthropology, Indiana University. Though always on the interdisciplinary fringes
of anthropology, I had never interacted consistently with political scientists until Elinor Ostrom invited me to the Workshop. The discussions of
common property, land tenure, and institutional economics with Lin and
Vincent Ostrom, Ron Oakerson, Louis D'Alessi (an economist who
commented expertly on several of my chapters), and Alan Schmidt were
fascinating and extremely useful. I am indebted also to Emilio Moran,
who first proposed the Bloomington visit, and who reviewed an early
draft of the manuscript based on his own comprehensive knowledge of
cultural ecology. Another perceptive reader, constructive critic, and major contributor to household studies was Rick Wilk, my long-time
friend, collaborator, and fishing buddy. He and Ann Pyburn welcomed
me to their house in both Bloomington and Orange Walk, Belize, to their
circle of courtesy kin, and to their groaning board of pundits. I owe a
great deal to Ester Boserup, whose published works and generous scholarly advice, including valuable comments on this manuscript in December 1991, have instructed and inspired me. None of my friends and mentors are, of course, responsible for my mistakes and misunderstandings,
which they did their best to correct.

A colleague and former student who made suggestions on the book
draft was David Cleveland, whose thoughts on sustainable agriculture
and indigenous crop varieties were particularly valuable. Mari Clark
shared with me her acute insights into Greek rural households. As a veritable novice in the thickets of sinological literature, I was saved from
some egregious blunders in the China chapter by the good counsel of
Susan Greenhalgh and John Olsen. Rhonda Gillett has listened with unflagging good humor to more obsessive disquisitions on smallholders
than anyone should have to endure. I owe a great deal to Glenn and Priscilla Stone, who rekindled my interest in the Kofyar, designed and carried
out much of the 1984–85 restudy, and were the best of companions in
those long Namu evenings under the Nigerian stars. This side trip to
synthesis did not mean that our real ethnography would be sidetracked.

Notions that percolate for a long time may become stale. I was fortunate that my matchless colleagues at the University of Arizona were unfailingly interested, often adding new observations or references to my
old brew of farm households, labor, and land. Special thanks are due

Jerry Levy, Norm Yoffee, Mark Nichter, Carol Kramer, Jim Greenberg, and Art Jelinek for their personal as well as professional concern on a long and sometimes rocky course. Perhaps my greatest source of satisfaction in the academic life has been the graduate students, many of whom learned, refracted, and reconceived the ecological position in chats and seminars over the years. Most recently Catherine Besteman, Jon Ingimundarson, John Higgins, Eric Poncelet, Catherine Tucker, John Gbor, and Dan Goldstein have kept the excitement and creativity lively. Hsain Ilahiane worked on the manuscript, tracking references in Chayanov, Lenin, and James Scott, while relating them to the moral economy of his own Moroccan family farm. Jim Spicer supplied some of the computer skills that I lack. The special contributions of Andrea Smith, who read the manuscript with discernment, prepared and endlessly revised bibliographies, pursued special topics in sustainability and Chinese land degradation, and found many of the illustrations, were invaluable.

As with my earlier volume on Swiss mountain ecology (Netting 1981), I relied on the editorial skill and strategy of Carol Gifford. This would be a different and a poorer book without her organizational suggestions and her unfailing confidence that I had something significant to say. Everyone that I have mentioned had some influence on the manuscript, though there are in it statements and themes with which they may heartily disagree. Appropriating the ideas of others without full attribution and misunderstanding the ideas one cites are often among the sins of synthesis, and if such occur herein, I am responsible.

Both Indiana and Arizona provided me with access to computers, and my long-delayed initiation to electronic writing took place with this book. Bearing with my penchant for penciled revision of the printout and my occasional recidivism to the pen, Gayle Higgins at the Workshop in Political Theory and Policy Analysis and Doris Sample in the Department of Anthropology at Arizona turned the manuscript into admirable copy with unbelievable affability and promptness. Marcia Lang at Arizona and Connie Adams at Indiana were unfailingly helpful and knowledgeable. Institutional support for the research assistance and technical services that allowed me to complete this manuscript came from the workshop directed by Elinor Ostrom and the Department of Anthropology under the capable and thoughtful leadership first of Bill Stini and then of Bill Longacre. I am very grateful to them all.

For the quiet and often invisible work and skill that transform an author's inevitably rough manuscript into a published book, I have my editors at Stanford University Press to thank. Bill Carver was supportive and enthusiastic in the early stages of the project. Paul Psoinos and Peter Kahn shepherded the book through to production. Glenn Stone generated some of the original figures, and Mike Maystead prepared all of the final figure art for publication. My copy editor, Peter Dreyer, managed to combine amazingly eclectic knowledge with keen literary judgment and precision in matters of English usage. Publishers, like smallholders, have reason to be proud of their craft.

The following tables and figures appear by permission of their copyright holders.

Table P1: Reprinted from Ester Boserup, "The Impact of Scarcity and Plenty on Development," *The Journal of Interdisciplinary History*, XIV (1983), 401, with the permission of the editors of *The Journal of Interdisciplinary History* and The MIT Press, Cambridge, Mass. © 1983 by The Massachusetts Institute of Technology and the editors of *The Journal of Interdisciplinary History*.

Table 2.3: Reprinted from *Proper Peasants: Traditional Life in a Hungarian Village*, by Edit Fél and Tamás Hofer, Viking Fund Publications in Anthropology, No. 46, 1969, by permission of the Wenner-Gren Foundation for Anthropological Research, Inc., New York.

Table 3.6: Reprinted from Gillian Hart, *Power, Labor, and Livelihood*. Berkeley: University of California Press, 1986. © 1986 by The Regents of the University of California.

Figures 3.1 and 3.2: Adapted from Carlstein 1982

Figures 3.4, 3.5, and 3.6: Adapted from Glenn D. Stone et al., "Seasonality, Labor Scheduling, and Agricultural Intensification in the Nigerian Savanna," *American Anthropologist* 92: 1, March 1990. Reproduced by permission of the American Anthropological Association. Not for further reproduction.

Table 4.2: Reprinted from M.J.T. Norman, "Energy Inputs and Outputs of Subsistence and Cropping Systems in the Tropics," *Agro-Ecosystems* 4: 355–66, with permission of Elsevier Science Publishers.

Table 4.5: Reprinted from Warren A. Johnson et al., "Energy Conservation in Amish Agriculture," *Science* 198: 376. © American Association for the Advancement of Science.

Table 8.2: Reprinted from A. K. Ghose, "The People's Commune," in A. Saith, ed., *The Reemergence of the Chinese Peasantry: Aspects of Rural Decollectivization* (London: Croom Helm, 1987), p. 38, with permission of the publisher.

Table 9.2: Reprinted from Ester Boserup, *Population and Technological Change: A Study of Long-Term Trends* (Chicago: University of Chicago Press, 1981), p. 19. © 1981 by The University of Chicago Press. All rights reserved.

Contents

30 pages of photographs follow p. 145

Tables and Figures

Tables

Figures

Smallholders, Householders

Julian Steward's anthropology had an earthy and common-sense orientation that regarded the exigencies of work and livelihood as among the most important of [the] determinants [of culture as caused and causal]. . . . Steward's theories also stressed the active and sensate over the symbolic and conceptual in social life, a logical enough outcome of his preoccupation with work and groupings. The approach was realistic, unetherealized, self-consciously tough-minded and dynamic. It found its subject matter in the more mundane aspects of culture, and it sought explanation in sinew and sweat. . . .

The heart of Steward's anthropology is the analysis of the ways the two givens, technology and resources, are brought together through human labor.

—Robert F. Murphy, "Julian Steward"
(1981: 175, 187)

Prologue: An Ethnological Essay in Practical Reason

A<small>NTHROPOLOGICAL</small> <small>THEORIES</small> of "practical reason" that are natural or ecological, utilitarian or economic like Steward's have in recent years been opposed by scholars who emphasize "culture" as symbolic or meaningful. In the words of Marshall Sahlins:

> The distinctive quality of man [is] not that he must live in a material world, [a] circumstance he shares with all organisms, but that he does so according to a meaningful scheme of his own devising, in which capacity mankind is unique. [This approach] therefore takes as the decisive quality of culture . . . not that this culture must conform to material constraints but that it does so according to a definite symbolic scheme which is never the only one possible. Hence it is culture which constitutes utility. (Sahlins 1976: viii)

This book on the practice of smallholder intensive agriculture by farm-family households relates elements of environment, technology, and human social organization in the tradition of cultural ecology pioneered by Steward (1938, 1955). It is not a study of "culture" in the widely accepted anthropological sense of a distinctive system of shared meanings and a symbolic organization of experience characterizing a particular society or social group. The focus on differences in ways of thinking expressed in language, beliefs, rituals, and myths, and interpreted from a wide range of cultural texts, has been and necessarily remains a central concern of the discipline. But anthropology has also been an empirical social science of practical reason, grounded in an Enlightenment faith that there are regularities in human behavior and institutions that can be understood as filling human biological and psychological needs under particular circumstances of geography, demography, technology, and history. These commonalities can be discerned cross-culturally in groups separated by space and time and displaying a splendid variety of cultural values, religions, kinship systems, and political structures.

The systematic comparison of practices and institutions that reoccur in different societies, and the analysis of how they function and change in a functional and comprehensible manner is *ethnology*, a somewhat quaint and old-fashioned designation that many contemporary anthropologists would never use to characterize their own work. What follows is not an attempt to interpret "culture," a project of eliciting and perhaps creating meaning so grand that only the artist or the literary critic would confidently attempt it. Rather it examines a limited set of social and economic factors that are regularly associated with a definable type of productive activity, despite considerable variation in a number of other "cultural" features that may themselves cohere internally in meaningful, patterned ways.

Smallholders: Characterizing a Type

Smallholders are rural cultivators practicing intensive, permanent, diversified agriculture on relatively small farms in areas of dense population. The family household is the major corporate social unit for mobilizing agricultural labor, managing productive resources, and organizing consumption. The household produces a significant part of its own subsistence, and it generally participates in the market, where it sells some agricultural goods as well as carrying on cottage industry or other off-farm employment. Choices of allocating time and effort, tools, land, and capital to specific uses, in a context of changing climate, resource availability, and markets must be made daily, and these economic decisions are intelligible in rational, utilitarian terms. Smallholders have ownership or other well-defined tenure rights in land that are long-term and often heritable. They are also members of communities with common property and accompanying institutions for sharing, monitoring, and protecting such resources. The existence of separate household enterprises, with a measure of autonomy and self-determination, in a larger economy with institutionalized property rights and market exchange, presents the likelihood of economic inequality, both among households in the community at any point in time and in the changing status of a single household at different times in its developmental cycle. But inequality is not equivalent to enduring class stratification *within* the farming community, and neither does it exclude socioeconomic mobility. The argument of this book is that these characteristics regularly co-occur, and that their systematic articulation and changing relationships can be reliably observed, described, and explained.

Not all food producers are smallholders. The characteristics put forward here do not apply to shifting cultivators practicing long-fallow or slash-and-burn farming where land is still plentiful and population density low, as in some parts of the humid tropics today. Nor does the designation *smallholder* fit herders, whether they be the nomadic pastoralists of East Africa or the ranchers of Texas. It does not match geographically, economically, or socially with the farming systems of dry monocropping of wheat, sugarcane estates, cotton plantations with slaves, or California agribusinesses. Smallholders practice *intensive agriculture*, producing relatively high annual or multicrop yields from permanent fields that are seldom or never rested, with fertility restored and sustained by practices such as thorough tillage, crop diversification and rotation, animal husbandry, fertilization, irrigation, drainage, and terracing. I am not talking here about amber waves of grain but about gardens and orchards, about rice paddies, dairy farms, and *chinampas*.

Even the casual observer has little difficulty in recognizing a landscape domesticated by intensive agriculture. The stepped stone walls and mirrored, ponded fields of Balinese wet-rice cultivation and the neat, fenced, manicured pastures of Dutch farmsteads bespeak high, dependable yields and diligent stewardship. But that these are, in fact, representatives of a distinctive type of land use regularly associated with specific demographic, social, and institutional factors may require something more than a leap of faith in practical reason. It is the virtue of Julian Steward's approach that consistent cross-cultural relationships can be demonstrated empirically despite striking variability in local environment, technology, culture, and politics. The common features form a definable cultural ecosystem with its own evolutionary patterns and probabilities of change.

The smallholder as depicted here may be what Max Weber (1949: 90) called an ideal type—that is, a "conceptual pattern [that] brings together certain relationships and events of historical life into a complex, which is conceived as an internally consistent system." As in Geertz's (1963) characterization of agricultural involution or Popkin's (1979) of the rational peasant, "the researcher posits a structured representation of a social category that singles out certain features and abstracts from others" (Little 1989: 194). The smallholding householders that I examine in this book are alike in that for all of them land is objectively a scarce good, agrarian production per unit area is relatively high and sustainable, fields are permanent, work takes skill and relatively long periods of time, decisions must be made frequently, and the farm family has some continuing rights to the land and its fruits. In these type traits, the smallholder differs in kind or in degree both from other food producers and from those who pursue other occupations. Drawing principally on ethnography, agricultural economics, and geography, I first describe what smallholders do and then attempt to account for the systematic commonalities of behavior and institutions that make a kind of sense according to the plebeian, but still powerful, canons of practical reason. There is no shared culture of meanings among the many disparate groups of smallholders, but the quest for functionally meaningful and coherent systems that transcend the distinctions of societies and regions is also part of the anthropological calling.

Why Study Smallholders? Some Subjective Reflections on Objective Research

It would be misleading and disingenuous to argue that scholarly work that styles itself "social science" arises from a single-minded search for timeless truths existing out there in the real world, or that those of us who essay this approach to knowledge believe that we shall discover natural laws of society, test hypotheses in some irrefutable way, or reliably predict future states of the system. Perhaps one of the attractions (and the solaces) of anthropology is that its deductive models are neither very compelling nor particularly intrusive, and that one is almost sure to learn something interesting and new by fieldwork (even if this is no longer always "exotica and trips"). The formal structuring of problems and hypotheses in the research proposal, and the (sometimes very different) relationships of data, argument, and theory in the finished product seldom overtly reflect the subjective experience or the sentimental journey that led the student in that direction. At the risk of ruminations and other evidences that I may be entering my anecdotage, let me ask how it was that I came to study agriculture, households, and land tenure, rather than, say, kinship terminology, the peasant view of the good life, or caste in India, as my esteemed teachers had. It might also be useful to try to reconstruct why, at this point, I should leave those ethnographic cases that I know at first hand for the much more hazardous terrain of global ethnological comparison and synthesis.

As a graduate student at the University of Chicago in the late 1950s, I read Julian Steward's *Theory of Culture Change* in Fred Eggan's course on ethnological theory and method, and I did a source paper on Fulani ecology for a seminar at Northwestern University with Jim Bohannan. But my proposal for my first field research, begun in 1960 on the Jos Plateau in Northern Nigeria, did not make such interests explicit. As I once admitted, "I did not have a carefully thought out plan of ecological study when I entered the field, and my findings came piece-meal in response to the elementary questions of why people lived where they do, what they did with their time, and how they got enough to eat. Many of my conclusions came in the analysis of quantitative material after leaving the field" (Netting 1968: 23). A generous interpretation of this choice of scholarly direction might be that I admired the apparent self-sufficiency of the isolated Kofyar, and their cultural vitality, and that I wanted to communicate to others some appreciation of the economy and "material culture" that supported this African society. Perhaps it is more accurate to say that I and my assistants had collected a lot of data in a fairly standardized form on household membership by name, gender, age, and kin relationship, and on the crops and domestic animals the household produced and consumed. I was impatient to begin my study before fully mastering the language, and I found that a household census, with its repetitive questions, relatively straightforward answers, and generally nonsensitive content, was a good way for me to get acquainted with people and practice my Kofyar. Perhaps a household survey also reflects a certain lack of imagination. I remembered the advice of Sol Tax, one of my professors at Chicago: "When you can't think of anything else to do, you can always census."

But judging from my field notes, I spent as much time attending divinations, recording folktales, exploring witchcraft beliefs, and drinking beer as I did talking about farming and observing work groups. Although Kofyar cultural concepts and behavior in such areas as gender relationships, politics, warfare, and contacts with the British colonial government have not been neglected (Netting 1964, 1969a, 1969b, 1972, 1974b, 1987), my core concerns have remained stubbornly centered on issues of work, agriculture, households, and rights to the means of production. One can count on the existence of activities and things that can be counted. The mundane, petty facts about residence, kinship, and crops that individuals can tell the interviewer with reasonable accuracy can be transmuted through numbers into statistics. From what people know and see can come approximations of mean and central tendency, classifications by age and sex and village of origin, information about differences (with, one hopes, some measure of significance) between subsistence cultivators and cash-croppers, correlations of household size and wealth. These are things we might guess at, but no one knows the answers accurately until you do the numbers. Moreover, unlike norms or ethical principles or aesthetic judgments, quantitative measures of behavior are not part of people's collective consciousness. Though individuals can and assuredly do make economic decisions about market exchanges, stored food, and labor expenditures, they generally do so without bookkeeping and exact calculations. They have little way of estimating changes in social behavior at the group level; indeed, there may be a vested interest in asserting a somewhat spurious cultural continuity and the strength of tradition (Murphy 1971). Statistical representation of a decline in fallow, an increase in age-specific female fertility, or a process of polarization in household incomes is not information that is available for people to apprehend or incorporate into their stock of cultural meanings. But these trends and changing relationships affect systems of farming, labor, and landholding, and they can be analyzed by the observer using the quantitative methods of practical reason.

It was in examining graphs plotting a regular association between field area and crop production in intensive farming at Kofyar homesteads, in contrast to the direct relationship between labor input and production on cash-crop farms worked by migrants, that I first became aware of differently patterned agricultural systems (Netting 1968: 135, 205). It appeared that farmers in the densely settled Kofyar homeland practiced permanent cultivation of small fields, with high yields per unit of land, as opposed to the *same people's* shifting cultivation of abundant land on the frontier, where fields were large and yields per unit of land were low. Moreover, traditional homestead cultivators had small households and nuclear or extended families, whereas migrant farmers had statistically larger households, achieved by increased rates of polygyny, and more multiple-family households. Household size appeared to correlate closely with land availability, and it varied with different labor needs (Netting 1965, 1968).

The household was not a static traditional grouping generated by fixed cultural rules of postmarital residence or the practice of polygyny (Goodenough 1956; Wilk and Netting 1984). Nor was it a predictable precipitate of a stage in the household developmental cycle of a social structure at equilibrium (Goody 1958). Because farm labor was largely mobilized and consumption organized in the family household, processes of household formation and fission might alter appreciably and quickly. The composition and structure of the household group, as it emerged from the figures of hundreds of household censuses, varied with changed circumstances of production. And this rapid adjustment was unaccompanied, as far as I could see, by changed cultural standards or expectations about household membership, marriage, socialization of children, or rights to land. Kofyar customary systems of meanings remained intact and did not constrain substantial nonrandom changes in social behavior.

Quantitative evidence of change was for me the genesis of recognition that the smallholder household had readily distinguishable characteristics related to a particular type of land use under a specific population regime. It also suggested certain limited ethnological comparisons to test the posited functional relationships. The Chokfem Sura, who lived near the Kofyar ancestral homeland in a similar plateau escarpment environment, practiced shifting cultivation and had a lower population density, with large, multiple-family households. One could also predict that as the Kofyar filled up their frontier land, they would revert to more intensive agriculture, with smaller fields, and that their recently augmented household size would begin to decline. These projections for change over time were in fact supported by a restudy of the Kofyar in 1984–85 (Netting et al. 1989; G. D. Stone et al. 1990).

Comparing and Generalizing: How to Recognize Smallholders

Like most people most of the time, the Kofyar have no means of reducing behavior to statistical terms, but they were quite ready to explain the patterned actions that the enthusiastic anthropologist had "discovered." Why shouldn't a young adult man remain in the parental household when his father provided bridewealth and a motorcycle from the family's new cash-crop earnings? Women recognized the value of an extra pair of hands and pressured their husbands to marry co-wives (M. P. Stone 1988). Successful farmers argued with me over the costs and benefits of cooperative beer-party work groups as opposed to wage labor. It may be that there is a strong streak of practical reason in certain areas of Kofyar life, just as there seems to be among other smallholders. It is also possible that the economically minded investigator asks questions that elicit pragmatic responses. But just as systems of meaning and behavior are not exhausted by a materialistic, ecological approach, so the utilitarian activities of production and reproduction are not solely culturally constituted or changed.

Could the Kofyar be nothing more than an interesting, but perhaps anomalous or idiosyncratic, ethnographic case? Smallholders are usually thought of as peasants with an intermediate technology of the plow and draft animals, living in a state, and subject to demands for tax or tribute from other elite groups in the complex society. The Kofyar practiced hoe cultivation in a rugged escarpment area of the Jos Plateau, where they had remained largely outside the political system and market economy of northern Nigeria's Hausa-Fulani kingdoms. Kofyar country was only made part of the British colonial state in this century, and they have retained a large measure of control over their own land and production system down to the present. They did not fit easily into the standard peasant mold.

The smallholder adaptation only became a generalizable category for me inasmuch as it appeared to encompass other examples of peoples practicing intensive agriculture and resisted conformity to the older and more conventional typologies to which these groups had been consigned. The most usual way of pigeonholing farmers is by contrasting technologies, often along an implicit evolutionary scale. Primitive farmers or horticulturalists use the hoe, the axe, the digging stick, and perhaps the sickle, and agriculturalists add animal draft power and the plow to these manual implements. The more developed technology captures nonhuman energy, presumably lowering human labor inputs and increasing agricultural production over larger land areas. Mechanization, energy from fossil fuels, and scientific methods of fertilization, plant breeding, and crop protection (using pesticides and herbicides) carry the same evolutionary process further. Smallholders with relatively simple tools farming small, often fragmented fields, and relying on traditional "prescientific" understandings of agriculture, are automatically relegated to a lower, and presumably earlier, farming type.

Scale and productivity are, however, slippery concepts. The bigger fields made possible by the use of nonhuman sources of energy do indeed save labor, and production per hour rises. But productivity as reflecting production *per unit of land* may in fact be *lower* under more extensive, technologically advanced systems. Because intensification refers to achieving and maintaining relatively high land productivity over time, it can be applied to farming systems with varying dependence on nonhuman energy. The Kofyar first claimed my attention because, with nothing more than iron-bladed hoes, digging sticks, and sickles, they achieved relatively high and reliable yields from small land areas, using compost manuring, intercropping, stall-feeding of animals, arboriculture, ridging for water retention and drainage, and terracing. The high labor inputs of intensive agriculture increase yields and reduce variability by conserving and enhancing soil nutrients and diversifying production.

If we include under the rubric *technology* the repertoire of skills, the folk knowledge and ethnoscience brought to the task, and the building and maintenance of intricate systems of irrigation, flood control, and drainage by means of hand tools, the evaluation of technology along the single axis of "labor-saving" becomes inadequate. Intensive techniques applied with care, and frequent monitoring of the field, garden, or orchard, also imply a *sustainable* agriculture that prevents the erosion and degradation that frequently accompany large-scale, extensive land use. Part of my reason for beginning research in alpine Switzerland during the 1970s was to see a system that had persisted for centuries in an easily damaged environment of steep slopes, short growing seasons, and low rainfall. Historical documentation attested to continuous use of irrigated mountain meadows, terraced vineyards and grainfields, forests, and high-altitude pastures with no evidence of erosion, declining soil fertility, waste of irrigation water, overgrazing, or deforestation. As in the Kofyar case, techniques of crop rotation, manuring, and controlling the tendency of worked soil to creep downhill were practical rather than based on a "scientific" understanding of hydrology and soil chemistry. Yet low-tech, highly effective methods maintained relatively dense, permanently settled local populations in a manner that both conserved and enhanced the production of existing natural resources. The modernist cant that traditional intensive cultivators must be taught how to farm with machines, purchased inputs, and scientific knowledge is directly contradicted by the land productivity, the reliability, the ecological sustainability, and the adaptability of these systems.

The fact that the Nigerian Kofyar and the mountain Swiss are both geographically and economically peripheral to the concerns of modern industrial nation-states stimulated my interest in smallholders who not only persist but play a dominant role in market production as well as subsistence. Though peasant smallholders have had an abiding presence in the north of Portugal, the Netherlands, parts of Germany, and Denmark, it is the ancient wet-rice societies of Asia where the type is most clearly and pervasively exemplified. There, with long-term, high-density populations in China, Japan, and Java, skill replaces scale (Bray 1986), renewable energy competes successfully with imported and mechanical energy, and household management demonstrates its superiority to both hired labor and collective farming. The great Chinese river valleys have proved more highly productive and more agriculturally sustainable, through cycles of intensification, than any comparable region on earth. Although the historic form taken by labor-intensive smallholder enterprises in the Chinese market economy (P. C. C. Huang 1990) may not be a model for emulation elsewhere, it does suggest the durability and amazing resilience of the smallholder techno-social type. As an ethnological comparison, the richly documented Asian cases best call into question the reigning hegemonic ideal of large-scale, energy-expensive, mechanized, specialized, scientific, capital-intensive, labor-saving agriculture enshrined by the West. Under certain circumstances of high population density and market economy, there *is* a viable smallholder alternative.

Myths of Modernization: Evolutionary Mystifications and Smallholder Persistence

Why have smallholders been ignored or regularly stigmatized as old-fashioned, resistant of innovation, inefficient, and a barrier to modernization? Almost from the beginning, my field experience tended to collide with and contradict conventional views of a unilineal development in agriculture and a static subsistence segment. No outsider had recently introduced the Kofyar to concepts of composting green vegetation with goat manure or preventing erosion by making rectangular ridges on top of bench-terraces. When the Kofyar summarily discontinued these practices in favor of slash-and-burn farming on the frontier, it was not a sign of some evolutionary regression but a reasonable reaction to abundant land and the desire to make labor more productive. Kofyar later bought and used chemical fertilizer as well as motorbikes and trucks for transport, but refused both the ox plow and the tractor, which got in the way of intercropping. They responded to growing land scarcity by reintensify-ing agriculture on smallholdings rather than taking the path of cultivating large-scale farms with hired labor. And the options they chose were fitted by trial and experiment to a savanna environment with seasonal, soil, and cropping differences that were in part new to them.

The Swiss smallholders had modestly revolutionized their alpine farm-ing system twice, once with the adoption of the potato as a greater and more dependable source of calories in the eighteenth century, and again when garden tractor-mowers for cutting hay replaced the scythe a few decades ago. This latter-day technology allowed agriculture to continue as a part-time activity along with employment in industry and the tourist trade. But the ancient peasant subsistence system had always coexisted with and mutually supported households whose income came in part from off-farm employment as everything from mercenary soldiers to chambermaids. The security of a diversified and intensive farming sys-tem maintained an astonishing proportion of village family lines from before 1700 through 10 to 13 recorded generations to the present (Net-ting 1981: 70–89), yet necessary cash and manufactured goods always came from outside the community (Netting 1984).

Perhaps the most stubborn and pervasive myth about smallholders is that their physical isolation in rural areas, their simple technology, and their modicum of self-sufficiency remove them from dependency on a market and the mentality of maximization, greed, private property, and inequality that is thought to be the market's inevitable accompaniment. Again, the evolutionary construct of the peasant mode of production, or of a precapitalist social formation where labor and resources are shared and reciprocity is unreckoned, did not seem to fit the intensive cultivators I knew. Scarcity was not an artificial, arbitrary creation of some elite but a condition of the ratio of population to land. Resources like irrigated fields or terraced vineyards, where the investment of labor and capital over years had built up and buttressed the productivity of the land, could not readily be loaned to others or periodically reallocated among village families. Ethnological comparison cross-culturally and through time as-sured me that intensive agriculture under circumstances of real, objective limitations on arable land makes primitive communism impossible.

Where money, legal titles, notaries, and courts exist, as in medieval Switzerland, land is bought and sold, and its price seems remarkably high. But even where market relationships are economically insignificant and no state legal system intrudes, as among the precolonial Kofyar and the Philippine highland Ifugao, households have clearly defined, very val-uable rights in real property, and land is heritable. With the assertion of continuing use, occupation, temporary exchange by loan or lease, and public litigation over disputed rights, an institution very close to private

property comes to exist, even if permanent alienation by sale seldom occurs. Individualized, socially recognized rights to scarce, highly productive resources and the improvements that increase and maintain their
yields are inherited along lines of close kinship or transferred in exchange
for other valuable goods. At the same time and place, land with low or
temporary production with little potential for intensification, as in marginal, long-fallow bush fields or rough grazing areas, may remain in
communal tenure with occasional redistribution or shared, controlled access (Netting 1969a, 1982a). The documentary evidence that the resident
families of the Swiss village had exercised private property rights in irrigated meadows, grainfields, gardens, and vineyards since the thirteenth
century while maintaining legally instituted common property in the
community alp and forest convinced me that there was no evolutionary
watershed separating an earlier stage of communal rights from a later
period of private property emerging with the market and the state (Netting 1976). Smallholder intensive cultivators hold land, and, all other
things being equal, it is the ecological factor of land use that most
strongly determines land tenure.

One implication of the scenario depicting small cultivators as low producers with poor technology, little market participation, and communal
tenure is that they are homogeneous in property and wealth (Redfield
1941, 1955). Even as smallholding peasants within a state, they are perceived as economically stagnant and politically inert, forming a mass "of
homologous magnitudes, much as potatoes in a sack form a sack of potatoes," as Marx notoriously put it.[1] The closed corporate community of
peasants systematically cuts back emerging inequalities of wealth by di-

1. Marx 1971 [1852]: 230. Marx found the French peasants of the nineteenth century a
conservative, antirevolutionary group who cared only for their selfish and narrow interests
in property: "The smallholding peasants form a vast mass, the members of which live in
similar conditions but without entering into manifold relations with one another. Their
mode of production isolates them from one another instead of bringing them into mutual
intercourse. The isolation is increased by France's bad means of communication and by the
poverty of the peasants. Their field of production, the smallholding, admits of no division
of labor in its cultivation, no application of science and, therefore, no diversity of development, no variety of talent, no wealth of social relationships. Each individual peasant family
is almost self-sufficient; it itself directly produces the major part of its consumption and thus
acquires its means of life more through exchange with nature than in intercourse with society. A smallholding, a peasant and his family; alongside them another smallholding, another
peasant and another family. A few score of these make up a village, and a few score of
villages make up a department. In this way, the great mass of the French nation is formed
by simple addition of homologous magnitudes, much as potatoes in a sack form a sack of
potatoes" (ibid.). It is noteworthy that the same characteristics of isolation, homogeneity,
and self-sufficiency that represent the strength and cultural integrity of the folk society for
Robert Redfield conveyed to Marx only stagnation, ignorance, and a bar to evolutionary
progress.

recting gossip and envy against the rich through an idiom of limited good (Foster 1965), redistributing use rights in the commons, requiring leaders to sponsor fiestas and host feasts (Wolf 1957), and relieving subsistence crises through forced generosity (J. C. Scott 1976).

Although such "leveling mechanisms" assuredly do exist, it is my impression that they by no means equalize access to resources within the rural community. What smallholders have, they hold on to with a tight grip, and they compete with vigor and craft for a scrap of garden, a larger herd of goats, or a new granary. A single family household may grow from relative poverty when an adult couple supports many young dependents to a large, prosperous group with several productive workers. Because families are at different points in their domestic developmental cycles, and because they do not all follow the same trajectory, inequality in wealth is the rule rather than the exception. There are few mechanisms of gifts among kin, religious charity, or communal sharing that effectively redistribute such important resources as residential buildings, livestock, and land.

Inequality among smallholders, as opposed to the quite profound differences among farmer-owners and merchants, government officials, professionals, and landless laborers, is present and measurable but not rigidly stratified. Over time, Swiss villagers showed considerable mobility, both up and down the economic scale, but they did not polarize into a class of landowning, wage-labor–employing managerial farmers, or kulaks, and an impoverished group of minifundistas and rural proletarians (McGuire and Netting 1982). Even without substantial charity, public redistribution, or institutional checks on accumulation, there seem to be economic factors acting to circulate wealth. The Chinese case shows rich farmers incurring transaction costs for recruiting and supervising paid labor while households provide more skilled and dependable workers willing to accept lower marginal returns on their work. The combination of higher costs, lower production per unit of land, and high land prices means that rich farmers can get better returns on their capital in commercial or other urban occupations outside of agriculture. Partible inheritance may also divide a big estate among many sons of the owner. On the other hand, intensive agriculture rewards management skills, conscientious work, knowledge of resources, and careful, long-term planning. Thus the more clever and industrious smallholders can potentially increase the size and wealth of their enterprises over time.

Population Parameters and the Smallholder

There is something of a paradox in the particular cast of thought and theory that I bring to the problem of the smallholder as an enduring social and economic type. Consideration, some of it quantitative, of systematic interrelations among factors of demography, technology, environment, economy, and social institutions, in search of cross-cultural associations and regularities in processes of change is an exercise in practical reason, but it does not fall neatly into place with the major paradigms of materialism. The most general orientation toward the functional interactions of what I have called effective environment, productive and protective technology and knowledge, and social instrumentalities (Netting 1965) was that of Julian Steward's cultural ecology. But despite his theoretical emphasis on causal change and evolutionary patterns, Steward (1938) was most persuasive in outlining a relatively simple hunter-gatherer ecosystem from his own superb Great Basin ethnography. The dynamic roles potentially played by change in population, technology, or environment appear only obliquely in comparisons between Owens Valley and other Paiute peoples or between Basket Maker and Pueblo settlements (Steward 1955). Where intensive cultivators appear in Steward's writings, they are pawns in the schematic play of hydraulic power politics of Karl Wittfogel's irrigation civilizations.

For me, a more precise and better-articulated model of agricultural change came from Ester Boserup. Reading her book *The Conditions of Agricultural Growth* (1965) only a few years after completing my dissertation on Kofyar farming gave me an electrifying sense of an inclusive and consistent pattern that logically accounted for both the regularities and the processes of contemporary change reflected in my data. The intensive, highly productive, permanent agriculture of the Kofyar homestead farm that I had described as occurring with dense local population, high labor inputs, and individualized land-tenure rights (Netting 1963, 1965) represented a subsistence type that apparently occurred worldwide as an adaptive response to population pressure. Boserup also asserted that if land increased in abundance, people would save labor by reverting to more efficient shifting cultivation—which was just what the Kofyar were doing on the Benue plains frontier.

My own variation on the Boserup theme was to emphasize the role of the small, nuclear or polygynous family household as the social unit that typically mobilized labor, pooled consumption, and exercised tenure over the intensively tilled farm. I saw household size and composition as correlated with and responsive to farm area, cultivation techniques, and especially the labor needs of the agricultural operation. It was easy for me to postulate that a dense local population, drawn to the Jos Plateau escarpment both for its desirable rainfall and oil-palm vegetation and for the protection it offered from slave-raiding neighbors, would create the higher demand for subsistence food that gave a selective advantage to intensive methods of agriculture. The numerically preponderant presence in the census of relatively small family households as the units of farm production and consumption, landholding, and residence suggested that a social dimension could be added to the original Boserup formulation that population pressure caused or made highly probable a more permanent and intensive system of cultivation.

The brilliant reductionism of the Boserup hypothesis seemed, however, to treat population growth as an exogenous factor rather than as a variable element in a local ecosystem. Under what circumstances of changing fertility, mortality, and migration did population increase? Were there environmental limits to agricultural intensification beyond which population could not grow, and could stability be achieved by social means or only through the harsh imposition of Malthusian checks? My attempt to reconstruct the demographic history of the Swiss peasant village of Törbel was an effort to examine the dynamics of a smallholder system in which documented population change could be seen as both caused and causal. The record reflected an alpine community continuously occupying a fixed agrarian territory, and a medieval population dense enough to require impressive irrigation works for intensive dairy / grain subsistence pursuits (Netting 1974a).

Törbel was not, however, a self-regulating ecosystem, delicately balanced in its mountain environment (Netting 1990).[2] In 1532–33, the Black Death eliminated many local families and opened places for in-migrants, and the Napoleonic invasion of Valais coincided with a dip in population (Netting 1981: 72, 118). The smallholder pattern, however, persisted as village population doubled between 1774 and 1867, well before the advent of modern medicine (ibid.: 95–97). These results convinced me that an exogenous technological change—in this case, the introduction of the potato as a food crop—could promote an increase in female fertility and raise the potential of village territory to support more people. There was equally good evidence that local institutions of land tenure, inheritance, marriage, and sexual control had operated to restrict fertility by encouraging relatively late marriage and frequent celibacy on the part of villagers while also promoting out-migration. Culturally specific ideals and practices of partible inheritance, monogamous marriage, chastity, and long lactation figured in the Swiss demographic regime, and other dense farming populations displayed functional systems different in operation but similar in effect. The sometimes remarkable persistence and continuity of the smallholder adaptation in this case appeared to lie

---

2. The homeostatic ecosystem with deviation-counteracting feedback loops was the favored model of biological ecologists in the 1960s and 1970s (Odum 1969, 1971; but cf. Worster 1990). As anthropologists looked to technoenvironmental relationships to expand their structural-functionalist formulations, the ecosystem became a major heuristic device (Geertz 1963; Rappaport 1968; Flannery 1968).

not only in the possibility of raising farm production to feed more people but also in indirectly controlling population growth itself. Despite doomsday scenarios of runaway world population growth, smallholder farms did not appear to be endlessly fragmented, their resources degraded, or their households impoverished.

The Smallholder Meets the Market

The greatest problem with modeling a viable system of rural population, land, technology, and labor has been the tendency to treat such systems as self-sufficient and independent. In fact, smallholders do not normally live in isolation from larger networks of economic exchange or political organization; indeed, the scarcity of their resources and their desire for goods and services they cannot produce at home necessarily involve them in important external relationships. Boserup tends to see a more complex division of labor, specialization, and trade as stimulated by the same population increase that fosters agricultural intensification. But it is also possible that market demand (Turner and Brush 1987) and the taxes or tribute of political systems that protect and extend the sphere of market activity may impel cultivators to produce a surplus considerably above their subsistence requirements.

If land is abundant, as it was originally on the Kofyar frontier, extensive or shifting methods may be used to raise production most efficiently. The original motivation for adding bush cash-cropping to existing intensive homestead farming was the desire to enter the market. As land availability and fertility declined on the frontier, the Kofyar reintensified their agriculture to maintain and even expand the amount of surplus food they could sell. It is true that population concentration along roads or in peri-urban areas often coincides with truck gardening or the intensive production of crops of high value, like fruits, dairy products, condiments, and flowers. But a unicausal model of smallholder intensive household farming systems that neglects either population pressure or market demands is inadequate to account for the prevalence of the type.

Just as smallholders are seldom solely subsistence cultivators, so the need to compensate for insufficient resources and turn unused agricultural labor to other productive purposes means that household members will generally pursue a variety of full- and part-time occupations. Processing and selling food; cottage industries like weaving, basketry, pottery, and knitting; and sidelines in trade, transport, and construction may all be potential sources of income for the farm family. The records of the Swiss village of Törbel showed that numbers of local men served abroad as mercenary soldiers in the seventeenth century, and jobs as muleteers, cheesemakers, herdsmen, mail carriers, cooks, waitresses, and factory workers have helped to support farm families for the past 150 years (Netting 1981: 97–108). Though the household may not be a full-time agrarian unit, the pooling of income from many sources, periodic cooperation to perform farm tasks, and the protection against risk that comes from a diversified economic base increase the resilience of the smallholder enterprise.

The salience that I give to the smallholder household is mirrored in the powerful characterization of the peasant household by the Russian economist A. V. Chayanov (1966 [1925]). The Chayanovian farm household is a subsistence unit whose workers expend only the effort or labor "drudgery" necessary to provide for the consumption needs of all household members. Although supported with impressive statistical data, this radically simplified characterization does not fit the case of intensive cultivators. At the most basic level, in Chayanov's model, the demand of more household consumers for more subsistence food is met by enlarging farm size or "sown area," an alternative not readily available on the land-scarce smallholding. Though the peasant household's activities in allocating labor and leisure conform in broad outline to the assumptions of neoclassical economics, they take place in an idealized context where there is no significant production for the market and no wage labor. Though Chayanov's ideas have been appropriated by many social scientists as a general characterization of peasant economy, they struck me on first reading as possibly applying only to shifting cultivators farming for subsistence in a land-abundant environment. The consumer/worker ratio bore little relation to the per capita production of each Kofyar worker, either on the homestead farm, where field size correlated with crop production, or on the big bush farms, where more effort was made by those who wanted to participate in the market.

Indeed, Chayanov draws systematic contrasts between his Russian case and that of peasants in Switzerland, where land is scarce and individually owned, where labor input per hectare is inversely related to farm size, and where holdings are unequal. Perhaps the major and still largely unacknowledged reason for the poorness of fit between the Chayanov model and most intensive cultivators, whether peasant or not, is that the Russian system was grounded in land-abundant, long-fallow cereal cultivation; a generally sparse rural population; periodic reallocation of fields in at least some communities; the former system of estate serfdom, where workers had little direct access to the market; and large multiple- or joint-family households under patriarchal direction. In almost every respect, traditional Russian farmers did not follow a smallholder pattern.

Unpaid household members can indeed produce subsistence when the employment of wage labor would be unprofitable, but Chayanov is merely specifying the conditions under which economic decision making took place. Workers increased their per capita labor sufficiently to feed household members, while minimizing the drudgery this entailed. I accept the motivational hypothesis that peasants are rational maximizers of personal or family welfare,[3] but I would insist that intensive cultivators calculate their interests over long spans of time, forgoing immediate benefits such as might come from cash-crop specialization in order to lessen risk in the short term (Cancian 1980). Savings in order to buy land and investment of effort and capital in land improvement are regularly made to secure the interests of future generations and of the elderly. A narrow neoclassical perspective may also deny the ability of peasants to take collective action for shared interests and manage common property at the village level (Popkin 1979). My own experience suggests that communities of farmers can support cooperative institutions for irrigation, grazing, and forestry and can protect their resources from some hypothetical "tragedy of the commons" (Hardin 1961). The dangers of free riders, theft, and mutual mistrust that economists derive from a postulated individual rationality (Little 1989: 34) can be mitigated by institutions for communication, monitoring, and sanctioning in active smallholder communities (Ostrom 1990).

Marx Against Smallholding

The tension between what I knew of the ethnographies of smallholder households and their societies on the one hand and the prevailing schematic, often polemical, categorizations of peasant cultivators on the other became particularly acute for me when I confronted the dominant concepts of Marxism and political economy. Neither the "lineage mode of production" nor the various descriptions of precapitalist and peasant social formations coincided with the systems of land use and social organization of the intensive cultivators I knew. The more aggressively materialistic and doctrinaire the political-economic assertions, the more rigidly evolutionary and abstract were the generalizations. In "primitive" societies, seen as both technologically rudimentary and representing an ear-

3. Those who, as I do, criticize a strictly "cultural" approach to understanding human behavior and institutions are said to be guilty of a simplistic and reductionistic economism. Economism "is the view that the moving forces in individual behavior (and thus in society, which is taken to be an aggregate of individual behaviors or some stratificational arrangement of them) are those of a need-driven utility seeker manoeuvering for advantage within a context of material possibilities and normative constraints" (Geertz 1984: 516). Sahlins makes a similar point when he refers to "the home-bred economizing of the market place . . . transposed to the explication of human society" (1976: 86). While I have tried to understand certain limited kinds of social behavior as they relate to work, household organization, and property rights in economic terms, I offer no comprehensive explanation of "society" nor of culture as "a moving and diversified frame of socially constructed meanings" (Geertz 1984: 513).

lier stage in cultural development, agrarian resources were supposedly not scarce, and neither did population pressure lead to competition. Rights to land were believed to be held communally, and the inequality that derives from private property had allegedly not yet emerged. The household as an important, semiautonomous unit of production and consumption tended to disappear.

In line with Lewis Henry Morgan and other nineteenth-century evolutionists, Marx believed that "in the most primitive communities work is carried out in common, and the common product, apart from that portion set aside for reproduction, is shared out according to current need" (Engels 1884, quoting Marx's letter to Vera Zasulich, as cited in Meillassoux 1972: 145). French neo-Marxist anthropologists (Meillassoux 1981; Terray 1972) have found in traditional African societies a lineage mode of production where descent-group work teams farm collectively, store the produce in communal granaries, and receive food allocated by elders. This may well typify large multiple-family households of shifting cultivators, and lineages may indeed allocate land and provide for territorial defense (Johnson and Earle 1987; Netting 1990), but I know of no cases anywhere where descent groups above the level of the household were the primary social units of production and consumption. Though intensive cultivators may have a variety of reciprocal exchange and cooperative labor groups, and though their communities often administer clearly defined rights in common property, households characteristically farm and eat separately, providing for their own reproduction, and protecting rights in valuable, heritable property.

If I had cherished any expectation that the Kofyar, like many hunter-gatherer groups, were primitive communists, because they had only recently been incorporated in a state or a market economy, their firm insistence on property rights in land and livestock and their autonomous corporate households rapidly disabused me. Given the lack of sound comparative data in evolutionary formulations and the "wistful romanticism" of the nineteenth century, it is perhaps no wonder that Marx "clearly failed to realize the complexity of rights over property, including property in land, characteristic of a primitive agricultural community" (Firth 1973: 36). It is less easy, however, to justify the equation of simple farming technology with lineage productive groups, communal rights to resources, and primitive egalitarianism in an anachronistic evolutionism that is still with us.

It can be argued that I have been self-deluded by ethnographic will-o'-the-wisps, projecting functional integration and timeless stability on contemporary groups of numerically insignificant cultivators who are technologically backward, economically undeveloped, and peripheral to the capitalist world system. For reasons of geographical isolation, folk cultural conservatism, or political-economic exploitation by the colonialist state or merchant capital, the Kofyar and the mountain Swiss may be seen as merely the detribalized or impoverished peasant remnants of previously autonomous, healthy societies. Although I must indeed plead guilty to consciously seeking out groups that have, or recently possessed, a modicum of self-sufficiency, a traditional low-energy tool-kit, and few direct relationships with dominant economic or governmental elites, I have come to feel that groups of smallholders with some essentially similar characteristics exist and persist in a wide range of social formations throughout the world. Their distinctive system of rural population density, intensive land use, household organization of production, and private property rights cannot be consigned to some evolutionary stage. Smallholders may use hoes or ox plows or tractors and live in rain forests or oases or temperate savannas. Their mode of agriculture or horticulture is not regularly associated with one set of political institutions, be it tribe or chiefdom or state (Sahlins and Service 1960; Netting 1990).

Marxist attempts to place smallholders unequivocally on one side or the other of some great historical divide separating use values and accompanying lack of accumulation, capital, and private property from exchange values with scarcity, inequality, wage labor, alienation, and capital (Firth 1975) seem to me to fabricate a Procrustean bed. Though households and village communities may appear inextricably bound up with the practice of intensive agriculture, more inclusive "relations of production" to absentee landlords or tax collectors or moneylenders seem more variable and less determinant (Attwood 1992: 42). I have avoided the "mode of production" designation as well, in part because of the resource abundance and production limited to basic needs implied by the lineage mode (Meillassoux 1981), the domestic mode (Sahlins 1972), the kinship mode (Wolf 1982), and the peasant economy (Chayanov 1966 [1925]), but also because smallholders flourish in such a variety of ideological and political contexts that links between infrastructure and superstructure become tenuous (Legros 1977; Friedman 1974).

It may be foolhardy of me, and it is certainly unfashionable, to question the singular role of capitalism in transforming peasants from small-scale, communal subsistence farmers to market-dependent, economically polarized rural people. But I have trouble finding those intensive cultivators of the Alps, the Low Countries, northern Iberia, or Scandinavia in Charles Tilly's succinct, magisterial characterization of European peasantry:

The peasant version of subsistence farming—in which land-controlling households devote a portion of their production to the market—expanded under the early phases of capitalism and state-making before declining under the later phases of the same processes. Capitalism reinforced private appropriation of the factors of production and gave priority in production decisions to the holders of capital. Thus capitalism challenged the collective use of the land, resisted the fragmentation of rights to the same land, labor, or commodities, and worked against the autarky of the household or village. By the same tokens, capitalism provided farming households with the means and incentives to dispose of a portion of their products for cash outside the locality. These features of capitalism promoted the conversion of a large number of peasants into agricultural wage-workers, pushed another large portion of the peasantry out of agriculture toward manufacturing and services, and gave a relatively small number of peasants the opportunity to become prosperous cash-crop farmers. (Tilly 1978: 408)

The admitted concentration of wealth and power in the factories of the Industrial Revolution, or even in the extensive agriculture of commercial East Prussian grain estates or Spanish sheep farmers of the Mesta (Tilly 1978: 410), may not reflect smallholder social processes under structurally different systems of organization and land use. We need also to explain why capitalist landlords were often unable to dispossess an existing smallholder peasantry. Land use *does* make a difference. Small-scale, intensive agricultural producers are not synonymous with "peasants," "precapitalist subsistence farmers," "petty commodity producers," or "rural proletarians," and smallholders demand their own explanations.

The literature of social science generally defines peasants not so much by what they *do* as by what they *don't* do, and by what is done *to* them. Marx found them a politically inert mass, lacking a consciousness of their own class status and unwilling to join the industrial proletariat in revolution. Modernization theorists have claimed that peasant conservatism and traditional values prevented the technological innovation necessary for land consolidation and economic development. A similarly universalistic dependency theory insisted that the social dynamics of agrarian societies are everywhere the same, varying only to the degree that production is oriented to external markets. Where external forces of world capitalism are overwhelmingly powerful, factors of regional and local ecology and history can have only negligible explanatory significance (cf. Attwood 1992: 12). Anthropologists have viewed peasants as politically and economically subordinate. "It is only when a cultivator is integrated into a society with a state—that is, when a cultivator becomes subject to the demands and sanctions of powerholders outside his social stratum—that we can appropriately speak of peasantry" (Wolf 1966: 11). Peasants "have very little control over the conditions that govern their lives," and the basic decisions that keep them poor and powerless are made outside their communities (Foster 1967: 8). Although I find the study of history as a material social process theoretically trenchant and illuminating, it does not seem to me that smallholders are adequately encapsulated in the "peasant" category of a political economy that primarily analyzes "social relations based on unequal access to wealth and power" (Roseberry 1989: 44). While not denying the elements of political and economic domination that affect many aspects of smallholder life, I contend that we must also examine the ecological relationships of population, agricultural technology, household organization, and land tenure that characterize a distinctive smallholder adaptation to local environment.

Julian Steward's strength was in part that his cultural ecology never tried to explain everything,[4] and the cross-culturally recurring elements of the smallholder pattern smack more of limited, middle-range explanations than of the technoenvironmental determinism of nomothetic cultural materialism (Marvin Harris 1969). But as I looked beyond my own field experience to the ethnographies and histories of peoples as different as the Ifugao, the Aztecs, the Chinese, and the Dutch, recognizable smallholders emerged from the obscurity of evolutionary stereotypes and overdetermined categorizations.

Modernization and Evolution: Smallholder Greening or Withering Away?

Evolutionary schemata combine conjecture about the past with an evaluative conception of the present and speculation about what represents "progress" in the future. It is intriguing that for both the socialists and communists of the left and the free-market capitalists of the right, the agreed-upon path to agricultural development has been the large-scale, mechanized, energy-dependent, scientific, industrialized farm. Smallholders have been universally stigmatized as unproductive, regardless of their yields per unit of land, on the grounds that (1) they use too much labor; (2) they do not produce a large surplus for the market; and (3) they do not make rational economic and scientific decisions about production and innovation. For most of my professional life, I have been content to remain within the conventional anthropological niche, attempting to understand human behavior in small-scale ethnographically specific societies with preindustrial economies. But the dominant evolutionary paradigms of agricultural change, both within and beyond the Third World, have increasingly seemed to be contradicted by the practice, and ultimately the logic, of the smallholder pattern.

Modernization theory that laid out apparently obvious stages of global agricultural development in the flush of optimism after World War II prescribed economic growth through applied science, capital investment, mechanization, a need-for-achievement mentality, and vastly increased

---

4. As Geertz points out, cultural ecology "forms an explicitly delimited field of inquiry, not a comprehensive natural science" (1963: 10).

labor productivity (Rostow 1960). It was believed that foreign aid and developed technology would inevitably (and rapidly) replace traditional, stagnant subsistence cultivation, freeing the poor and "underemployed" rural masses for the urban industrial sector where they belonged. I envision the emblem of this movement as a tractor triumphant on a field of Iowa corn. Its flaws, at least in Africa, had something to do with the Western hubris that ignored the existence of working indigenous solutions to the problems of farming an alien environment. Local ethnoscientific knowledge of soils, rainfall, crop varieties, and pests was not appreciated by outsiders and could not readily be duplicated on experimental farms and in laboratories (Paul Richards 1985). No one seemed to consider the fact that bigger fields, even with machine plowing, might require more seasonal weeding and harvesting labor than a typical household could muster (Baldwin 1957), or that shifting cultivators might be working less and enjoying higher returns per hour than intensive farmers (G. D. Stone et al. 1990). Furthermore, the costs of high-tech irrigation systems and manufactured inputs for rice or wheat in West Africa were far in excess of what the disappointing yields would ever cover (Pearson et al. 1981; Andrae and Beckman 1985).

Where modern Green Revolution technology of high-yielding rice varieties, chemical fertilizer, and improved irrigation was most effectively adopted, Asian systems of intensive cultivation were already operating and the scale-independent innovations were accessible to smallholders. Agricultural economists also began to detail the advantages of household labor in comparison to the opportunity and transaction costs of hired agricultural workers (Binswanger and Rosenzweig 1986; Pingali et al. 1987). The spontaneous and very effective effort of the Kofyar to produce a surplus for the market, using hoe technology and mobilizing labor by household and reciprocal means, suggested that a mechanical modernization model of development was not adequate to understanding this process (Netting et al. 1989; G. D. Stone et al. 1990).

The template of large-scale, monocropped, labor-saving agriculture dependent on fossil fuels was applied with equally uncritical abandon to systems of production in the socialist countries. There, however, the vision of radically reformed farming brought with it an ideological stress on the communal organization of production and the abolition of inequality founded on private property rights. The history of land reform after the Russian Revolution, followed by Stalin's forced collectivization campaigns, the violent seizure of grain, and massive rural starvation suggested the depth of peasant resistance to communist economic policies. The notorious failure of collective farms to allocate the factors of production efficiently and to provide incentives for responsible skilled-labor inputs has been a major cause of the long-term Soviet agricultural crisis (Shanin 1990: 188–205).

Even in the midst of those huge spreads of dry wheat, invariably depicted with ranks of combine harvesters under fair summer skies, the individual household plots of collective members remained significant suppliers of the nation's food. Tim Bayliss-Smith (1982) describes a 3,144-hectare collective farm in the Moscow district with only 7 percent of its 1,700 arable hectares in private plots. Yet over half of all household labor time was devoted to these fields and gardens, achieving yields per hectare that were six times those of the collective area. Energy returns on each calorie of input were estimated at 11.2:1 as opposed to 1.09:1 on the heavily mechanized collective farm. Although the private plot was not technically owned by the user, it resembled a smallholding in being near the house, receiving constant care and attention, being heavily manured, and supporting diversified plant and animal production. Decisions concerning these small operations were made by the household, with women playing a more important role than men. Such management differs decisively from that of the collective, where "any response by farmers to signals from their environment must, in all important respects, be made with reference to instructions from a remote bureaucracy, instructions which cannot foresee all the local vagaries of weather, disease, and soil conditions" (Bayliss-Smith 1982: 97). The inherent contradiction between smallholder efficiency and communist ideology has at last been recognized by Gorbachev's call for the freeing of Soviet farmers from the state-run system of collective agriculture. "Comrades, the most important thing today is to stop the process of de-peasantization and to return the man to the land as its real master," Gorbachev has been quoted as saying (*New York Times*, October 14, 1988). The return of private property in some form is a foregone conclusion.

When I began to think tentatively about the extension of some general characteristics of smallholder farm households to cultures with long traditions as complex civilizations, the case of China was both attractive and problematic. Even the casual student (and I am no sinologist) is aware that Chinese agriculture uses methods of double-cropping, controlled irrigation, fertilization, and terracing in a highly productive system of wet-rice agriculture, and that China supports an unusually dense rural population. Given a continuous, documented history, I inferred that the tools and techniques of intensive agriculture and perhaps some indications of labor organization and land tenure could be investigated over time. Recent work on Chinese agricultural history (Hsu 1980; Chao 1986; Anderson 1989), on sustainable energy input/output measurements (Wen and Pimentel 1986), and on the distinctive features of comparative Asian wet-rice economies (Bray 1986) suggested the exciting possibilities for such an investigation. It was the massive, politically inspired, and centrally planned institution of agricultural collectives in the People's Republic of China in the 1950s that really piqued my curiosity. If an archetypal system of intensive cultivation could indeed be carried on by extra-household work teams, if harvested goods and income could be shared equally, and if private property could be replaced by communal control of pooled land and livestock, my smallholder speculations would be just another romantic anthropological Just-So story. Marx's primitive communism might in fact have a dramatic modern analogue, and there might already exist a bountiful, egalitarian future at work on the farm. According to many popular and scholarly reports, to propaganda, and to theoretical analysis, collective agriculture had succeeded.

It was only in the mid-1980s that I began to hear of an official retreat from communes, brigades, and work teams and their replacement by a national agricultural policy of *baogan daohu*, or the household responsibility system. Households were given use rights on collectively owned land, and they contracted to fill quotas for delivery of certain products at fixed prices to the state in return for the right to dispose of their entire surpluses on the free market (Perkins and Yusuf 1984). An administrative change initiated in 1978 encompassed the majority of Chinese villages by 1982 and was all but universal by 1984 (Smil 1985; Hartford 1985). Linking economic rewards directly to output and encouraging household initiative, innovation, investment, efficiency, and risk-taking explosively raised production and the rural standard of living, effecting what has been called "the most far reaching and orderly socioeconomic transformation of the 20th century" (Smil 1985: 118). An unparalleled experiment in the social engineering of agriculture had been reversed, and I resolved to follow the smallholder story to China.

Any discussion of smallholder household farming worldwide raises expectations that it cannot fulfill. The tendency of many Americans is to categorize the issue as the familiar one of the family farm, whose demise at the hands of agribusiness, a national government of subsidies and controls, a volatile land market, the agricultural-research establishment, and international commodity trading has been heralded for decades. Values ranging from the agrarian democracy of Thomas Jefferson to the virtues of raising hard-working and pious children in the salubrious country air have been enlisted in this debate, and we are now entering a decade in which questions of the mythic economies of scale (Strange 1988), stewardship of the land (Berry 1987), and sustainable systems (Francis and Youngberg 1990) will be rightfully (and righteously) publicized.

I once thought that the plenitude of North American land, its frontier history of relatively sparse rural population, its marvels of technology

TABLE P 1
Agricultural Output in Six Countries, 1880 and 1970

	Hectares per male worker		Output per hectare		Output per male worker		Fertilizer (kg/ha) (1970)	Workers per tractor (1970)
	1880	1970	1880	1970	1880	1970		
United States	25	165	0.5	1	13	157	89	1
England	17	34	1	3	16	88	258	–
Denmark	9	18	1	5	11	94	223	2
France	7	16	1	4	7	60	241	3
Germany	6	12	1	5	8	65	400	–
Japan	1	2	3	10	2	16	386	45

SOURCE: Boserup 1983: 401. Used with permission.

NOTE: *Hectares* refers to agricultural area—that is, to all land on farms. *Output* refers to output of both crops and animal products (excluding fodder consumed by the farm animals). This output has been recalculated in wheat units, equivalent to one ton of wheat, by Yujiro Hayami and Vernon Ruttan (1971). *Workers* includes adult male workers, but not women and children. Kilograms of chemical fertilizer are measured in fertilizer content per hectare of arable land—that is, agricultural area minus pasture and fallow.

and science, and its cheap power made most considerations of small-holder intensive agriculture here, and by extension in any modern state, either trivial or anachronistic. I was wrong. Quantitative comparisons among leading industrial nations show that the logic of population density and agricultural intensification distinguishes types and defines trajectories through time (see Table P 1). Since 1980, U.S. male workers have greatly increased the areas they farm and their output, but output per unit of land has not kept up with that of Denmark, France, and Germany, and is only one-tenth that of Japan. Surprisingly, fertilizer per hectare gives the same picture of extensive U.S. as compared to intensive European agriculture. Although productivity grew in all modern nations from 1880 to 1970 (Ruttan 1984), Japan and Germany have raised land output, while the United States has pushed up labor output (Fig. P 1). Secular trends in the relationship of population to arable land, the cost of energy, and the dangers of erosion, chemical pollution, and declining fertility point to more intensive and sustainable methods of land use on an Asian or European model as an inescapable necessity even for the United States. We ignore the proven advantages of household labor, management, ownership, and inheritance at our peril.

My purpose in this book is not to describe U.S. agriculture or prescribe some illusory smallholder panacea. The adaptations that will surely come with changing energy prices (Pimentel 1973), government subsidies (Strange 1988), water and soil depletion, and new models of alternative agriculture (National Research Council 1989) have legions of able and passionate expositors. The large class of self-provisioning family farmers that existed in the United States before World War II has almost disappeared in a sea of specialized commodity production, and "by the

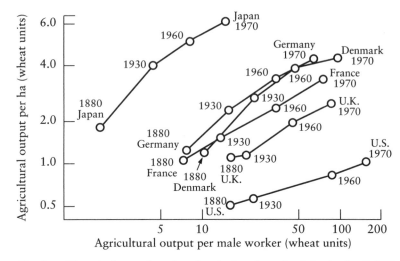

Fig. P 1. Historical growth paths of agricultural productivity in the United States, Japan, Germany, Denmark, France, and the United Kingdom, 1880–1970. (From Ruttan 1984: 109)

1980s the farm household had become virtually isolated from the farming operation" (Adams 1988: 467). Where facets of the old smallholder system remain, as among the Amish, in the diversified, family-descended, "yeoman" operations of Illinois (Salamon 1985), and in part-time and retirement farms in Georgia (Barlett 1987b), I shall note them. But it is in the Third World, with its sweeping dynamics of population growth, market expansion, and agricultural change, where an understanding of the smallholder pattern may be most relevant.

Though I never became the applied anthropologist and secular missionary that I expected to be when I entered graduate school, my concerns with the cultural ecology of cultivators, and more generally with social change, have brought me into repeated contact with students of development. From the vantage point of an occasional consultant, I looked at the economics of rice production in Senegal and the Ivory Coast, the proposed dams on the Gambia River, and the small farmers in northern Portugal on the eve of EEC membership. Perhaps even more important was my long-term view of the indigenous development of the Kofyar as they entered a market economy with almost no outside planning, extension, new technology, or credit (Netting et al. 1989). Reports of the death of the smallholder in a modern high-tech, large-scale world have proved to be vastly exaggerated. Indeed, scarcity of rural resources and national demands for food production create just those circumstances in which agriculture intensifies and the household organization of production demonstrates its comparative advantage. But in the shadow of proliferating industry, bureaucracy, and education, smallholders often become invisible or embarrassing.

The fact is that Asian smallholders, with an assist from Green Revolution methods, have astoundingly kept ahead of food demands, side-stepping the sinkhole of involution in Java, and liberating unimagined effort and productive enterprise in China. Where state policy has allowed agricultural prices to escape the controls of marketing boards, and where inputs like fertilizer are no longer left to the (mal)distribution of parastatal firms in Africa (Bates 1981), rural people have usually responded with a deluge of food. Where conditions of population density and market relationships are conducive to intensive, sustainable production, we can reasonably expect smallholders to rise phoenixlike from the ashes of thousands of collective farms. Even the indolent estates and deforested ranches of Latin American oligarchs may someday become the scene of much higher yielding gardens and orchards (Anderson 1990) if they are portioned out to solvent, experienced peasant households by effective land reform. I believe that intensive agriculture by landowning smallholder households is economically efficient, environmentally sustainable, and socially integrative. This book is an attempt to use the evidence of social science, the logic of practical reason, and the personal conviction of a garden-variety ethnographer to show how the smallholder works.

CHAPTER ONE

The Technology and Knowledge of Intensive Farm Practices

FARMERS MAKE NUTRIENTS, water, and sun available to the plants and animals they nurture, and they do this as simply and efficiently as they can. The tools they use may be nothing more than axes, pointed sticks, ropes, and knives, but their knowledge of their environment and the species they grow for food and fiber is complex and systematic. Hunter-gatherers may strew a few seeds onto their nitrogen-rich trash middens and harvest what has sprouted when they pass the camp site again. Paiute seed collectors even dammed tributaries of the Owens River in California to encourage production of the wild grasses (Steward 1938). More serious food producers need only to clear away competing vegetation so that the sun can reach their cultivars. Fire can then convert the nutrients stored in the slashed brush and felled trees to ash, making the nitrogen, phosphorus, potassium, and other mineral elements that food crops need readily available. A sharp dibble stick punches holes for the seeds, and if the burn has been good, competing weeds and insects may be wiped out, at least for the year or two during which the swidden field produces. Permanent cultivators may not necessarily have a more elaborate tool kit, but they have to know more about agriculture and expend more time, energy, and skill than those with the luxury of more abundant resources.

In intensive agriculture the task is not so much to tap naturally existing sources of plant and animal nutrients, water, and sunlight as to increase their supply to support more biotic growth, to maintain the proper conditions over longer seasons and more years, and to replenish and regulate the supply of those elements that are exhausted. The signs of permanent agriculture in a landscape are familiar and unmistakable. They involve (1) moving and manipulating soil to feed and foster plant growth and to control erosion, as in deep tilling, ridging, and terracing of permanent fields; (2) regulating water by increasing its supply through irrigation or removing an excess through drainage; (3) restoring or increasing soil fertility by systematic manuring, usually involving the stall-feeding and fencing of livestock, production of crops for fodder and green manure, and the collection and processing of household wastes; (4) diversification of production with a wide variety of cereal, legume, tuber, vegetable, fodder, and tree crops that are interplanted, rotated, and scheduled according to existing microenvironments and seasonal conditions, and with a range of large and small domestic animals; and (5) protection of plants and animals from growth-inhibiting competition or predation by weeds, diseases, insects, and other pests through guarding, fencing, and reducing exposure.

These practices, though often displaying remarkable ingenuity, clever management, and disciplined diligence, do not reflect unique inventions with a particular geographic spread and a history of diffusion. When people are faced with the problem of farming continuously on sloping terrain, they make terraces, whether they live in widely separated parts of Africa (Floyd 1965, Ludwig 1968), the Americas (Donkin 1979), or Southeast Asia (Spencer and Hale 1961). It is true, on the other hand, that the plow has not been infinitely reinvented, and the introduction of its wheels, its moldboard, and the traction furnished first by eight oxen and subsequently by a tractor can be placed historically with some accuracy (Lynn White, Jr., 1962). But worldwide the goals of intensification in modifying the physical environment to achieve higher, more reliable, and more sustainable production show a compelling similarity of ends gained by an amazing multiplicity of means. Intensive agriculture as practiced by smallholders has no single origin or history, and we are deluded if we tie its multistranded development to any model of unilineal technological development. Digging stick, stone hoe, wooden and then iron plow, and the John Deere tractor provide an evolutionary scheme that is obvious and satisfying, but they do not reflect stages of intensification. One can use each of these tools either in a manner that economizes on resources and energy or for food production that is extensive and wasteful. Intensification, as I shall suggest, is alive and well, and it does not require a technological breakthrough to introduce it to a waiting world.

The Nigerian System of Permanent Subsistence Cultivation

The agricultural technology of the Nigerian Kofyar (Netting 1968: 55–107), with its short-handled Sudanic hoes and metal-bladed axes, knives, and sickles, still strikes outsiders as backward, primitive, and inefficient. But without the plow, animal traction, or any of the modern mechanical or chemical inputs that distinguish Western labor-saving cultivation, the Kofyar have managed to produce their own subsistence from scarce, permanently tilled land resources for generations. Their shovel-shaped hoe blades can cut deeply into the soil or scoop and pile dirt into the ridges and mounds that make a bold relief on their fields. Where topsoil is thin and can be leached by the tropical rains of the savanna, old ridges are broken in the middle, with half the earth turned over into the furrow to the left and the other half to the right. The new ridges, 20 to 40 cm high, both bring to the surface lower soil levels and concentrate the soil and decomposing organic manure as a substratum for the crops planted on the ridge or already sprouted in the previous season's furrow.

Ridges, oval mounds for sweet potatoes, and conical heaps for yam tubers all elevate the growing plant from water and allow drainage after heavy wet-season rains. The Kofyar traditionally carried their ridging a step further, hoeing the homestead field of millet and sorghum after the early millet harvest in July or August. The eight- or nine-foot millet stalks were buried under a pattern of rectangular tie ridges around enclosed basins, giving a waffle-like surface to the field. This elaborate tillage procedure (1) banked up and supported the sorghum plants in the same field, which would not be ripe for another four or five months; (2) removed weeds and created beds for the cowpeas that had just been planted; and (3) hollowed out enclosed basins that would catch the water from the second peak of the rains, allowing it to soak in near the roots of the crops and preventing the water from carrying the precious topsoil away.

Ridges alone are excellent erosion protection where the land is reasonably level, as on the plain below the Jos Plateau, but on the slopes of the escarpment and in the highland valleys, a cleared field could suffer significant damage in a single six-month season of 50-inch rainfall. The Kofyar meet this challenge by constructing terraces with dry-stone walls along the contours of their hills. Benches of level, stabilized earth are then formed into parallel or rectangular ridges for the crops (Netting 1968: plates IIIA, B, and C). Large terraces with oil palms in kitchen gardens around homesteads, and the smaller, more haphazard walls in bush-field areas that may be periodically fallowed all suggest considerable amounts of planned, organized labor. The Kofyar now farming them do not claim to have built all of their own terraces, and the serried hillsides are probably the work of generations. Terraces that I saw being built were often constructed by individuals moving an old wall and hoeing the soil down to make a wider surface. It was a part-time or a dry-season activity, and there were no social mechanisms beyond the village exchange-labor and beer-party groups to provide large groups of workers. But terraces do require vigilant maintenance, and when a wall develops a fissure or collapses, the unanchored soil may begin to move rapidly down-slope, endangering the benches below. Intact terrace systems that I saw being farmed in the 1960s were scarred by erosion in 1984 after the owners had deserted their hill farms for larger, more level fields in the Benue Valley. Once the pressure that keeps dense populations in the hills is relaxed, whether by removing defensive needs or lowland diseases, the extra labor of terrace building and maintenance may become superfluous, and an orderly artificial landscape may dissolve into silt and flooding.

Rain-fed Kofyar agriculture has not yet been faced with the necessity of irrigation, though water control has always been a problem. Ridges and basins may trap water, but a low or marshy section of the same field may threaten waterlogging. Here the farmer constructs mounds in parallel lines along the contour, and the depressions between the mounds disperse the water and gently drain away the excess without allowing entrenched gullies to form. Some springs are now tapped to moisten little fenced gardens of pumpkins, grown for their leaves to feed pigs in the dry season. A cracked pot with a slow leak may be rigged to supply drip irrigation to a single tomato plant in a corral wall. An entrepreneurial Kofyar farmer on the frontier heaped up earthen dikes around a shallow valley to provide controlled dry-season water and wet-season drainage for his commercial banana orchard in the streambed. Other Kofyar take advantage of slow-draining natural swamps to grow rice. Many households are now paying to have wells dug for domestic water, increasing the potential for some dry-season irrigation. A knowledge of the uses of irrigation is not lacking, and it may yet become a more important tool of Kofyar intensification.

Perhaps the most unique element in Kofyar agricultural technology is their means of restoring and in part creating fertile soil, which involves the transferring of organic material from the wild, the fields, and the household to domestic animals, and the collecting of composted manure that can maintain the fertility and soil structure of permanently cultivated plots. During the nine-month growing season, the Kofyar confine their goats to a circular, stone-walled corral about 3.5 or 4.5 m in diameter, standing at the entrance to the homestead cluster of round, thatched huts. The goats are individually tethered and fed with grass cut in the fallowed fields around the village, or, in the latter part of the dry season when the bush has been burned, from the leaves of tree branches cut for the purpose (Netting 1968: plate VIIB). They may also consume crop leaves and residues. A daily chore, often for women or children, is fetching the goat fodder and hauling their drinking water from the stream. The animals are fed more than they eat, and a layer of compost or mulch one to two meters deep accumulates over the course of the year, while dung and urine also build up in the stable hut where the goats are kept at night. Just before the coming of the rains in March, this manure is dug out, loaded into baskets, and dumped in piles on the nearby homestead fields, which resident family members also use for elimination. In addition, during cultivation, weeds are turned under, and millet stalks are systematically buried under ridges. Every traditional homestead has a miniature hut near its entrance where ash from cooking fires is stored. Both fuel wood from the bush and sorghum stalks contribute to the ash, which is applied by the handful to peanut plants. Fertilization processes thus both recycle organic matter directly from the farm back to the soil and add to it by bringing grass and forest products to the homestead and transforming them through the digestive tracts of animals and people, as well as by fire.

The major strategy for combining high production per unit area with risk reduction and sustainability in Kofyar agriculture is diversification. This entails the careful fitting of crop complexes and techniques to existing microenvironments, the interplanting of mixed crops, and systems of crop rotation. In the early days of my field research, I was struck by the fact that most Kofyar homesteads had interplanted millet, sorghum, and cowpeas as their primary crops, while in Bong village, where I lived, the crops were maize and coco yams, with lima beans on trellises along the paths. After months of puzzlement on my part, a Kofyar farmer patiently explained to me that Bong's placement near the edge of the scarp meant that it had more rain and mist than neighboring villages. Maize and the elephant-ear coco yam or taro tubers flourished best with extra moisture, while the savanna cereals of millet and sorghum were adapted to drier, sunnier locations. These conditions could vary within a kilometer, and people maximized homestead production by matching their crop association to the environmental potential. Similar observations led Kofyar migrants on the Benue Valley plains to plant coco yams in the midst of a millet/sorghum field where an African locust bean (*Parkia filicoidea*) or other economic tree cast dense shade. Differences between the deep, manured soils of the homestead fields and the thinner, sandy soils of periodically fallowed bush fields also necessitated a different choice of crops. Acha (*Digitaria exilis*), a domesticated relative of Bermuda grass with tiny white seeds, could reach maturity in the stony bush fields that would not support larger grains, and peanuts would produce there as well. Bush fields in several locations also decreased the risk of crop failure or loss to marauding baboons in any one place.

Planting several crops together in mixed stands rather than monocropping is a frequent device of both intensive and shifting cultivators. Millet and sorghum, unlike maize, are drought-resistant and can go temporarily dormant during a dry spell. But the fact that millet ripens in about four months, while the sorghum planted in the same field at the same time takes almost eight months to reach maturity, means that they may be differently affected if the rains fail. With poor early rains, millet yields go down, but sorghum can still produce effectively. The grain crops, combined with cowpeas near the ground, also provide a stratified cover, protecting the field surface from pelting rain and direct sunlight that can contribute to soil degradation. Interspersed occasional plantings of pumpkins, okra, bitter tomatoes, and tree cotton had a similar effect, and also interfered with the spread of insect pests that often feed on specific crops and multiply most rapidly when they can move through pure stands.[1] Kofyar homesteads are also dotted with trees that produce important food items, such as palm oil, locust beans from the carob, and leaves that can be harvested for goat fodder. Tilling manured fields under the trees increases their production, while the trees further stabilize terraces and bring up nutrients with their root systems that would otherwise be lost to the crops. Studies among the Hausa have demonstrated that interplanting, though it may decrease the yield of any single crop component from that achieved in a pure stand, raises the total productivity of the field, reduces its variability and thus the risk of total crop failure, and allows labor to be applied to several crops and their specific needs at the same time, thus increasing labor productivity (Norman, Simmons, and Hays 1982).[2] Tending to several densely interplanted crops at various stages of growth is not difficult with hand tools, but the Kofyar explain their reluctance to use draft animals and tractors by pointing to the need to cultivate relatively widely spaced, even rows of single species if mechanical means are to be used effectively. They claim that if they monocropped, total yields would in fact be reduced per unit of land, risk would increase, and money costs would be higher.

Both crop rotation and interplanting with nitrogen–fixing legumes can aid in the maintenance of soil fertility. The Kofyar use cowpeas, various beans, and peanuts for this purpose. On bush fields, they may grow an early crop of peanuts, and when they harvest the peanut plants by hoeing them up, they simultaneously ridge the field for late millet transplanted from a nursery. The extra work of starting the millet in a seedbed and moving the plants in time to take advantage of the later rains means that two crops can come from the same field in a single season. In the next year, the field may be planted with the less-demanding acha, before reverting in a third year to the peanut–late millet succession. When opening new fields from the forest on the Benue plains, the Kofyar often planted the roughly cleared and burned area with sesame (beniseed), to be followed in the next year by yams, which require very fertile soil, and then by the traditional millet/sorghum crops. Systems of rotation combined

1. Intercropping is a primary means both of intensifying land use and of reducing the risks of harvest failure in Africa (Richards 1983). "When exposed to stresses such as drought, flooding, insect attacks, disease, and high temperatures, total crop yields will decrease relatively little in mixed plantings of diverse traditional varieties compared to monocultures of genetically uniform industrial varieties," note David Cleveland and Daniela Soleri (1991: 286).

2. Planting cowpeas among sorghum plants, as the Kofyar do, or intercropping maize and beans with large-leaved squash vines, using a widespread New World pattern, creates a "living mulch" that suppresses weed growth. Such an association can reduce weed growth as much as several hand weedings (Altieri and Liebman 1986).

with interplanting and manuring may reduce and finally eliminate fallowing, allowing a field to be permanently productive.

Like all types of intensive cultivation, the Kofyar farming system is oriented to reducing the threat of competition and predation. The farmer's success in generating a fertile, friable medium for plant growth and in regulating access to water and sunlight means that unwanted natural plants (i.e., weeds) are also encouraged. Weed removal, either by uprooting them with a hoe or pulling them up by hand, is an extremely important task of the early growing season. Weeding may be done in the homestead field as many as three times, and the labor is too great to be consigned to women alone, as is the case with many shifting cultivators. Adult men weed, and indeed most jobs on the farm are shared between the sexes with considerable equality. It also requires a discerning eye to distinguish between a shoot of young acha and competing grasses. Intractable weeds like imperata (*Imperata cylindrica*) are signs of declining fertility and the need for more vigorous manuring of the field. Plant parasites like striga (*Striga hermontheca*), which grows on sorghum roots, can only be combatted by deep tilling.

The Kofyar do not fence their fields, but they either confine livestock like goats and pigs during the entire growing season or closely herd sheep and cattle. The small homestead field, often encircling or directly adjoining the living compound and its courtyard, can be supervised so that birds and antelope or rodents can be driven off. Dogs are on guard, and chickens are protected from hawks by dense stands of hemp adjacent to the house. Homesteads are usually clustered away from forested areas, and large game that might damage the gardens cannot move through the settled area without attracting attention. After the crops have been harvested, the fields are bare of underbrush, and grain is either bundled on the courtyard drying rack or sealed in mud-walled granaries like giant pots inside the huts. Livestock and crops are too hard-won for the Kofyar to slacken their vigilance, and they cannot make up for agricultural losses by relying on the scanty wild foods in their area.

A Swiss Alpine Dairy Farming System

In the very different physical environment of the Swiss Alps, German-speaking peasants practice an agro-pastoral mode of livelihood with a technology very different from that of the Kofyar, but similarly aimed at the intensive use of scarce resources (Netting 1981: 10–41). Their reliance on dairy cattle fed over the long winters on hay harvested from natural meadows means that working the soil is confined to grainfield and garden areas (Fig. 1.1). Gently sloping areas with occasional stone terrace walls were formerly plowed in June and August in the alternate years of fallow

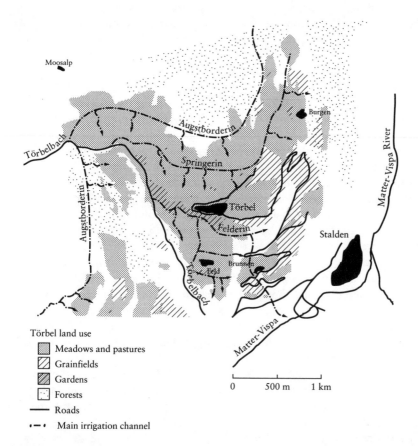

Fig. 1.1. Land use and irrigation in Törbel, Switzerland. (From Netting 1974a: 68; and 1981: 4, 11)

and then sown broadcast with winter rye. The wooden-wheeled ox- or mule-drawn plow used until the 1920s had an iron-pointed share that scratched the soil to a depth of four or five inches and a simple moldboard that turned the furrow. Smaller or steeper plots are still tilled for potatoes or broad beans with a long-handled broad hoe or a bent-tined fork. To combat erosion in these canted fields, the Swiss fill pack baskets with dirt from the bottom of the slope and laboriously carry it back up to the top every spring. Tiny vegetable gardens planted with turnips and cabbages are made of well-hoed, raked earth in raised beds. The most extensive terracing is on stony slopes near the river, where vineyards are tended with hand tools.

Because the village of Törbel in Valais Canton has only about 20 inches of rainfall a year, making it one of the driest areas in Switzerland, the technology of irrigation has for centuries been the key to intensive production. Meadows that can produce two crops of hay, plus some grazing, constitute 56 percent of the arable land. Without regular artificial watering, only one meager hay cutting would be possible. The system of irrigation still in use appears unbelievably simple. Water is tapped by damming a mountain stream and channeling it through stone-lined ditches, which one can step across with ease. Movable steel plates in concrete boxes (replacing the old wooden sluice gates) direct the water into subsidiary channels, and from one of these the irrigator captures the flow with a temporary dike of mud and stone. As the water travels along the meadow contour in a feeder ditch, thin stone slabs are used to block the trench, causing the water to spill over the edge and cascade downhill. The stone is moved so that each transverse section of the meadow can be soaked in turn. It is a task best done by two people working together, so that one can relay the message that the water has reached the foot of the slope. The only tool used is a mattock, and these, along with spades, are wielded by groups of the men with fields in the area served by a single off-take channel, who cooperatively clean and repair the ditches in the spring. In the past, some meadows were reached by conduits of split, hollowed-out logs, and today rubber hoses and rotating sprinklers may water potato fields that could not be irrigated in the past. When water had to be brought in from some distance, canals were constructed with considerable engineering skill. Törbel, evidently having reached the limits of its own water supply, purchased rights in the Embd brook in 1270, and with the neighboring village of Zeneggen built the ten-kilometer Augstbord water channel, which is still called *die Niwa* (the New One). This perhaps doubled the irrigated meadow area of the village, carrying it up to almost 2,000 meters. There were also devices for collecting small sources of spring water. These amounted to little more than an excavated pool with a central drain that could be opened with a log plug whenever the pond had filled to capacity. Each one was operated, like the stream-fed systems, by a society of users who had their own rules and constitution.

Though such relatively small-scale gravity systems of water distribution are not mechanically complex, the organization of water sharing is intricate and well regulated (Stebler 1922: 67–83; Netting 1974a). For maximum use to be made of the water, it has to be distributed day and night on a predictable schedule and at intervals sufficient to maintain growth of the grass. Each privately owned meadow carries with it rights to water on a specific day of the 16-day cycle. The duration of a water-use period is based on the movement of sun and shadows on the surrounding landscape—for example, *Tagaufgang* (sunrise), *Dreifurren* (shadow reaches the three terraces), *Mittag* (midday), and *Schattiwasser* (shadow reaches the river). The owners of neighboring meadows may rotate their periods so that one will not always fall during the daytime and another at night, as well as to compensate for the changing lengths of the daylight periods as summer days become shorter. Close coordination is necessary to ensure that no irrigator takes more than his share, and that someone is always ready to take over the water flow at the proper time. If the water were allowed to flow uncontrolled through the system, or if an undetected break developed in a canal, a landslide of water-soaked earth might severely damage the down-slope fields. The society of users managing the Augstbord canal hired a *Wasserhüter* (water guard) to patrol the channel every day for breaks, and users who stole water or did not adequately control it were subject to judgment and punishment. Irrigation water was so valuable that it was known as *das heilige Wasser* (holy water), and it could not be wasted.

Organic material to maintain the fertility of Törbel meadows and gardens had to be collected, processed, and distributed in an equally painstaking manner. The Swiss systematically appropriated the humanly inedible products of their fields and forests, concentrated and enriched them in composted livestock manure, and returned them to the soil. Over half the farmed area of Törbel was in irrigated hay meadows, from which the average householder cut with a scythe and packed into barns more than 2,200 kg of dried, tightly tamped-down grass (Netting 1981: 39). The dung of the cattle, sheep, goats, and mules that were fed on this hay for about seven months each year was retained by their bedding material, which was frequently renewed in their stalls.[3] Dry rye straw remaining from the threshing of the major cereal grain was the preferred stall bedding, but until recently, when baled straw trucked in from elsewhere in Switzerland became available, there was never enough for the purpose. Whole families went to the communal forests above the village and into steep ravines to scrape up the litter of conifer needles, fallen leaves, moss, and humus and bring this back in large pack baskets to their barns. Eight to sixteen cubic meters of this *Waldstreue* might be collected, and some observers felt that the practice might be detrimental to normal forest growth (Stebler 1922: 98). The stall bedding retained the liquid portion of animal wastes and helped to prevent loss of its nitrogenous components.[4] Manure removed in the frequent cleaning of the stall was neatly stacked in front of the barn, where it formed a visible gauge of the

3. An estimate of total weight of animal excreta and stall bedding per year (in metric tons per 500 kg live weight) is horses 12, cows 15, sheep 9.75, pigs 18.25, and hens 4.75 (Wilken 1987: 49).

4. Animal dung and urine that is not contained and composted loses nutrients as the organic material dries up and blows away. Absorptive stall bedding such as straw provides

farmer's prosperity. In the fall it was loaded into mule panniers or pack baskets and deposited on the meadows. After the winter snows had melted, the residue was raked up, dried, and returned to the barn for bedding—a triumph of labor-intensive recycling. The composted manure not only promoted the growth of the hay but is also reputed to keep down broad-leaf weeds and other undesirable plant species (Netting 1981: 47). To increase the fertilizing potential of cow manure, the Swiss even rearranged the dung left by the herds on the alpine pasture over the summer. As part of their required labor contribution to the maintenance of the communal alp, Törbel villagers swung pitchforks to break up and scatter the cowpats more evenly in the fall.

Manuring of potato fields benefited not only this important tuber crop but also rye, which was often rotated with potatoes in the same field. Wine grapes also require fertilization, and this meant carrying loads of composted manure to vineyards far below the village. Törbel still maintains a communal vineyard for the wine used on civic and ceremonial occasions, and each household is responsible for furnishing a given amount of manure, which a village official checks off when it is delivered.

The diversification of agricultural production in mountain regions responds powerfully to altitude, which affects such climatic factors as exposure to sun, winds, evaporation, humidity, precipitation, and, above all, temperature (Viazzo 1989: 16). Because of the steep slopes of the Visp Valley, in which Törbel is located, the village territory includes lands from about 800 meters above sea level to a mountain peak almost 3,000 meters high (Fig. 1.1). Tiny, gravelly terraces near the river can support grapevines, but above 800 or 900 meters, there is not enough warmth to produce wine grapes. Rye and some barley, oats, and wheat could be grown in the past on glacial terraces up to about 1,600 meters, and potato fields might occupy the same areas, as well as smaller garden patches as high as 1,900 meters. Vegetable gardens clustered below the village church at 1,500 meters, and the dominant irrigated hay meadows stretched all the way from near the river to the edge of the forest at approximately 2,000 meters. The forest and the grassy basins and slopes above the timberline at 2,200 meters have a growing season too short for crops, and they are devoted to summer grazing by cattle, with sheep finding sparse forage as high as 2,600 meters. Vertical zonation does not neatly divide up the environment by altitude, because cliffs, gorges, and areas of thin soil intersperse wastelands or rough grazing among the meadows, and the north-facing or "shadow sides" of the mountains have fir and spruce at lower altitudes than the warmer, larch-covered "sunny

---

carbon and air for microorganisms that can store the nitrogen in their bodies (Cleveland and Soleri 1991: 171).

sides." Much of Törbel's territory is inclined toward the south, and its favorable location seems to have supported an agricultural population since the Late Bronze Age, before the Christian era (Netting 1981: 8).

In order to use this wealth of contrasting microenvironments intensively and at the same time provide secure and relatively complete household subsistence, Törbel farmers needed access to resources at several levels and barns, granaries, and dwelling houses allowing convenient processing and storage of foodstuffs and fodder. An independent peasant family required vineyards, grainfields, meadows, gardens, and pastures in the appropriate zones, and there were good reasons for having several differently situated parcels of each type. Because grass in the lower meadows allows earlier spring grazing and matures first as hay, it is useful to own some of these. Successively higher meadows mean that haying can be spread out as the season advances, and the family may work up the slopes from July through the beginning of August before returning to the lowest meadows for a second cut. The labor-intensive reaping, raking, drying, and back transport of the great bundles of hay is thus not concentrated in a narrow window of time. Meadows may also be sited in several of the three irrigation areas, which have somewhat differing water availability, and different years may threaten grainfields with frost, wind damage, or arid conditions according to their location. Risk reduction as well as labor scheduling were thus prime considerations in the dispersion and parcelization of a family's holdings (Bentley 1987).

The nature of alpine crops and indigenous plants, as well as limitations of temperature and season imposed by altitude, encouraged diversification that utilized significantly different ecological niches and several spatially distinct spots in each zone. Instead of intercropping and complex rotations on particular fields, the Swiss used grasses that flourish best in pure stands, such as planted rye and self-seeding meadow grasses. To take advantage of the less abundant and more widely dispersed forage of the higher altitudes, they moved their grazing animals to the summer alp and herded them to take advantage of the immediate potential of specific pasture areas. They relied jointly on animal products that could be consumed immediately, like milk, or stored, like cheese, butter, and dried meat. The highly diversified subsistence base also provided vegetable products such as cereal grains, potatoes, broad beans, sauerkraut, schnapps from the fruit of apple and cherry trees near the houses, and wine from the vineyards, all of which could be stored. Climatic variation was buffered by the very different environmental requirements of crops, the mobility of livestock, and the technology of storage that conserved cheese, grain, and wine over several years in cellars and locked log structures set on mushroom-shaped pillars to protect them from vermin. Even the high, rugged forested areas were a long-term resource, specialized in the production of firewood for warmth and cooking and of timber for construction. That a single community could encompass and afford access to this diversity of products meant that a relatively dense population in an area of constrained biotic potential could maintain a remarkable degree of security.

Despite the appearance of difficult, marginal subsistence potential in the Alps, agricultural change could increase the potential of this area and allow more intensive exploitation. Grain crops are not well adapted to high mountains, and even hardy winter rye may yield poorly if spring is late and an overcast, rainy early summer retards ripening. There is evidence that climatic conditions of this type led to widespread harvest failure in the latter part of the eighteenth century. It was at this time that the Swiss, along with other European farmers, were able to broaden their subsistence base by the adoption of the potato, an American tuber crop domesticated in the Andes, where environmental conditions resemble those of the Alps. In Törbel, potatoes could produce almost 15 times as much as rye from the same area by weight and 3.3 times as many calories (Netting 1981: 163). Moreover, growing potatoes did not mean that grainfields had to be sacrificed. Previously, rye had been planted every other year in a field, but it was found that potatoes could occupy the fallow period and thus raise the total production of the land. They could grow in poorer soil and at higher altitudes than rye, could occupy small patches of land too steep to plow, and resisted the hailstorms that devastated grain crops. Potatoes, which are rich in carbohydrates, could supplement a diet that already had good sources of proteins and fats in its dairy products. The new food crop, which rapidly became a regular item of daily consumption, increased the support capacity of the existing land, and in Törbel and similar European alpine communities, it may have been instrumental in sustaining a doubling of population over the next hundred years (Netting 1981: 164–68; Viazzo 1989: 212–14, 269).

Protection of agricultural plants and animals among the Swiss was less a matter of inhibiting the growth of competing organisms, as in the tropics, than of sheltering livestock from the elements and keeping domestic animals out of the crops. The lush individual meadow plots, which averaged only about 1,300 square meters each, were too small to divide up with fences or hedgerows. There were a few wooden fences along heavily used paths, but animals were usually either closely herded when they were in transit to other barns or pastures or tethered to a peg and moved about in a meadow of their owner's. Today, temporary pasturage is defined by portable electric fences, powered by batteries, which are strung up to enclose small areas. Because of the long winters, all livestock must be confined to barns or stables for warmth and proximity to their stored fodder. Substantial log buildings with roofs of local slate ring the Swiss villages and hamlets of Valais and dot the meadows. Sometimes a barn with its haymow will be joined to a cabin for temporary residence when the farm family must care for cows at some distance from the settlement. A farmer might have access to five different barns, but in many cases only a fractional share, averaging two-fifths of the structure, would be owned. In the past, groves of tall trees were left standing around stone-walled corrals on the alp to give the cows protection from late spring and early fall storms.

Valuable crops may be subject to specific pests, and today farmers use a variety of chemical sprays. Potatoes are menaced by the Colorado beetle, and the grapes must be sprayed periodically against mildew and insect larvae. In the nineteenth century, the phylloxera epidemic ruined many of the Törbel vines, and the vineyards had to be replanted with American root stock. There is a large traditional pharmacopoeia for treating animal diseases, and in the past, the death of one of the average family's two adult milk cows would have been a severe economic hardship. Today scientific breeding has produced animals with higher milk yields, and veterinary medicine is readily available. Though dogs were too expensive to maintain in former times, many families had cats to keep down rodent infestation that threatened stored foodstuffs. In the long-domesticated Swiss landscape, there have been no large animal predators for many years. Theft of livestock was also unlikely in the small, isolated, face-to-face community.

Wet-Rice Farming as an Intensive System Par Excellence

Perhaps the epitome of intensive agriculture is represented by the irrigated wet-rice systems of Asia. No brief survey can do justice to the technological achievement, the ecological sophistication, and the variety of such systems, but a few examples may illustrate the principles and some of the specific practices of cultivation. The smallholder who cannot gather the grain of a wild cereal grass from a swamp or a seasonally inundated river valley, and who lacks the forested tropical uplands to grow dry rice by slash-and-burn methods, must domesticate and control an environment specifically adapted to this plant. The underlying soil may be quite poor in plant nutrients, and sources of organic replacement may be limited, but managed water can supply these deficiencies and has done so over centuries of use. Clifford Geertz (1963: 31) likens this means of converting natural energy into food to "the fabrication of an aquarium," a felicitous image that recalls the first model of a balanced ecosystem that many of us learned about in elementary school science textbooks. On a wet-rice terrace or pond field, the base of impervious clay, the low banks or bunds around the edges, the precisely leveled surface, the water inlets and the outlets, are all designed to retain water at measured depths for accurately timed periods in accordance with the needs of the developing plant. The "rice-growing brew," though it may be from rainfall, serves its fertilizing function best if it conveys dissolved nutrients and silts from a river or some other external source (Hanks 1972: 37). Irrigation water not only restores crop-depleted nutrients to the soil annually but also promotes the fixation of atmospheric nitrogen (up to 50 kg /ha) through the symbiotic association of blue-green algae and fern that it supports (Altieri 1987: 76). In the words of L. M. Hanks:

A flooded field is short on oxygen . . . , yet oxygen becomes available through the bacteria that break up the organic products of fermentation. In this dank airlessness nitrogen is converted by other bacteria into ammonia rather than more familiar nitrates. In the presence of ammonia the brew tests more nearly neutral or even [more] alkaline than acid, so that phosphorus becomes available to plants as ferrous and manganese phosphates rather than the more familiar phosphorous acid. Unlike roots in dry soil, the rice roots at the bottom of a flooded field serve mainly to anchor the plant, while the higher rootlets drink in the necessities of growth. (Hanks 1972: 33–34)

The chemical and bacterial decomposition of organic material, including the remains of harvested crops, contributes to this process.[5] Dissolved nitrogen can be brought in by the water, especially if there is a slow flow through the pond field (Seavoy 1986: 156), and the rice rizosphere can also fix considerable amounts of atmospheric nitrogen (Ruthenberg 1976: 184).

Where irrigation is practiced, the high water table prevents the vertical movement of fluids, thus limiting nutrient leaching (Altieri 1987: 76). Standing water also fills a protective function, shielding the soil surface from high temperatures and the direct impact of rain and high wind that could induce erosion. An inundated field restricts the growth of many weed plants that would compete with the growing rice for space and nutrients. With water as the crucial variable determining rice productivity, the timing of its application is also important. "Paddy should be planted in a well-soaked field with little standing water and then the depth of the water increased gradually up to six to twelve inches as the plant grows and flowers, after which it should be gradually drawn off until at harvest the field is dry" (Geertz 1963: 31). Small fields permit the maintenance of an even depth of water and varying it on a schedule that may stagger crop maturity so that the critical transplanting and harvesting operations need

5. Under dry-land conditions in the tropics, the nitrogen in organic matter is changed during decomposition into ammonium, rapidly oxidized to nitrate, and then quickly leached or denitrified into a gaseous form that is lost to the atmosphere (Bayliss-Smith 1982: 72). Waterlogging slows the decomposition rate of organic matter and the rate at which nitrogen oxidizes, making nitrogen available to the growing wet-rice crop.

not be performed on all fields at the same time. Even temperature can be regulated by moving water along a short course, thus retaining warmth, or over a longer distance that allows it to cool (Hsu 1980: 119). Creating and precisely controlling a liquid microenvironment for rice not only optimizes and stabilizes crop yields but may in fact improve soil quality under permanent land use over the long run of 50 to 100 years (Ruthenberg 1976: 184).

On both irrigated and rain-fed fields under a variety of crops, Asian intensive cultivators used a wide variety of complex and laborious fertilization techniques. Texts from the Han period in China (206 B.C. to A.D. 220) show the existence of an elaborate soil classification of three grades and fifteen types, along with the knowledge of how their agronomic qualities could be improved (Hsu 1980: 94). Green manuring by cutting, burning, and soaking weeds or by turning them under is attested by the fifth century B.C. Terra-cotta models of Han pigpens with adjoining privies show how pig manure and human feces were collected by pipes and channeled to a lower clay plate for drying and later distribution to the fields in powdered form (Hsu 1980: 97). The most careful fertilization was devoted to the seedbed, as in the instructions of a medieval Chinese agronomist quoted by Francesca Bray:

In autumn or winter the seedbed should be deeply ploughed two or three times so that it will be frozen by the snow and frost and the soil will be broken up fine. Cover it with rotted straw, dead leaves, cut weeds, and dried-out roots and then burn them so that the soil will be warm and quick. Early in the spring plough again two or three times, harrowing and turning the soil. Spread manure on the seedbed. The best manure is hemp waste, but hemp waste is difficult to use. It must be pounded fine and buried in a pit with burned manure. As when making yeast, wait for it to give off heat and sprout hairs, then spread it out and put the hot fertiliser from the centre to the sides and the cold from the sides to the centre, then heap it back in the pit. Repeat three or four times till it no longer gives off heat. It will then be ready for use. If it is not treated in this way it will burn and kill the young plants. Neither should you use night soil, which rots the shoots and damages human hands and feet, producing sores that are difficult to heal. Best of all the fertilisers is a mixture of burned compost, singed pigs' bristles and coarse bran, rotted in a pit. (Bray 1986: 45–46)

The products of other farm activities, such as silkworm excrement, bonemeal from slaughtered animals, and pond muck containing decomposed plants and waterfowl droppings, were all systematically collected for manuring. Old straw thatch and cooking-fire ash entered the midden, and even carbon-impregnated brick stoves and adobe walls were broken up and returned to the soil (M. C. Yang 1965 [1945]: 240). There was also a thriving market from as early as medieval times for organic materials from external sources. Night-soil collectors called regularly on urban

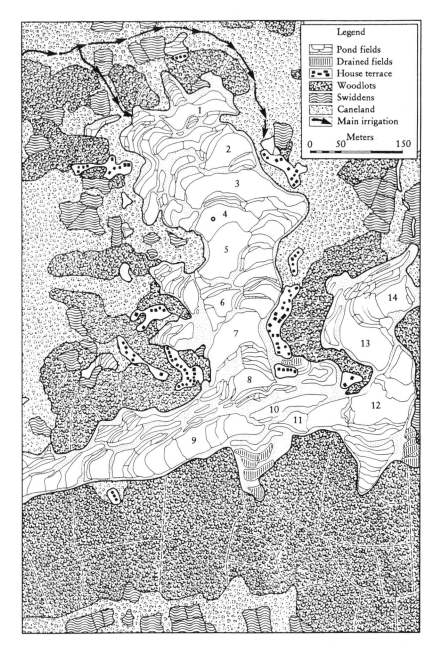

Legend
- [⬚] Pond fields
- [⫼] Drained fields
- [▪-▪] House terrace
- [▨] Woodlots
- [〰] Swiddens
- [∷] Caneland
- [➤] Main irrigation

Meters
0   50   150

Fig. 1.2. Land use and settlement in the central portion of Baynīnan, an Ifugao district community in highland Luzon, Philippines. (From Conklin 1967: 115)

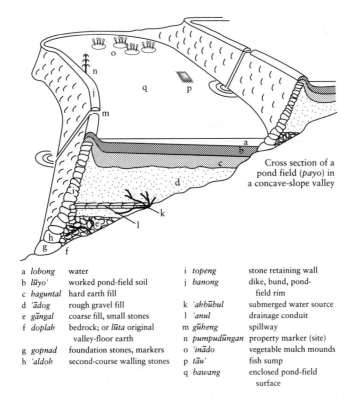

Cross section of a
pond field (*payo*) in
a concave-slope valley

a	*lobong*	water	i	*topeng*	stone retaining wall
b	*lūyo'*	worked pond-field soil	j	*banong*	dike, bund, pond-field rim
c	*haguntal*	hard earth fill			
d	*'ādog*	rough gravel fill	k	*'ahbūbul*	submerged water source
e	*gāngal*	coarse fill, small stones	l	*'anul*	drainage conduit
f	*doplah*	bedrock; or *lūta* original valley-floor earth	m	*gūheng*	spillway
			n	*pumpudūngan*	property marker (site)
g	*gopnad*	foundation stones, markers	o	*'inādo*	vegetable mulch mounds
h	*'aldoh*	second-course walling stones	p	*tāu'*	fish sump
			q	*bawang*	enclosed pond-field surface

Fig. 1.3. Ifugao pond-field terrace composition. (Adapted, with permission, from Conklin 1980: 16)

households before dawn, and there was trade, sometimes over considerable distances, in oil cake, fish meal, waste from making bean curd, mollusc shells for lime, and river mud (Bray 1986: 49).[6] Modern manufactured chemical fertilizer fits naturally and inevitably into this process, but it may not have the same effects on soil tilth and texture. This vigorous historic recycling effort was undoubtedly a key element in the energy efficiency and sustainability of Chinese agriculture (Wen and Pimentel 1986).

6. A similar process of intensification was compressed into a relatively short period in Japan, where a general shift from dry- to wet-rice farming took place in the eighteenth and nineteenth centuries, weeding was introduced, irrigation was extended, new rice varieties were diffused, and commercial fertilizers such as dried fish, oil cake, and urban night soil were used to supplement barnyard manures and ash. Few of these technological changes were the result of inventions, and most were known techniques that spread from the localities where they had been developed. See Smith 1959: 86–97.

The most spectacular examples of landscape transformation to accommodate wet-rice production are the terrace systems in upland areas of China, Japan, Java, and Bali. Harold Conklin (1980) has mapped, illustrated, and described the pond fields of the Ifugao in central Luzon, in the Philippines, and his discussion of terrace construction shows how massive, finely engineered structures are created with simple hand tools by small groups of workers (Fig. 1.2). Sloped dry-stone walling is built up along the contour of a hill and the terrace is filled with successive layers of small stones, rough gravel fill, hard earth fill, and worked pond-field soil that can be flooded (Fig. 1.3). Rock, gravel fill, and earth are transported hydraulically by temporary canals and ditches and raised conveyor flumes of hollowed logs, with water impounded and solid materials excavated and sluiced down from higher slopes (Conklin 1980: 16–17). Springs that emerge from the hillside behind the terrace are drained by underground conduits of bundled canes protected by flat slabs of rock. Spillways allow water to be drained down the face of the supporting wall to terraces below. The field surface is slanted slightly downward toward the uphill-side margin of the terrace to prevent total loss of ponded water if there is a washout. When the field is intentionally drained, an excavated sump confines the mud fish that have been stocked and allowed to mature in the seasonal ponded environment. Walling stone can be quarried and split from boulders or outcrops with metal and wooden wedges, assisted by fire and pry bars (Conklin 1980: 18). The time spent in the original construction of terraces and water channels is greatly added to by the frequent need to repair damage caused by landslides, seepage, and rainstorms (Conklin 1980: 29). Wet-rice production in terraced pond fields exemplifies the skills and labor intensification that support Ifugao populations of some 250 per km² of cultivated area (Conklin 1980: 51).

Technological Change in Chinese History

There is a temptation when describing Chinese agricultural intensification to recount marvels—the clever techniques for capturing and creating organic material that restore soil fertility, the practical and sophisticated understanding of biological processes, the disciplined and skillful application of labor. But even the fascinating history of agrarian technical virtuosity describes a series of spurts rather than a straight line of development through time. The artifacts, images, and writings of a literate civilization allow us to reconstruct the course of technological change in agriculture. The introduction and wide adoption of iron tools in the pre-Han era (before 206 B.C.); of the moldboard plow during the Han period; and of steel for the edges of plowshares, hoes, and sickles under the Tang dynasty (A.D. 618–907) meant in each case that farmers could till larger areas, fallow could be shortened, and individual productivity could rise (Chao 1986: 194–96). However, labor-saving innovations in cultivating implements virtually all took place before the fourteenth century. "Biological," as opposed to "mechanical," technology involves new crops and varieties, fertilization, and cropping techniques (Hayami and Ruttan 1971), many of which may be labor-using innovations that give higher output at the cost of reduced returns per unit of labor (Chao 1986: 21).[7]

Kang Chao (1986: 198) claims that there was a turning point in the twelfth century when population pressure compelled the Chinese to adopt multiple-cropping, reclaim land and extend land use to inferior or difficult terrain, and increase output by fertilization and the introduction of higher-yielding varieties. Eugene Anderson (1989), on the other hand, finds the seeds of the "first Green Revolution" in the 400–100 B.C. period, when the appearance of the moldboard plow, advanced harrows, seed drills, and oil- and flour-milling machinery was accompanied by the introduction of grapes and alfalfa, techniques of grafting and intercropping, the pretreatment of seed with fertilizer and insecticide, and the lining of storage bins with pesticide plants. Intensive pig rearing and aquaculture of several crop species were also developed, along with the dryland farming techniques of rolling snow into soil and dust mulching. Farming populations had already become concentrated around cities, and their strategies of intensification were powered in part by the drive to participate in a growing market-dominated economy (Anderson 1988: 109).

From its beginnings in China, agricultural intensification was encouraged by the state and diffused by a literate elite. The Han emperor Wen propagated the ideology of agriculture as the basis of the state, symbolizing this commitment by himself plowing a ceremonial furrow, while the empress raised silkworms. In more material terms, peasant taxes were kept to a low 3 to 7 percent of the harvest, and large estates were discouraged. Wen "recognized that small farmers are particularly prone to intensify" (Anderson 1989: 141). From the ninth century, woodblock printing allowed the publication of practical agricultural treatises with pictures of tools and appliances. Wang Chen's book in 1313 had 136,000 characters and almost 300 illustrations (Elvin 1973: 116). Methods of steeping seeds in a decoction made of boiled bones, scooping mud from creek bottoms

---

7. The failure to continue technological development in China because the labor of the fast-growing population was cheaper than machinery or capital-intensive innovations has been called the "high level equilibrium trap" (Elvin 1973) and likened to the involution that Geertz (1963) has characterized in Java (Anderson 1988: 88, 100). The question remains as to whether the biological innovation and intensification of labor sketched in the Boserup model inevitably led to a decline in average rural welfare and the immiseration of the population.

for compost, and applying quarried lime to fields were recommended. A thirteenth-century description of Han-chou mentions swarms of boats carrying away rubbish and night soil from the city and the organized business of ordure carriers who daily emptied the tubs that people set outside their doors (Elvin 1973: 120). An urban life without sewers or fouled water sources and a farming system dependent on recycling valuable organic material had combined in an ecologically symbiotic relationship unmatched in the Western world.

The processes that not only maintained fertility in rich alluvial land but literally created an agricultural medium where it had not previously existed were described in vastly appreciative detail by F. H. King (1911), a pioneer soil chemist from the University of Wisconsin. Irrigated, leveled fields south of Canton produced two crops of rice in the summer before being ridged for a winter crop of leeks and vegetables. Night soil imported 90 miles by boat from the city was diluted with water and applied at the rate of 16,000 gallons per acre to sustain the multicropping (King 1911: 73). Composting reached a pinnacle of baroque organic elaboration in the practice of (1) shipping stable manure from the city to the country; (2) unloading the manure and saturating it with mud scooped from the canal bed; (3) cultivating a field in nitrogen-fixing clover; (4) digging a pit in the field and filling it with alternate layers of composted manure and cut clover; (5) fermenting the mixture for 20–30 days before distributing it over the field for the following rice crop. Excavating canal-bottom mud had the additional effects of preventing the waterway from silting up, adding organic matter to the soil, raising the level of the field, and giving it better drainage (King 1911: 74). King observed a man who had carried about a ton of mud from a 10-foot-deep, tidally exposed canal bottom, using flat pans and a carrying pole, all before 10 in the morning (1911: 169).

Time as well as space was carefully economized in the Chinese intensive system. Interspersed rows planted at different times—for example, wheat nearing maturity with Windsor beans two-thirds grown and cotton just planted—provided for the "fullest possible utilization of every minute of growing season and of the time of the family in caring for the crops" (King 1911: 266). Using a one-acre seedbed for rice plants that would occupy ten acres left nine acres free for 30 to 50 days, giving extra time for a crop there to mature (King 1911: 11). Double- and even triple-cropping could only be achieved with careful leveling of the field to provide a constant and accurately controlled depth of water. The saturated field was plowed, weeds were buried, and harrowing produced a smooth, liquid mud. Transplanting had to be rapid so that the rice plants did not wither between the nursery and the field. Equally important was the precise spacing of the seedlings to ensure the maximum number of plants with enough soil and water to bring each one to fruition (Rawski 1972: 13).[8]

Other crops received similar care. Formerly millet-based agriculture in North China was intensified with the arrival of New World crops such as sweet potatoes and peanuts in the late sixteenth century (Anderson 1988: 79). M. C. Yang (1965 [1945]: 19) describes procedures of selecting peanut seeds, sprouting them in warm water, planting them with fertilizer, flattening the field with a stone roller, weeding twice, cutting the vines, and then plowing the field into a fine powder, from which the peanuts are sifted out. In alternate years, sweet potatoes are grown in the same field. The tubers are first buried in damp sand on a warm brick bed. The shoots are transplanted to a nursery bed with heavy fertilizer and kept wet. The vines are then cut, bundled, and transplanted onto ridges, with a pint of water applied to each plant. To prevent small roots from sprouting, the vines are turned from one side of the ridge to another after every rain. New earth must be applied to the ridge tops as they wash down. Even the processing of the harvest is laborious. Women slice the tubers thin and put them out to dry in the sun, while boys camp in the field to protect them (ibid.: 20–21). After describing the practices of a farmer who supported 20 people from a half-acre garden of cucumber vines trained up trellises, King points out the alertness and efficiency with which the cultivator treated each plant individually. "Forethought, after-thought, and the mind focussed on the work in hand are characteristics of these people" (King 1911: 205).

The multiplicity of tasks in a highly developed system of intensive agriculture, the premium on effective execution, the necessity for planning, scheduling, and managing of farm operations—all of these are elements of production based on skill as opposed to scale. Francesca Bray's discussion of the rice economies of Asia makes this crucial distinction, noting that "Eurocentric" models of historical change in agriculture have always equated progress with the increasingly efficient substitution of alternative forms of energy for human labor (Bray 1986: 2–3). Where land was plentiful and labor scarce, as in the underpopulated New World and Australia, production could be increased most rapidly by land-extensive methods and mechanical technology. The same relationships of land and population that encouraged adoption of the McCormick reaper and the tractor had, at an earlier period, spread the eight-ox, heavy wheeled plow and the horse collar in medieval Europe (Lynn White, Jr., 1962). China's situation of land shortage and abundant labor, already evident in the Han era, gave the advantage to innovations like quick-ripening rice and the

8. In Java, men pull the seedlings from the nursery and women plant clumps of two to six seedlings 20 to 40 cm apart depending on soil fertility and variety grown; 12 to 20 people working as a team can plant 1 ha a day (Seavoy 1986: 157).

techniques of irrigated multicropping. Larger amounts of highly skilled labor were required, but this did not necessarily mean that the productivity of labor was drastically reduced (Bray 1986: 5). Agricultural implements might even become simpler as cultivation techniques became more complex. "When rural populations are dense and opportunities for alternative employment few, technical changes that absorb labour and reduce agricultural underemployment are preferable to those which increase output at the cost of reducing the labour force" (ibid.: 4).

The Ethnoscience and Specialized Technologies of Indigenous Intensification

The practices that characterize and define intensive agricultural systems of course form a continuum with other means of food production. Swidden or shifting cultivators may plant a great diversity of tuber, legume, grain, and tree crops in their burned clearings (Freeman 1955; Conklin 1957, 1961), achieving similar effects of soil protection, shading out competing vegetation, and restricting the spread of insect pests. They may allow leguminous trees to grow there as a means of restoring soil fertility, and they may fence their temporary plots, as do the Tsembaga in highland New Guinea (Rappaport 1968). But none of the normal techniques, such as weeding, are pursued as assiduously or combined with more permanent investments like terracing or irrigation as is the case with intensive farming. Smallholders with a permanent, limited land base must *do* more and probably also *know* more. Ethnoscience refers to local knowledge formalized into systems of classification that can be studied by linguistic methods. The Philippine Hanunóo swiddeners distinguish 430 different cultigens and recognize 10 basic and 30 derivative soil and mineral categories (Conklin 1954). The Tzeltal in the Chiapas highlands of Mexico make fine varietal distinctions in maize and other crop plants (Berlin 1973), and it seems clear that these extensive and orderly lexicons have been elaborated for utilitarian purposes (Hunn 1982).

The Kofyar classify soil by color, texture, and moisture content, accurately judging its organic content, its suitability for specific crops, and the difficulty of working it with a hoe when wet or dried and hard (Netting 1968: 82). Leached, unfertilized bush-field soil becomes lighter in color and increasingly sandy, and the Kofyar explain this change by saying that the water has soaked out its goodness. The natural plant cover and succession give the informed local cultivator an excellent sense of the underlying soil status (Nye and Greenland 1960) and prevent the mistakes that inexperienced settlers may make in an unfamiliar environment like that of the Amazon rain forest (Moran 1981: 108–15). Whereas shifting agriculturalists judge those inherent characteristics of soil that influence

immediate crop production, intensive cultivators who improve soil fertility go a step further to classify managed soils in terms of workability,
response to fertilizers, and moisture absorption and retention where irrigation is practiced (Wilken 1987: 35). Not only are the inventories and
hierarchical organization of knowledge about soils, climate, vegetation,
water quality, and manures detailed and practical (though naturally without the theoretical understanding of these processes contained in geology,
chemistry, and the other physical sciences), they are also organized by
word and experience into folk ecosystems. French alpine farmers group
specific naturally occurring plants and animals with their characteristic
habitats of altitude, slope, sun, and water into biotopes that have particular human uses (Meilleur 1986). In these situations, cultural knowledge
transmitted by language and shared by members of a social group is fo-
cused on and elaborated around those elements of the local environment
that are critical to the continuity and sustainability of an intensive agroecosystem.

The very amplitude and complexity of these systems of knowledge
suggest why they are often only partially documented, and the observer
is left to infer what people understand from their practices and the physical evidence of farming behavior. In some striking cases, this involves
the creation and maintenance of a cultivable surface where none existed
before. In swamps and seasonally flooded areas, raised fields may, at the
cost of considerable labor, divide the wet from the dry. The chinampas
near the present site of Mexico City represent large-scale land reclamation through drainage dating back to the time of Teotihuacán, perhaps
2,000 years ago (Coe 1964). Strips of land perhaps 300 feet long and between 15 and 30 feet wide were heaped up from excavated channels along
the sides and at one end (Coe 1964; Chapin 1988). These spring-fed canals
that formerly drained into Lake Xochimilco provided not only a high
water table for plant growth but also aquatic vegetation and bottom
muck for fertilizer, fish, and waterways that served for canoe transportation. Stakes interwoven with branches and reeds formed the banks of the
raised fields originally, and willow trees reinforced the margins. The plat-
form strips are raised a foot or two above water level, and if the chinampa
grows too high by the repeated application of mud, some of the surplus
soil must be removed (Coe 1964).

Corn, beans, chili peppers, amaranth, tomatoes, and squash have always been grown on chinampas (Gomez-Pompa and Jimenez-Osornio
1989), but higher-value vegetables, flowers, spices, and herbs destined for
the market are also produced in quantity (Wilken 1987). The wild plants
that grow on chinampas have been selected for food, medicines, fodder,
pesticides, green manure, mulches, and forage, and it is claimed that the
concept of "weed" does not exist in this farming system (Gomez-Pompa

and Jimenez-Osornio 1989: 237). Besides the carp, salamanders, frogs, and ducks from the channels, the *chinamperos* utilize turkeys, chickens, pigs, cows, and ducks, which are fed crop wastes and forage plants and provide manure for the fields.

One of the most ingenious procedures in chinampa cultivation is the creation of muck seedbeds as nurseries for eventual transplanting. The canals collect deposits of eroded soils, decomposed aquatic plants, household waste, and residual animal manure (Wilken 1987: 85). This muck, dredged from the deep canal bottoms with long-handled scoops, is taken to the plots in canoes and there strained by hand to remove clay lumps and stringy plant remains. Once formed into rectangular seedbeds and solidified, the muck is cut with a knife into squares and seeds are planted in holes poked in the center of each cube. The nursery bed saves field space during germination and initial growth, allows convenient watering with sprinkling cans, and provides a fertile medium for the developing root systems. When the small plant and its entire cube is transplanted to the chinampa field, the rich muck is effectively distributed over the entire chinampa surface (ibid.). Since up to seven crops a year may occupy the raised field in rapid succession (Coe 1964), starting the plants in seedbeds allows the limited field space to be more fully utilized (Wilken 1987: 257). It also promotes selection of the strongest and most resistant seedlings (Gomez-Pompa and Jimenez-Osornio 1989: 237). As land is filled in for urban building, water sources are appropriated, and more chinamperos work in tourist and other industries, intensive agriculture is shrinking, but there are increased efforts by Mexican botanists and horticulturists to describe and perhaps introduce the system of drained fields into other areas. The high, sustained production achieved by controlled drainage, muck fertilization, animal manure, seedbeds, transplanting, plant diversification, and the canals forming protective moats around the platforms is an indigenous model of wetland use worthy of emulation.

On Ukara Island in Africa's Lake Victoria, a similarly intensive regime developed, in which the major challenge was not the building up and fertilizing of drained land but the creation of organic material to support the continuous cropping of very limited land. Presumably protected in the past by their island location, some 16,000 Wakara in 1965 made a living from 30 square miles, living at densities of 671 per square mile of arable land (Ludwig 1968: 91). They cultivated rain-fed millet, sorghum, Bambara nuts, and cassava, interplanted or in rotation, as well as irrigated rice. To maintain the fertility of their fields, they kept dwarf, shorthorn Zebu cattle, goats, and sheep in pit stalls occupying half of the dwelling huts. The livestock were fed in part on leguminous plants, especially *Crotalaria striata*, which was grown in the millet field and also used as green manure on the wet rice. The leaves of this plant were used as stall bedding, and the manure was supplemented with household refuse, including ashes and night soil.

Additional fodder for the livestock came from artificial water meadows, either dammed upland stream valleys or four-foot-deep pits, 5 by 100 yards, close to the lakeshore. Flooded by groundwater or seepage from the lake, these pits could produce four cuttings of the quick-growing elephant grass a year (Ludwig 1968: 115). Weeds and stalk residues from the millet fields, as well as branches from privately owned trees and shrubs of 39 different kinds, were also fed to the domestic animals. Perhaps three to four hours' labor a day went into the procurement and transportation of fodder, feeding the animals, and driving them to the watering place. On the way back to the fields, people frequently carried 15-kg baskets of manure, fertilizing at the annual rate of 3.6 or 4.5 metric tons per acre (ibid.: 120).

Since the island is quite hilly, the Wakara have independently developed "nearly all conceivable pre-machine methods of erosion control" (Ludwig 1968: 121). Along the contour, they cultivate rectangular raised mounds, and they have terraces with earth and stone walls up to two meters high. Erosion gullies are blocked with stone dams planted with grass, sisal, or euphorbia. The larger rivers are regulated with levees along their lower courses to prevent seasonal flooding. Elevated paths with embankments along the sides are used to take the animals from the homesteads to watering places. The slow-growing, deeply-rooted *Crotalaria* green manure crops also tend to anchor the soil and counteract leaching. Given the high labor demands of stall-feeding, manuring, and erosion control in this system of intensive land use, it is not surprising that Wakara islanders have migrated in recent years to less densely populated mainland areas, where shifting cultivation brings a much higher return per hour of work (ibid.: 133).

Garden-Variety Intensification

Perhaps the most universally familiar example of small-scale agricultural intensification is the garden, and one way to think of the technology of the smallholder household is to consider that smallholders' relatively simple tools, highly skilled but often manual labor, and detailed instrumental knowledge of the local environment all suggest those of the good gardener. Indeed, non-farmers both in the city and in the country may be the most avid gardeners of all, and their motivations may range from producing a few vegetables for the table in a kitchen or dooryard garden, to specialized market gardening or truck farming, to the recreation afforded by flowers or ornamental plants. The beauty and bounty of a luxuriant garden form an apt image of the earthly paradise, whether it be an imagined Garden of Eden or the walled, irrigated orchard of fruit trees and vineyards celebrated in Islamic literature and art (Watson 1983). While the form, location, size, and types of crops grown are varied, household gardens are often seen as "secondary sources of food and income, while field production, animal husbandry, wage labor, professional services, or trading are the major sources of support" (Cleveland and Soleri 1991: 2). Yet though gardens may be physically enclosed, their economic and conceptual boundaries are seldom as distinct.

Gardens can be distinguished from arable cropping by a number of features that are usually, but not always, found together. Hans Ruthenberg (1976: 118) lists (1) production of small amounts of produce from a great number of different food crops, (2) small plots, (3) proximity to the house, (4) fencing, (5) mixed or dense planting of a great number of annual, biennial, and perennial crops, (6) intensity of land use, (7) land cultivation several times a year, (8) permanence of cultivation, and (9) cultivation with hand implements. As I have suggested in the Nigerian, Swiss, and Chinese cases, these characteristics may also be extended to an entire farming system. Gardening can indeed become the dominant branch of the holding where there is a marked shortage of land or where there is a town in the vicinity (ibid.). A number of different crops may be planted in raised beds (Omohundro 1987), in rows at uniform intervals, or randomly, and gardening techniques are common whenever cultivators need to maximize total output per unit of area.

Guatemalan dooryard gardens make efficient use of both horizontal and vertical space with two dozen or more different economic plants in less than 0.1 ha (Wilken 1987: 250). A four-tier arrangement ranges from tall trees such as mangoes, papayas, capulins (*Prunus serotina*), and guajes (*Leucaena esculenta*), to trees of medium height like bananas, peaches, avocadoes, pomegranates, and citrus, to high and low field crops of maize, shrub beans, tomatoes, chili peppers, squash, flowers, and medicinal and cooking herbs. Vines of beans and chayote climb maize stalks and trees. The garden plot is fenced with additional productive and ornamental plants. Turkeys and chickens patrol between the plants and eat insects (ibid.).

An Ibo compound garden plot of 400 m² contained, at the height of its vegetation period, 1 oil palm, 2 coconut palms, 30 other useful trees and bushes, 307 yams, 230 coco yams, 8 manioc plants, 46 maize stalks, and 566 vegetables (Ruthenberg 1976: 89). Trees and bushes were trimmed before the major planting season to allow sunlight to reach the lower crops. Goats were confined during the main cropping season and fed with palm leaves, crop residues, and household refuse. Their dung and mulch materials were collected in pits for manuring the garden. Though the compound garden occupied only 2 percent of the household farmland, it produced half the farm output (ibid.). Regardless of season, the Ibo garden always has something to harvest for food or to sell.

Javanese *pekarangan* house gardens are multistoried orchards: a canopy of coconut palms, durian trees, and mangoes shelters bananas and coffee, with taro and cassava at ground level (Seavoy 1986). Though most studies of Indonesian agriculture emphasize wet-rice and extensive swidden cultivation, gardens make up 15 to 75 percent of the cultivable land area, and they may provide more than 20 percent of household income and 40 percent of the households' caloric requirements (Stoler 1981: 243). Dominated by perennials rather than annuals, and by woody rather than herbaceous growth, gardens "provide sustained yields and cause minimal environmental degradation under continuous use" (ibid.: 244). Tree roots bring underlying minerals into the topsoil, fallen leaves provide a protective mulching cover and introduce more humus into the soil, and the tree cover prevents erosion (Terra 1954). Firewood, bamboo, and other house-building materials come from the garden, and leafy vegetables supplement the human diet with vitamins and protein. Animal fodder, medicinal herbs and roots, and handicraft materials are also garden products. Low-income households may be particularly dependent on the leafy vegetables, spinach, long beans, cassava, and fruit trees of the garden (Stoler 1981: 250), and these foods in part compensate nutritionally for the lower consumption of rice by the poor. Households with less cropland per head devote a higher proportion of their land and more effort to gardens. Ninety percent of households in one village had access to gardens through ownership, rental, or sharecropping (ibid.: 248). Javanese house gardens occupy only 8.4 percent of their owners' total working time, but they provide a higher return to labor than other agricultural activities.

Gardening can be a very commercial operation, incorporating market crops and livestock production. Using 0.6 ha in Singapore, three brothers grew leafy vegetables on 127 beds, getting eight crops annually (Ruthenberg 1976: 212). They compensated for irregular rainfall with five man-made ponds, which provided water for the vegetables, water hyacinth for pig fodder, and fish. The pigs were fed purchased concentrates and a mash prepared from the vegetable wastes of the garden. Diluted pig manure, chicken droppings, mud from the ponds, compost, prawn dust (from a local shrimp industry), and chemical fertilizer were used to maintain the garden's fertility (ibid.: 213). This vegetable and pork-producing smallholding garden was built on the recycling of plant nutrients, and the energy sources imported into the system were chiefly feed for the pigs rather than purchased fertilizer.

Diversified gardening on small land areas with heavy, sustained production is a microcosm of the practices and virtuosity of intensive cultivation. The intimate association of gardens with residences, compounds, or kitchens reinforces the role of the household as a labor, management, and consumption unit that derives substantial benefits from what may be a very limited set of resources. Continuous individualized husbandry of plants and animals involving practices such as intensive tilling; carefully timed planting; transplanting; multiple weeding; elaborate systems of fertilization, with the recycling of plant and animal wastes; fodder production; and selective harvesting qualifies agriculture as horticulture—that is, gardening.

Smallholder Tools and Know-How

Evolutionary models of human development have often characterized progressive stages by the technology of food production and processing, from the chipped stone implements of the hunter-gatherer to the ground stone and pottery of Neolithic cultivators and onward to the more "efficient" means of energy capture in the Iron Age and the Industrial Revolution. Agricultural typologies have contrasted digging-stick and later hoe-culture practices (*Hackbau*), mostly by women on small plots, with male-dominated plow agriculture (*Ackerbau*) and horticulture (*Gartenbau*) employing irrigation and fertilizers (Hahn 1919; Curwen and Hatt 1953). The basic distinction that I am making between extensive and intensive modes of food production does not rest on different types of tools or on technological change as the engine that powers agricultural evolution. Digging stick, hoe, plow, and combine harvester may each be part either of extensive, shifting systems of land use or of more intensive, permanent cultivation. The critical elements in the process of intensification are knowledge of the local environment and the specific requirements of domesticated plant and animal species, combined with a tool kit of practices for soil manipulation, water control, nutrient conservation and restoration, and the protection of cultigens that act to increase and sustain biological production. Intercropping, rotation, diversified grain-tuber-vegetable-tree crop production, stall-fed livestock, terracing, irrigation, and manuring can all be carried out with different physical means and with various cultigens.

Evidence of intensive, permanent farming systems is present on almost every continent, encompassing a great many culturally autonomous groups in prehistory and in the contemporary world. The same general type of land use occurs among the Nigerian Kofyar, who were until recently largely cut off from the market and the state; among the alpine Swiss in subsistence communities dating from the Middle Ages; and in the heartland of Chinese imperial civilization. Wet-rice paddies in Asia and the raised fields of Mexican chinampas achieve similar high, continuous yields in the tropics. And gardens everywhere exemplify the same techniques and strategies in miniature. Smallholders meet similar ecological/economic problems with a multiplicity of means and understandings. Technological invention and scientific discovery are not the crucial causal factors in the course of agricultural intensification.

CHAPTER TWO

The Farm–Family Household

Sᴍᴀʟʟʜᴏʟᴅᴇʀs ᴀʀᴇ almost by definition householders. The people who do the work of intensive agriculture, who consume and exchange the products of the farm, who have continuing rights in the property, and who live together on or near their fields are almost always the family of a married couple and their children, along with other kin and possibly permanent employees. The household is a social group so ubiquitous in human society that it is easy to take it for granted, and its forms are so varied—monogamous and polygynous, patrilocal and matrilocal, nuclear and extended, with and without servants—that no universal common functions or activities seem to exist. But wherever we go, there appear to be recognizable domestic groupings of kin with a corporate character and an identity that is recognized in the use of terms like *family*, *house, hearth,* or *those who eat from a common pot.* The household may be seen as "the next biggest thing on the social map after the individual" (Hammel 1984: 40); "the smallest [social] grouping with the maximum corporate function" (Hammel 1980: 251); or a task-oriented, culturally defined unit (Carter and Merrill 1979).

My first experience with households in a non-Western society was the census of Kofyar homesteads that I began when I embarked on field research in northern Nigeria. Because little clusters of thatched, round mud huts inhabited by close kin and surrounded by well-defined farm plots belonging to the family were the most prominent feature of the physical and social landscape, I used them without a second thought as the units of my survey. Membership in these *lu* (house) or *koepang* (homestead) groups seemed relatively stable, changing with birth and death, marriage and divorce, or when a son and his wife left to set up housekeeping on their own, but aside from these events and a few little children sent to live with grandparents or aunts to "run errands," household makeup showed considerable continuity through time. Though people had important relationships as members of patrilineal lineages, citizens of autonomous villages, kin of mother's brothers in other settlements, dwellers in named neighborhoods, drinking friends, and voluntary labor-group members, the household remained the fundamental social unit. The household occupied and maintained physical premises, its organized and cooperative labor produced the food that was stored, prepared, and eaten there, members had joint rights to important resources, tools, and livestock, and dependent children and elderly people were cared for in the household context. I knew that elsewhere in West Africa, spouses might not share a residence, and that family members might work at different jobs with little pooling of food or money (Fortes 1949; Guyer 1981, 1986; Woodford-Berger 1981), but Kofyar households did not strike me as problematic. They were visible, countable, seemingly salient social units that organized a great deal of important everyday behavior.

Though the morphology of households, their shape and composition, may show great differences cross-culturally or even among economic, ethnic, and occupational groups in the same society, there do seem to be certain activities that form a core of what the domestic group actually does. Households are engaged in some combination of production, distribution (including pooling, sharing, exchange, and consumption), transmission (trusteeship and intergenerational transfer of property), biological and social reproduction, and co-residence (shared activity in constructing, maintaining, and occupying a dwelling) (Wilk and Netting 1984: 5–19; Wilk 1991). Though these activities tend to overlap in the household, various societies may remove many tasks, such as earning a living, the formal education of children, or even cooking meals, from the household context.

Smallholder agriculture helps to restore or maintain a common focus on the domestic functions, concentrating the time and effort, the rights and duties, of household members so that the density and significance of shared, cooperative activities are increased. The household does farm work, but not as separate and independent producers (even though some members may also have individual plots). Neither are household members equivalent to hired laborers, serfs on a manor, slaves on a plantation, specialized employees of an agribusiness, or members of a collective, and in fact these agricultural institutions seem poorly adapted to a regime of intensive cultivation. People from the family may work for wages extramurally or hire labor to help on their own farm, but they are neither proletarians totally dependent on an employer nor usually the owners of farms they do not work.

The household is the scene of economic allocation, arranging collectively for the food, clothing, and shelter of its members, and seeking to provide for these needs over the long term with some measure of security against the uncontrollable disruptions of the climate, the market economy, and the state. The corporate continuity of the household unit is apparent not only in the reproduction of family members and the inter-dependency between generations but also in ongoing property rights in the means of production. Valuable and scarce land, whether owned or held as tenants or sharecroppers, may absorb the energies of the house-hold, provide its livelihood, and condition its future. When subsistence, investment, and wealth are embodied in fields, livestock, trees, imple-ments, and buildings, a center of economic gravity is formed that at-tracts and influences the movements of the household members who hold and inherit these assets. Without new members entering the household through birth, marriage, adoption, or contract, the labor to use these re-sources would not be adequate. The coordination and management of the diversified smallholding enterprise is greatly facilitated by the intensive interaction, division of labor, and socialization in the intimate context of a common residence. Securing the benefits and paying the considerable costs of smallholder agriculture requires an emphasis on the social unit that most effectively combines production, distribution, transmission, and reproduction, and that is the co-resident household.

The "family farm" is one of those indistinct but positively valued icons of the modern world that conveys images of hard but honest toil, close kin who share the work that results directly in their wholesome daily bread, virtuous thrift and piety, the warmth of a protective, lifelong fam-ily circle, and the independence of a self-sufficient unit freely cooperating in a kind of agrarian democracy. It is a romantic view untinged by pov-erty and conflict with the state, by crop failure, greed, parental domina-tion, and marital dissension. It exalts beliefs and attitudes seen in Hesiod's *Works and Days* and characteristic of farmers in many cultures: "the peas-ant's emphasis on agricultural industriousness as a prime virtue . . . sup-ported by . . . security, respect, and religious feeling" (Redfield 1956: 113). Like the noble savage in the natural world, Jean-Jacques Rousseau's Swiss alpine farmers enjoyed the abundant fruits of their own efforts, liv-ing without masters in the little republics of their villages, in harmonious households where children and parents made decisions together for their common welfare. In this impressionistic mirror, country people in an-cient Boeotia and contemporary Maya milpa cultivators look very much alike and equally opposed to urban alienation, commercialism, and hier-archy (Redfield 1956). Though not denying that rural people may share certain views of "the good life," I shall be examining smallholders among whom constraints on agricultural land and capital regularly condition not only land use and technology but also the household organization of pro-duction. I contend that an intensive cultivation strategy, requiring greater, more disciplined labor of high quality, sustained use and im-provement of resources, and the establishment and transmission of property rights, increases the number and importance of economic activities carried on in a household, and hence the centrality of the household as a social institution.

A frankly economic approach to the household is not meant to consign cultural concepts of gender, age, class, or even ideal household type to some epiphenomenal realm (Yanagisako 1979, 1984; Guyer 1981; Löfgren 1984; Laslett 1984). Rather, it is designed to address particular problems arising from cases of scarce resources in the physical environment, the finiteness of earthly space, and the limitations on human time; it hypothesizes that individuals, acting together in households, make rational choices about the costs and benefits associated with various alternatives (Sawhill 1977). A preoccupation with work and the social groupings most directly involved in getting a livelihood has characterized ecological anthropology since the writings of Julian Steward (1938, 1955; Murphy 1981: 175). I share with Richard Wilk the working hypothesis "that agrarian social formations including the household, are constituted partially as work groups that motivate and apply the proper combination of labor, knowledge, and leadership to each task in an efficient (if not the most efficient) fashion" (Wilk 1991: 85).

Certain characteristics of small agricultural producers, their family farms, and their households that have been used to define peasants in general become even more pronounced among intensive agriculturalists:

The produce from the farm meets the basic consumption needs of the peasant family and gives the peasant relative independence from other producers and from the market. This makes for relative stability in peasant households which, in crises, are able to maintain their existence by increasing their efforts, lowering their own consumption and partially withdrawing from any market relations they may have. . . . The family farm is the basic unit of peasant ownership, production, consumption and social life. The individual, the family and the farm, appear as an indivisible whole. . . . Family labor is an essential requirement for conducting a farm adequately. . . . Family solidarity provides the basic framework for mutual aid, control and socialization. The individualistic element of personal feelings is markedly subordinated to the formalized restraints of accepted family role behavior. Forming the basic nucleus of peasant society, the life of a family farm determines the pattern of peasants' everyday actions, interrelationships and values. Together with the mainly natural economy, it makes for the segmentation of peasant society into small units with a remarkable degree of self-sufficiency and ability to withstand economic crises and market pressures. (Shanin 1971: 240–43)

Though there is general agreement that "peasant enterprises are familial," Donald Attwood (1992: 17) has incisively observed that peasants are not the individualistic economic men and women of neo-classical economics, nor are they the communal conformists of Marxism and dependency models. "The family enterprise is coming to be seen as the crucial middle term, one which avoids the atomism of the first position and the determinism of the second" (ibid.: 17).

Though family solidarity, values, and personal feelings may be difficult to measure and compare, the attributes of household membership and management that positively affect farming behavior can be suggested. Long-term experience with the farm's specific mix of soils, terrain, climate, tools, crops, and livestock, plus the passing on of this knowledge to other family members, emphasizes the tie between the household's productive and reproductive roles. Agriculturally specialized human capital is created and supported in the domestic setting. The organization and completion of farm tasks also suggests the advantages of the household over other institutional means for mobilizing labor on a smallholding with relatively low-energy technology. Smallholder households everywhere have had to perform the same basic tasks to ensure the survival of the household members over time. As Clark Sorensen points out in his compelling study of Korean intensively cultivating households, traditionally, "food for subsistence had to be provided, the house built and maintained, food produced by the household members processed and stored, domestic animals cared for, meals made and served, clothing provided and maintained, and the young who will provide the continuity of the family, born, raised, and married" (Sorensen 1988: 133). In the farming community, no other social unit is as salient and significant as the household.

The Household as a Repository of Ecological Knowledge

The knowledge that the intensive cultivator brings to his or her activities is not merely that of a generalized environment and the appropriate farming systems to utilize it. Certainly a collection of facts, an agricultural lexicon, and a set of basic skills are widely shared within the society, but the farmer who makes the most effective use of resources is the one with specialized knowledge of the specific microenvironments on the smallholding. Soils, with their special qualities of moisture retention and drainage, may change along a slope. A Sierra Leonean rice cultivator may practice shifting cultivation of an upland area with particular rice varieties and have a schedule of activities that differ from those used farther down the soil catena in the permanently farmed valley-bottom wetland (Paul Richards 1985: 50). Conscious experimentation with new crops or interplanting techniques in specific locations (ibid.: 149) contributes to practical ecological knowledge, which is supplemented by the recollection of past yields under a variety of seasonal conditions.

Swiss farmers delighted in regaling me with stories in which the names, appearances, milk production, calving rates, and personality quirks of long-dead cows figured prominently. Household heads had grown up tending vineyard terraces and meadow parcels, and they worried that their children might never learn the complex irrigation schedule of the far-flung family plots. Ibo women who carry on the intensive horticulture of compound gardens surrounding their courtyards may cultivate 18 to 57 plant species, including tubers, legumes, grain, and fruit trees, as well as raising dwarf goats and poultry. They may practice more multiple-cropping, plant more carefully, and have more knowledge of crop varieties than their husbands (N. D. Hahn 1986). Detailed ecological knowledge and experience with farm resources and operations over long periods of time create a system of high information content and considerable distinctiveness at the level of the individual farm (Leaf 1973). If plots of land are different and weather variability is high, experience specific to those plots has a high return. Economists note that the specificity of this experience makes it profitable for generations of kin to work together on the same plots and makes sales of land to non-kin less likely than bequests of land to offspring (Rosenzweig and Wolpin 1985). Knowledge and experience in the use of the land results in higher production and greater returns than would accrue to the same land cultivated by a stranger applying strictly standardized techniques.

Members of a household may share a long attachment to a farm, and the fund of ecological information so vital to the agricultural endeavor is transmitted through observation, imitation, and instruction that accompanies more general processes of socialization and enculturation in the family. Utilitarian knowledge of the environment, such as what trees are good for firewood, which grasses are best for pasturing small stock, how to get clean water from a riverbed, and how to distinguish weeds from crops are learned early by children in the Sudan (Katz n.d.), but increased schooling and lack of opportunity to find farmland may result in "deskilling" and a rapid decline in agricultural information. The key relationships that order the social world of the family—parent/child, older/younger, female/male, brother/sister—simultaneously both structure social and moral expectations and provide a conduit for environmental knowledge, task skills, and modes of labor organization. Firm concepts of authority, of responsibility, and of property seem to be especially important aspects of the training of intensive agriculturalists.

How the Household Works: Labor, Social Organization, and Meanings

Given the labor load that accompanies permanent agriculture, the need to perform daily chores dependably, and the seasonally specific bottleneck demands for concentrated effort, it is no wonder that a work ethic pervades the lives of smallholders. This is learned and reinforced in the household. When I showed the Kofyar a drawing of a boy watching a group of people hoeing and asked them what they thought was happening, they all replied that the boy in the picture was on an errand and had politely paused to greet the workers. When I insisted that the boy was in fact refusing to take part in the task, the response was that this was impossible: "If a child of ours does not work, we do not feed him!" I later saw this penalty for laziness imposed on an adolescent boy, and an adult male who preferred hunting to field work was relentlessly criticized.[1] Kinship rights and duties may be constituted in part by production and consumption activities within the household. The Kabre, intensive cultivators in the interior hills of Togo, have clear expectations of the generational division of labor:

Kabre say it is their parents' feeding of them as children which obligates them to work for their parents, both before and after marriage. Put otherwise, the parents' initial labor creates a debt that ties parents and children into a life-long reciprocal relationship. Thus, the process of producing and consuming in the household creates long-term dependencies not only between spouses but also between parents and children. Indeed the Kabre see "kinship" as deriving not from blood or genealogy but from the production and consumption process. . . . My "father" and "mother" are the people who fed me when I was young, not those whose blood I share. (Piot 1988: 273–74)

A folk model of social relationships that seems based on "bloodless" economic interdependencies and the flow, albeit very long-term, of goods and services along kinship lines seems to contradict the conventional wisdom of anthropology. In the following sections, we shall consider the radical dichotomies that social scientists have devised to separate rational calculation of advantage from altruism, the domestic mode of production from capitalism, and the moral economy, or economy of affection, from the impersonal market and political economy. In examining the smallholder household as simultaneously a corporate, enduring social group and a farm enterprise, we need to consider household activities and their meanings in a more integrated, systemic perspective. Nancy Folbre has reminded us that differences between household and market production are differences of degree rather than kind, that households and firms share common goals, and that they respond in similar ways to economic constraints. "Economic self-interest seems to penetrate even the most intimate domains of family life. This does not imply that household decisions can be explained in purely economic terms. It merely suggests that

1. The boy accused of indolence was a temporary resident in a Kofyar household, and his experience growing up in a Chokfem community of shifting cultivators may not have prepared him for the more stringent Kofyar labor demands.

the boundary between self-interest and altruism does not necessarily co-incide with the threshold of the home" (Folbre 1986a: 33).

The same structure of cultural understandings about the appropriate division of labor in a society or the social rules pertaining to postmarital residence and the rights and duties of kinship statuses can be approached through the concept of implicit contracts. Labor economists have developed this concept of contracts whose provisions are unwritten and not formally binding to account for situations in which employers and employees violate the precepts of neoclassical economics by refusing to optimize prices and wages respectively over short periods of time. For instance, employers who make large investments in training workers implicitly agree to increase workers' wages and job security as they gain experience, and employees accept lower wages during the training period and stay with the firm afterward, thereby achieving an upward trajectory of rewards. "It is the structure of incentives on both sides that embodies the implicit contract" (England and Farkas 1986: 22). The metaphor of the implicit contract has been used recently to analyze the long-term incentives and costs of marriage and other enduring cohabitational arrangements, of divorce, and of the "relationship-specific investment" of rearing children (ibid.: 22–23, 42–70). Unlike the flexible and impersonal transactions of the spot market or the legal provisions of a contingent claims contract, implicit contracts involve

(1) informal understandings rather than explicit contracts regarding the obligations of the relationship, and (2) enforcement only through the disinclination of either party to incur either the bad reputation that would come from violating the understandings, or the costs of search and new relationship-specific investment entailed in starting over. . . . By virtue of this informal understanding, transaction costs are decreased, a modicum of security is obtained, and options remain open for continued tinkering with the specifics of the agreement. Because they involve elements of both rational self-interest and social convention, such implicit contracts provide a vehicle for the integration of economic and sociological interpretations of behavior in both households and employment. (England and Farkas 1986: 47)

Because kinship, family, and household relationships are so interwoven with residence, labor, and the property of the joint enterprise, small-holder households can be seen as a system of functional implicit contracts.

Treating family relations as a kind of transaction or exchange illuminates the ways in which the implicit contracts of the household both resemble and depart from the more formal and explicit contracts of the business firm. Yoram Ben-Porath observes that the household transactions (1) extend over a long period of time, but the duration is not specified in advance, (2) include a wide range of activities, though not all terms are specified explicitly, (3) involve no explicit balancing of exchange and tolerate large outstanding balances without specifying when and how these balances are to be liquidated, and (4) embed the family contract in the identity of the partners, so that it is specific, nontransferable, and, to some degree, nonnegotiable (Ben-Porath 1980: 3).[2] The continuity through time, high value, and scarcity of the smallholding; the complex cooperative production of the farm family; and the kinship identities of household members in a reproductive group with considerable duration ensure that a great many important exchange transactions will take place according to the understood contracts of family-household affiliation.

The shape and the activities of farm households in the Swiss alpine community of Törbel derive from the cultural precepts and commitments of implicit contracts that form one, but by no means the only possible, social pattern for solving the problems of smallholders in a mountain environment (Viazzo 1989). In terms of residence, a newly married couple and the parents of each partner expect that a neolocal household will be established in an apartment dwelling separate from that of the parents. Apartments occupying a floor or a section of a large log chalet, or small freestanding houses, are bought and sold, inherited, or rented as units, and their cost and comparative scarcity may in the past have occasionally delayed the formation of new households. The autonomy of the household, which has its own dwelling, barns, storehouses, cellars, livestock, and diverse mix of fields, is emphasized by the separate postmarital residence correlating with the establishment of an independent enterprise.

In the late nineteenth century, 3 percent at most of Törbel households

2. Use of the term *implicit contracts* is not meant to exhaust the multiplex meaning and functions of social and psychological relationships within the household or the family. Rather I am emphasizing the economic aspects of these institutional relationships in the smallholder context. The emic view of household members may well be that their exchanges reflect the ideals of a perfect community in which the doctrine of "from each according to his abilities and to each according to his needs" actually holds. But as Sahlins points out (1972: 194–95), such generalized reciprocity consists of transactions that are "putatively altruistic," and "the expectation of direct material return is unseemly" or, at best, "implicit." "The material side of the transaction is repressed by the social: reckoning of debts outstanding cannot be overt and is typically left out of account" (ibid.: 195). Yet though there is no explicit quid pro quo, terms are not explicit, and relationships are not impersonal as is the case with contracts, reciprocity of this type seems to stand midway between amoral exchange and altruism, and it is consistent with self-interest. The relationship is one that Ronald Oakerson, following Hobbes, calls a covenant, as opposed to a contract. "A covenant . . . represents an agreement as to a set of principles or norms to govern future conduct. It draws as much upon moral reasoning as upon economic calculation. This in turn requires an ability to take on the perspective of others, and to consider the interests of others, as opposed to a *purely* self-interested orientation" (Oakerson 1983: 7). To see the material and economic side of long-term intrahousehold exchange and cooperation is not to deny the presence of unselfishness or the power of love.

ever contained more than a single married couple, and these few instances seem to have arisen because an elderly couple needed daily assistance or a wealthy family with an only child had brought in a spouse to provide additional household labor (Netting 1979). Unmarried children remained in the parental household, continuing to care for aged parents, and often forming an independent farming unit of celibate brothers and sisters after the deaths of the senior generation. Solitary individuals, often widows, bachelors, or spinsters, might continue to occupy separate premises and maintain their own farms, even when they took meals with a younger relative. The residential household was usually coterminous with the major farming unit. Work by household members in their joint farming operation was not reimbursed in wages, but the individual was entitled to support—food, clothing, and shelter—for as long as he or she remained co-resident. Parents brought up dependent children, trained them in farm and domestic work, and provided for their elementary schooling, while children were expected to work to the best of their ability under parental direction on the farm or, as representative of the household, on communal labor parties for the village.

Adult offspring who remained at home might assume authority gradually as their work contributions increased, or they might continue under parental direction until a parent's retirement or death. Personality factors and domestic negotiations made for considerable leeway in power relationships and labor management, both between spouses and between parents and adult children. The generality of implicit contracts neither ensures harmony nor avoids conflict. There was no question, however, of the necessity of carrying out the annual round of tasks on the farm. Both sexes and all ages from the age of about seven on took part in raking hay, tending and foddering livestock, milking, irrigating the meadows, and grape harvesting. Gender differentiation was manifested in male plowing, manuring, making cheese or herding on the alp, kneading bread, cutting and transporting firewood, pruning the vines, wine making, threshing, and slaughtering, while women gardened, fed the pigs, cooked and kept house, performed the spinning, weaving, and sewing, and took care of children. Women reaped grain with a sickle, while men used a scythe, but almost all tasks could be handled by members of the other sex when necessary. When men were working away from the village, women managed the farm on their own, and single adults could be self-sustaining. The best security for old age was to have adult children still resident in the village, and the value of this implicit contract to provide care for dependent elders may be reflected in marital fertility that shows no signs of family limitation (Netting 1981: 143). Children both worked on the farm and provided support for the aged, and there was the further expectation that unmarried children working at nonfarm occupations would send all but a subsistence minimum of their earnings back to their parents. Only when they married could they retain their entire wages (Netting 1981: 172). Over the duration of the family cycle from the marriage of the founding couple to their deaths or retirement, wealth flows went disproportionately from children to parents, an economic process that has been linked to the value of children and the demography of high marital fertility (Caldwell 1982).

If, on the other hand, implicit contracts of children were largely limited to their labor while in the parental household, and if their marriages and establishment of separate households effectively ended most of this cooperation, economic ties might be less powerful.[3] In the Swiss case, extremely valuable, heritable property in land of various types with attached water rights, in buildings, and in livestock enters significantly into both implicit contracts and the legally defined rights of household members. Men and women acquire by partible inheritance equal shares in the parental estate. Because the resources necessary to establish a viable household are not transmitted until the death or voluntary retirement (which may be at a very advanced age) of a parent, marriage may be delayed. There is no dowry or other provision for giving part of the farm to a son or daughter at marriage, so married children may have no rights to farmland, while their celibate siblings remain in the parental household and continue to be supported by the smallholding. Land that is inherited by a husband or wife is used jointly but it is only pooled when it is passed to their offspring. If a spouse dies without children, the land may be returned to the next of kin rather than remaining with the surviving husband or wife. Medieval documents show that family land could not be sold without the written permission of spouse and children. Siblings are implicitly committed to divide the parental estate equally, and they do so by agreeing among themselves on equal shares of the entire property and then drawing lots for the shares (Netting 1981: 173). Out-migrant family members can sell their portions at market rates, but they are expected to give their siblings rights of first refusal.

Rights to use such important communal property as the summer grazing alp, the forest, and the wastelands are held by only those village households descended in the male line from citizens. In-marrying wives, but not husbands, share in this privilege of citizenship; in the past, the village resources were such an important supplement to smallholder properties that only households with communal rights continued to reside in the community (Netting 1981: 70–89). Genealogical records show that, over the past 300 years, twelve Törbel patrilines have maintained

---

3. Among the Kekchi, who are shifting cultivators with little scarcity of land, frequent changes of residence, and easily constructed houses, labor-based implicit contracts between parents and children are relatively short-term (Richard Wilk, pers. comm.).

unbroken continuity, nine lines have died out or moved away, and not a single new male surname has been established in the village (ibid.: 71–76). When the access to a reliable livelihood through household labor and property is tied so closely to a farm and its resident kin group, we may predict that a web of implicit contracts will emphasize the interdependency and economic welfare of the group over a duration that may approach several lifetimes. Studies of European historical demography show that farmer families with significant rights in land tended to remain in the neighborhood for generations, while cottagers, landless laborers, and craftsmen were much more mobile as well as being considerably poorer (Schofield 1970; Skipp 1978). The smallholding, and the accumulated knowledge, skills, and labor investment that it embodied, served as an anchor, grounding the household in scarce resources that were difficult or expensive to replace. The best guarantee of household stability and reproduction was the intensively tilled farm that sustained it.

Engendering a Division of Labor

The need of the smallholder for labor that was diversified, yet coordinated, and the use of more work time per capita than in other agricultural systems (see Chapter 3), means there must be social mechanisms for mobilizing a small pool of workers effectively. The gender division of labor may allocate tasks in a complementary fashion within the household. The familiar example of males plowing and driving draft animals while females transplant, weed, garden, and tend domestic livestock is widely represented cross-culturally (Burton and White 1987). This does not mean that women lack the physical abilities and skills to plow, and even with their somewhat smaller body size and strength, they may do so when there is a pressing need (Segalen 1983). Kofyar men whose wives threaten divorce and leave the homestead right at the beginning of the rains, when labor needs are at their highest, must grind and winnow grain, cook meals, water the goats, and care for children (Netting 1969b). People sometimes dislike seeing the conventional division of labor set aside, even temporarily, and they may ridicule those who take on a task associated with the opposite sex, like a Kofyar woman climbing a palm tree to harvest its fruit, but the multiplicity of jobs and their frequent urgency in intensive cultivation suggests that there will be a de facto blurring of work roles. There are, however, undeniable rewards to specialization, and an increase in field work for men may see women participating less in direct crop production and more in food processing at home, care of stall-fed animals, and household tasks.

Massed and relatively undifferentiated work during labor bottlenecks, such as Kofyar planting after the first rains or Swiss hay raking, may be done by entire household groups, including adult males, females, and children. In a comparison of extensive, low-labor systems of agriculture with intensive plow or irrigated cultivation, total daily work hours for women went from 6.70 to 10.81, while men's work hours rose even faster, from 5.15 to 9.09 (Ember 1983: 285). Women may specialize in time-consuming daily activities, like the gathering of fresh mulberry leaves for silkworms, or they may be removed from agriculture entirely into cottage industry. Characterizations of the Chinese rural household as early as Han times refer to female weaving for the market as coordinate with male farming and equally necessary to the economic viability of the household enterprise (Hsu 1980). Kofyar beer brewing, an almost wholly female occupation, is tied both to farm work, because beer parties enlist the large groups of volunteers needed for yam mounding and other co-operative tasks, and to the cash economy, because women brew for sale during the dry season and on market days.

The existence in the intensive-cultivator household of demands for simple labor to be performed en masse, of specialized complementary tasks, and of both agricultural and nonagricultural subsystems, combined with the domestic operations of reproduction, enforce a heavy interdependency between adult males and females and suggest the likelihood that the conjugal family will most frequently meet these demands. People in a farming society do not cultivate solely as individuals, and the omnipresence of households as productive units derives from the gains of specialization in various tasks, gains that exist because of age- and sex-specific differences in aptitude, strength, knowledge, and experience (Binswanger and McIntire 1987: 81). More labor and skills, more diversified tasks, better coordination and management of work activities— these correlates of more intensive agriculture on a limited land base can be provided more effectively by the smallholder household than by single individuals or larger, nondomestic institutions.

Being a Farm Kid Is Not Child's Play

Unlike a business, a household recruits much of its labor by marriage and by birth, both involving long-term social and economic commitments. The bearing and raising of children takes place on the farmstead, where their training and early labor contribution are linked directly to the exploitation of the resources that feed them and the land they may eventually inherit. Given the demand on the intensive smallholding that labor be of high quality, responsible, and plentiful, children may be seen as an unquestioned good. Although both the Kofyar and the Swiss originally had little land to accommodate increasing numbers of households, they agreed that children were a great benefit to their parents, and that any sane householder wanted more working offspring rather than fewer. As Adam Smith (1776, cited in Becker 1981: 97) remarked about colonial America, "Labor is so well rewarded that a numerous family of children, instead of being a burthen is a source of opulence and prosperity to the parents. The labor of each child, before it can leave their house, is computed to be worth a hundred pounds clear gain to them."

Children in Java and Nepal are producing economic returns equal to the costs of their upkeep by the age of 9, and by the time they are 12, they have in effect repaid the total costs to their parents of raising them to that age (Nag et al. 1978). Bangladeshi children begin economically useful lives around age 6, gathering fuel, fetching water, carrying messages, and caring for younger children (Cain 1980). Boys care for cattle at the age of 8 or 9, begin crop work at 11, and take on the skilled and difficult task of transplanting rice at 14. Girls are processing rice and preparing food by ages 9 to 10. Children reach the full adult labor input of over nine hours a day by age 13, and they are already putting in half of this time by the ages of 7–9 (ibid.: 237). Male children of landed households working on their own fathers' farms have a higher rate of return on their labor than the children of the landless, who put in more total hours for wages. Between the ages of 10 and 13, boys become net producers, in that the average daily product of their work in calories exceeds their average daily consumption (ibid.: 243). Home handicrafts and marketing may absorb the time of children that cannot be expended on the farm. A Philippine survey showed children contributing 15 percent of market income and putting in 30 percent of the farm family's productive time (Folbre 1984). Indeed, children may be even more continuously occupied with a simple operation like twisting and winding rope than they would be in farming, and villages where this domestic craft predominates in Guatemala show correspondingly larger family size and earlier marriage (Loucky 1979).

Though there are certainly costs in food, housing, and especially mothers' time in raising children to the point where they are producing economic benefits for the household, these expenses are less for farm than for urban families, and the children are much more productive, accounting for the high demand for children under these circumstances (Becker 1981):

Even in peasant villages of high population density, households with a relatively large number of children appear to ensure themselves a lengthy period of economic "success" during the latter phase of their development. The duration of this period depends both on the parents' ability to produce children who *survive* . . . and on their ability to retain control of their children's labor by postponing their dispersal from the household. (Nag et al. 1978: 300)

Those Swiss whose children married late or not at all, or who received remittances from a single child even after he or she left home, indicate the extent to which the productive potential of children can be retained by their parents. Even younger children have a considerable value for necessary tasks of low marginal return like scaring birds, herding, and water carrying (Cain 1980: 245). Such light but time-consuming work frees adults for activities with higher output. It appears that intensive agriculturalists have more children than extensive farmers, perhaps by means of limiting postpartum sex taboos and changes in diet and nutrition that decrease birth spacing (Ember 1983).

Higher fertility also improves the chances of old-age security. In a Javanese peasant village with 121 elderly couples or individuals, an average of 2.5 children out of a living sibship of 4.0 were resident in the village. Only in 13 cases did elderly people have no living children present, and 88 lived with various combinations of sons and daughters (Nag et al. 1978: 299). Land is increasingly divided equally among male and female heirs (in contrast to the traditional system, where females received only half a male share), and there are various arrangements by which land may be transferred to adult children during the parents' lifetimes in such a way that the parents are assured of either income or food (Nag et al. 1978: 298–99). Again property mediates a relationship in which children provide care and sustenance for their aged parents.

Household Help and Hired Hands

It is only recently that agricultural economists have begun to devote concerted attention to the comparison of household and hired labor. They have noted that smallholders tend to have a more intrinsic grasp of the agronomic attributes of their land, and that the household allows efficient transmission of this knowledge and the requisite skills, but they have also dealt with the supervision and motivation problems in the application of labor (Binswanger and Rosenzweig 1986, Rosenzweig and Wolpin 1985, Binswanger and McIntire 1987, Ellis 1988). Part of the reason why largeholders and smallholders obtain differing yields from the same land area is that hired labor on big capitalist farms is paid more and has higher opportunity costs than household labor. Large commercial farms characteristically use less labor per ha than small farms dependent on family labor (Hayami and Kikuchi 1982). These differences are often accounted for by variation in transaction costs, "the negotiation, supervision, and enforcement costs associated with contracts" (Alston et al. 1984).

The costs of searching for and recruiting or hiring labor may be quite high where there is easy access to land and the technology is simple. A worker's output on his own plot would be at least as large as on an employer's, and the employer has to spend time finding and then supervising the worker. With these extra costs, the employer cannot compensate the worker for the output from his own plot forgone by the worker (Binswanger and McIntire 1987: 76). As the Kofyar moved into the plentiful land of the Benue Valley frontier, young men worked only briefly for established farmers, learning the lay of the land and acquiring some seed yams before striking out on their own. With land almost literally for the taking and labor as the scarce resource, individuals used the workers they could count on: themselves and the members of their households. There were no landless Kofyar who had no choice but to work for wages, and the preferred option of all the migrants was to establish homesteads of their own. When household labor was inadequate to cultivate a millet / sorghum field or to expand the area devoted to the yam cash crop, Kofyar preferred to stage a beer party and enlist the help of neighboring Kofyar households. Farmers justified this choice to me by saying that (1) neighbors could be collected in large groups of 40 to 80 members, while wage laborers coming from other ethnic groups on the plateau arrived in little groups of 3 or 4, (2) Kofyar could arrange reciprocal beer parties for particular dates, whereas you never knew when the transient laborers would be available, (3) the wage costs of ₦ 5 per 100 yam heaps plus food costs, increased by the workers' negotiation for meat, oil, and other expensive items, added up to much more out-of-pocket cash than the millet for brewing and the unpaid work of the women who made the beer, and (4) the work might not be well done, and the laborers might leave before it was completed. In other words, the cost of searching for local workers was minimal; moreover, they both constituted a larger group than the available pool of outside wage laborers and were more dependable and less expensive. Kofyar neighbors, like household members, were familiar with the careful operations of intercropping and ridging, so that the quality of their work was higher and they did not require monitoring.[4] Indeed, the owner of the field might not even have to be there when the volunteers marked out the sections to be done by individuals and went to work. Neighbors also know that their work carries with it an obligation by their host to reciprocate, so they can count on experienced help when they in turn hold a beer work party.

4. The new institutional economists, following Douglass North (1981), emphasize the role of transaction costs, including the crucial dimension of labor quality, which supplements the factors of price and quantity conventionally considered by neoclassical economics. A person's common sense, intelligence, knowledge, skill, and especially trustworthiness are at least as critical to the profitability of the employer as how many hours he or she works (Ensminger 1992). Household labor is likely to be of high quality and dependability, with less cheating and shirking, which increase transaction costs.

Household members do have substantial "committed" costs, because they must be fed for the entire year, and younger members must be supported and socialized until they are old enough to contribute to the labor pool. The sick and infirm have an implicit guarantee of basic maintenance. But all able household members are there when the exigencies of climate or crop growth demand work, they can put in very long hours, and they ask for neither cash nor extra benefits for doing the job.[5] Because tasks are learned and training given as a part of growing up, family members generally need little further instruction. Working with spouses, parents, or siblings means that household members are often doing the same thing at the same time, at a pace with which they are familiar and in congenial company. The structure of incentives also favors the household, because individual members receive as consumers what they produce, and they have a continuing corporate interest in farm resources and in the returns on the labor and other investments they make:

Family members, being residual claimants, have more incentives to work, for given supervision, than hired wage workers. Moreover, more information is available on the family workers over the long-term, hiring or search costs are minimal, and family laborers share in the risk. Of course while the incentive problem is smallest for family members, it is only incompletely solved. When marginal product cannot be precisely identified and effort cannot be discerned, shares in profits will be independent of effort. Thus the larger the number of claimants (the greater the family size), the greater the incentive problem for each individual family worker, as shares are diluted. Nevertheless, the problem will always be less than for the time-wage, hired workers. As a consequence, as the size of the labor force grows, any fixed amount of resources devoted to supervision will have smaller effects on each worker; labor costs will rise with farm scale. (Binswanger and Rosenzweig 1986: 520)

Information about the habits, character, knowledge, skills, and reliability of farm workers is costly to acquire, and for a farm manager to investigate the background and abilities of an itinerant day laborer would not be worth the trouble. The next best alternative is to keep the worker under surveillance, but the time of an overseer is also expensive, especially when a variety of field activities in widely separated areas are taking place simultaneously.[6] A kin-based household reduces these transaction

5. Polly Hill (1982) regards it as axiomatic, on general grounds, that Hausa farmers prefer family labor to hired workers, and they are usually anxious to retain the services of their married sons. Though "fathers would never stop to calculate (if they could) the relative costs of family and hired labor," the preference for the household source derives "from its superior exploitability and instant availability, as well as from its greater reliability, so that it requires no supervision" (Hill 1982: 92).

6. If productive tasks are diverse, in terms of a variety of operations in different locales, and also simultaneous, in that they cannot be done in a linear fashion by one person (Wilk

costs and the uncertainty inherent in hiring outsiders. The co-residential group "is a reasonably effective 'insurance company' in that even an extended group is sufficiently small to enable members to monitor other members—to prevent them from becoming lazy or careless, and in other ways taking advantage of the protection provided by their kin. The characteristics of members are known and their behavior is easily observed, since they live together" (Becker 1981: 238). A household head may schedule tasks and actively cooperate in many of the group activities, but there may be little overt direction and few commands, because members know what they are supposed to do. Some knowledge may be unequally distributed—for example, the women who plant wet rice in Gambia may be the household experts on the plant varieties and keepers of the seeds (Weil 1973)—but such specialized information is respected by males and forms an important part of the division of labor.

There are also certain areas of individual choice and flexibility regarding when and how some of the myriad tasks are completed. I remember returning to a Kofyar village after a trip of several hours with the local chief. When we came in sight of his homestead, he pointed out to me with delight that his three wives had continued working on the farm, even though he had not been there to help them. It appeared that he could not compel or control their participation, other than by setting a good example and sharing in the effort (Netting 1969b). Household members may also devote some time to their personal fields from which the proceeds do not go into the household pool. Priscilla Stone (1988b) has evidence that Kofyar women in the cash-crop area spend up to one-fourth of all their agricultural labor time on their individual fields or those of other women, and their value as key producers in the household labor force may encourage their husbands to grant them usufruct land rights and time off from both the homestead farm and household exchange-labor groups (M. P. Stone 1988a; Netting et al. 1989). Where production depends not only on amount of labor but on the individual skills and conscientious application of the experienced individual worker, coercion and micromanagement by a boss can be counterproductive. The household as a small, enduring, self-reproducing kinship unit can mobilize disciplined and responsible labor for intensive agriculture in a way the wage-labor farm cannot match.

---

and Netting 1984: 7), we might expect to see large households engaged in a complex division of labor. This is the "practical reason" for Fijian multiple-family households, where members may be tending distant yam and taro gardens, harvesting coconuts near the shore, and fishing or collecting shellfish from the reef at the same time (Sahlins 1957). The products are pooled and consumed in the household, but the tasks need not be directly supervised.

An alternative means of enlisting labor is by paying a daily wage. Even after recruitment and training expenses, however, the transaction costs of supervision and enforcement associated with explicit contracts of this kind may be high:

A daily paid laborer has no incentive to work hard, unless supervised closely via direct observation of his effort or via monitoring or inspection of his output. Incentives to work hard may be improved by providing share contracts (piece rates at harvesting or for earth digging, or crop sharing tenancy contracts). Since the worker receives only a share of the full marginal product of his effort he will still not work as hard as an owner-cultivator, unless again he is supervised or monitored in other ways, and/or is penalized in terms of loss of future repeat contracts. (Binswanger and Rosenzweig 1986: 507)

Tasks that are vital to a farming operation, but have relatively low returns to labor time, like frightening birds away from ripening rice or herding goats, are better performed by unpaid family members with a low opportunity cost, whose eventual product returns in part to them and who claim a long-term share in the benefits of the diversified smallholding. If Little Boy Blue who sleeps under a haystack while his cows are in the meadow and his sheep are in the corn is a paid employee, he can only be fired. If he is the son of the owner of the livestock damaging neighbors' crops, he may be subject to punishments that only an enraged parent can mete out.

There are, of course, occasions when family labor, despite the absence of hiring costs and its incentive advantages, is not adequate to the labor demands of the smallholding. Where there is a marked seasonality of farm operations and labor demand, it may be optimal to hire supplementary workers for the peak periods. A permanent force of household laborers that could meet all bottleneck requirements might be less productive during the slack period while they are waiting for their peak employment. Conversely, relying on hired labor alone to beat the peak would create problems in a smallholder area where certain crucial operations may be synchronically timed across farms and yields may be substantially reduced by delay. Therefore, the optimal permanent household labor force is likely to be greater than the minimal or off-peak quantity of labor demand, and family members may seek slack-period off-farm employment (Binswanger and Rosenzweig 1986: 520). The same smallholders who hire extra workers may themselves work for wages in agriculture or cottage industry, or as migratory laborers at other times.

In northwestern Europe, the traditional pattern of life-cycle servants was used to even out household labor demands. Land-poor or cottager families could reduce consumption by sending out children as live-in servants to larger farmers, and many young single men and women, even from affluent landed families, served a kind of apprenticeship as farm-hands or domestics before marrying and taking up their own farms (Kussmaul 1981). Although they received low wages, these hired people lived in the household, ate and slept with the family, and were treated as temporary junior members who would eventually themselves be small-holders. When growing rural population, more commercialized farming, and increasing stratification between the landless and propertied peasant households developed in Scandinavia, there were pressures to remove "lazy" and "uncouth" servants from the family circle at meals, to provide separate and inferior sleeping arrangements, and to treat them as a separate proletarian class who could not aspire to own land or marry into farmers' families (Löfgren 1974). An ethic of household participation by some non-kin workers was succeeded by the social and economic separation of family and wage-earning outsiders.

Even within various forms of agricultural contract, such as wages, shares, and fixed rents, there are differences in supervision costs. Piece rates, such as the Kofyar standard payment of ₦5 per 100 yam heaps, provide a superior incentive compared to time rates. They are only feasible, however, for operations where worker-specific output is easily measurable and where quality of output does not decline with increased speed (Binswanger and Rosenzweig 1986). Paying by area weeded may not be effective, because a worker might easily miss weeds or damage crops in hastening to complete the field. The varied tasks and emphasis on labor quality in intensive agriculture make it difficult to measure individual output and assign piecework rates.

Supervision costs tend to decrease the more closely effort is linked with reward. For this reason, supervision costs are greatest under wage contracts, since the reward—the wage—is based on the amount of time a worker spends rather than on his output. On the other hand, supervision costs are least under a fixed-rent contract because the landlord receives his payment for land regardless of the length and intensity of work effort. . . . Since workers under sharecropping receive only a share of their marginal products, sharecrop contracts occupy an intermediate ground in the supervision cost spectrum. (Alston et al. 1984: 1124)

If a managerial, landowning farmer is producing crops that are labor-intensive and require valuable inputs like fertilizer and motorized equipment, there will be considerable incentive to monitor both labor and other inputs directly. Careless use of work stock, equipment, or land might result in its depreciation, and incorrect plowing, pruning, fencing, or irrigation might affect future output. However, the costs of constantly monitoring the wage workers might well be prohibitive. Fixed rent and sharecrop contracts reduce the need for direct supervision of labor and increase the benefits to labor.

TABLE 2.1
Ibo Family and Non-Family Farm Labor

	Okwe (low density)	Owerre-Ebeiri (high density)
Man-hrs of family labor / household	734	577
Man-equivalents available for farm work	3.9	2.5
Man-hrs of family labor / man-equivalent	188	231
Non-family labor / household	470	200
Proportion of non-family labor	39%	26%
Labor cost in naira / household	₦37.17	₦14.48
Non-farm income	300	721

SOURCE: Lagemann 1977: 90, 109

The self-supervision characteristic of smallholders is also present to a lesser degree among intensive cultivators who sharecrop or rent. Share-crop contracts secure labor better than wage contracts because the opportunity cost of leaving (the expected harvest payment to the worker) exceeds that of wage hands who are paid on a time basis, bear no risk, and require considerable supervision (Alston et al. 1984: 1126). Under the terms of fixed-rent contracts, the landlord does not even need to monitor output, and the tenant bears the risk of crop failure. In return for lower transaction costs to the landlord, the renter retains the net proceeds of his energy and skill after deducting a fixed payment. Where the renter's labor and management inputs are particularly important, as on a small, intensively tilled farm, the higher returns to labor of renting may be an important incentive, the renter's expected income will be greater than the sharecropper's, and the landlord will have fewer enforcement costs in ensuring that the renter honors the length of his contract. With the generally lower cost of family labor, a large landowner will find it more profitable to rent out land rather than to manage hired workers (Rosenzweig and Wolpin 1985).

Though low fixed rents and heritable long-term contracts may lower transaction costs and raise returns to tenants, it appears that maximum incentives are still present when a household administers its own labor on a smallholding on which it has rights to the output it produces. Hiring labor is not, however, confined to large, and usually more extensively tilled, farms. Intensive cultivators may also employ non-family workers, especially when the farm household is deriving substantial income from nonagricultural tasks and there is a pool of landless peasants, microproprietors, or seasonal migrants on which to draw. Ibo farmers in low- and high-population-density villages make considerable use of non-family labor hired on a contract or exchange basis (Table 2.1). Household members continue to do many of the more sensitive, skilled tasks, such as burning, planting, staking the yam vines, and thinning young plants, and almost all harvesting is carried on by the family. Non-family labor is used for heavier jobs like clearing (52 percent of total labor), ridging and mounding (44 percent), and weeding (43 percent), with a total of 37 percent of all agricultural labor, excluding tree crops, coming from outside the household (Lagemann 1977: 225). Okwe, a low-density village with bigger farms and more fallowing, uses fewer man hours per man-equivalent in the household but considerably more non-family labor than does land-scarce Owerre-Ebeiri (Table 2.1). Where farms are small, the proportion of non-family labor to total labor is smaller, and the total labor cost per farm may be less than half that in a village where the average farm is bigger. Cash expenditure for hired labor was significantly related to farm size in both Okwe and Owerre-Ebeiri (Lagemann 1977: 89), suggesting that farmers with relatively more land supplement their household labor supply. Where there is significant household income from other sources, as in Owerre-Ebeiri, the cost of hired labor was found to be significantly related to nonfarm income ($R = 0.77$) (Lagemann 1977: 89). Smallholders continue to perform more intensive tasks themselves, but they may rely increasingly on hired labor for the work of shifting cultivation on bigger farms and also in circumstances where the time of household members is going more heavily into various nonfarm occupations.

The extent of hired labor, purchased agricultural inputs like fertilizers aand pesticides, and paid machine services may be quite high without displacing the smallholder household from its central role. Egyptian cultivators of irrigated wheat, clover, beans, cotton, and maize are so vitally involved in a market economy that Nicholas Hopkins (1987a, 1987b) designates them petty commodity producers rather than peasants. Farmers hire a tractor and driver for plowing; purchase irrigation water from diesel-pump owners; pay harvesters, loading crews, specialized winnowers, and threshers; and rent camels or tractor-drawn wagons for transport, as well as tractor-powered drum threshers. They sell wheat, fodder straw, and cotton to different sets of merchants. The household's adult males sow, fertilize, and manure their crops by hand and carry on the various irrigation tasks. Women take care of the stall-fed water buffalo and cows found in 64 percent of the households and make the cheese that is the most common food, along with bread, in the village (Hopkins 1987b). Sheep, goats, poultry, and rabbits are raised in most households, and children cut clover for the animals. Paid crews of children pick cotton and destroy cotton worms. This variegated mixture of household and hired labor, animal power, and machine services is centrally organized by the individual farmer

and only he and other household members are really in a position to follow one crop through all the different steps needed to produce it. Thus the managerial function remains located in the household. Hired hands and hired machines are paid for their labor, and they are not concerned with the quality of the end-product. Hence even though the household only provides a small part of the labor, it retains its centrality because of its managerial role. The devolution of the managerial role to the household level minimizes the need . . . for the hierarchical control of labor. (Hopkins 1987a: 165)

The household can efficiently organize, recruit, and control labor because of the small size and shifting composition of the work gangs, and because key daily tasks are articulated through the sexual division of labor. Despite the frequency of cash employment and unequal commercial relationships with machine owners and richer patron households, household members are engaged in a dense interaction of cooperation and exchange, articulated not through money "but through cultural understandings about appropriate roles" (Hopkins 1987a: 166).

Household Harmony?

Any generalized, functionalist model of the relationships between smallholder agricultural production and household organization risks overemphasizing positive or adaptive features and tends to attribute an idealized, ahistoric stability to the smallholder household. Conflict and change are neglected, whereas universalized economic principles are given explanatory power over a broad range of different cultures and environmental circumstances. The feminist critique of using the household as a key unit in social analysis is based in part on the lack of neat congruence of production, consumption, and reproduction within the co-residential domestic group. Those who focus uncritically on the household are also censured for ethnocentric attitudes presuming sharing and "joint utility functions" within a household where there may, in fact, be great inequality, patriarchal dominance, and exploitation of women and the young (Guyer 1981, Folbre 1986b, McMillan 1986). A highly integrated and complementary division of labor, implicit contracts that provide for long-term reciprocities in care for children and the elderly, and enduring relations to crucial productive property do not guarantee a spirit of harmony or the continuity of the corporate group. As Nancy Folbre (1984) points out in her criticism of the neoclassical economists' concern with household behavior as motivated by efficiency, (1) altruism in the family coexists with conflicts of interest over the distribution of goods and leisure time, (2) individual shares of family income are determined in part by individuals' bargaining power within the household, and (3) the relative bargaining power of men, women, and children changes in the course of economic development.

The gender roles of Chinese male farmers endow them with superior decision-making power and economic privilege, while their jurally dependent wives and children are subordinate. Sons, who will stay on the family farm and care for their parents, are preferred; their schooling and marriage are paid for, and they are given more personal autonomy. Daughters, whose shorter-term implicit contracts last only until they marry and move away, begin domestic chores at the age of five or six, receive less education, and are expected to remit outside incomes to their parents (Greenhalgh 1988). Daughters are socialized to feel that they are personally worthless and owe everything to their parents. An earlier pattern of bringing up child brides in the homes of their parents-in-law emphasized intergenerational authority at the expense of a strong emotional relationship between the spouses, who were raised together. Women lacked power in the household until later life, when they might exercise control over their sons and especially daughters-in-law (A. Wolf 1968).

Japanese rural households with a stem-family organization (a senior couple and their co-resident married son and heir) put the son's incoming spouse at a severe disadvantage. She lacked a long-standing relationship with other members of the household and local community, and she was not familiar with the household's political workings and financial resources. The new partner had to "prove herself in order to gain access to resources" (Cornell 1987: 151). Because the junior couple could not become master and mistress of the house and farmlands until the older generation were dead or household division occurred, they might be well into middle age before gaining the status of household heads. This "subservience of adults to their elders" gave the system "a strong patriarchal flavor" (ibid.), and norms of respect could not alone prevent friction and covert hostility within the household.

The control of property and the economic considerations of inheritance that established the viability of a newly formed household also engendered conflicts over marriage. In the Swiss Alps, the parents' preference that their son or daughter marry a spouse with wealth in land and cattle might run contrary to the desires of the younger person, sometimes resulting in lifelong celibacy for the child (Netting 1981: 175). An Austrian folk song cited by Lutz Berkner (1972: 403–4) portrays the anguish of a peasant son who is balked in his wish to form his own household.

> Father, when ya gonna gimme the farm,
> Father, when ya gonna sign it away?

> My girl's been growin every day,
> And single no longer wants to stay.
> Father, when ya gonna gimme the farm,
> Father, when ya gonna gimme the house,
> When ya gonna retire to your room out of the way,
> And dig up your potatoes all day?

Turning over property to an heir often entailed a written contract stipulating in detail the annual amount and type of food, the rights to garden, firewood, and a cow, and the lodging to which the retired couple were entitled. That a legally enforceable contract here formalized the implicit contracts of members of a domestic kin group suggests the existence of competition and a major opposition of interests within the household.

The recurrent drama in the developmental cycle of the farm family is often the fission of household units, especially where multiple-family households are common. Though Kofyar seldom expressed anger in the father/son relationship, and the household head commands the respect and obedience of his children, a married son may take exception to his father's division of the grain harvest among the various wives. Notwithstanding that the son may have been contributing most of the farm labor, his appropriation of the right to allocate the harvest might precipitate a split with his father and oblige the son and his wife to move off the parental homestead. There are undoubted benefits to a large, multiple-family household in the plains cash-cropping area, but a Kofyar man who has worked under his father's direction and received the bridewealth allowing him to marry and begin to have children may decide to set up his own independent household. Although he has had the benefits of the larger labor force, the accumulation of capital necessary for bridewealth, and the insurance against sickness and bad crop years provided by household storage and wealth accumulation (Binswanger and McIntire 1987), he may feel that his current contributions to farm output are not balanced by the payments (consumption from the common household store and cash for personal use) that he receives.

The Salience of the Smallholder Household and the Organization of Consumption

The disputes and negotiations over marriage, rights to resources, authority, and co-residence that we have been discussing take place between individuals, and it is possible to regard the household in which they take place as what Peter Laslett calls "a knot of individual interests" (Laslett 1984: 354). Other corporate groups, the lineage or clan, the village community, the business firm, the church, or the state, may seem much more directly involved with production, consumption, and accumulation than is the household. British social anthropology in the tradition of A. R. Radcliffe-Brown saw social groups as constituted around the jural relations between persons, with the prototype in the domestic domain being dyadic links involving husband/wife, parent/child, and sibling/sibling relations (Saul 1989: 350). The boundaries of the household may fade in Africa, where spouses often own property separately and retain rights to their own earnings, and where what they accumulate devolves differently for men and women (Goody 1976). The joint conjugal estate found in Europe does not exist in Africa (Saul 1989: 350). It has been claimed that household and kinship institutions as part of social structure have an a priori existence—they "comprise endogenous, independent variables that are not reducible to exogenous forces and conditions," such as those of ecology and economics (Fortes 1949: 289). I am arguing to the contrary that even in Africa, when the household is part of a system of intensive, permanent agricultural production, with labor provided by smallholders and continuing rights to land and other scarce resources, the functional relationships of the household and the farm enterprise will be direct, powerful, and readily discernible. The household's form and functions are certainly responsive to social norms and cultural rules embedded in regionally specific, historically persistent systems of marriage, descent, and the rights and duties of kinship. But there are also significant lines of causality leading from the material conditions of livelihood to the size and composition of the household. I want to demonstrate that the smallholder household is a particularly flexible and responsive social grouping, "sensitive to minor, short-term fluctuations in the socioeconomic environment and a prime means by which individuals adapt to the subtle shifts in opportunities and constraints that confront them" (Netting 1979: 57).

The smallholder household is not only the primary institution for mobilizing and administering labor on the farm; it also sees to its own reproduction and long-term security by organizing consumption. The primary distinction between a commercial farming enterprise, even when it is family operated, and a smallholding is that smallholders consume at least part of what they produce. An important part of what they eat comes from their own fields, orchards, and barnyards. Though some food may be purchased and a substantial part of their agricultural produce may be sold for cash, there remains a subsistence component necessary to household maintenance that is self-provisioned. The integration of smallholders into the market economy is never so complete that they are fully dependent upon it (Ellis 1988: 9). The persistence of a peasantry in Europe over centuries of warfare that laid waste the countryside, of confiscatory exactions by feudal lords and state governments, and of economic booms and depressions is a testimony to the worth of a home-controlled productive base (Greenwood 1974).

Smallholders consciously decide to provide part of their own household subsistence rather than becoming entirely specialized, market-oriented producers of agricultural commodities. The Embu of Kenya diversify both their cropping and livestock, as well as off-farm enterprises:

> The Embu view a good farmer as one who grows enough maize, beans, potatoes, sweet potatoes, cassava, and bananas to feed the family without purchasing such staples and without relying on income from wage labor to purchase goods that could be grown at home. This local ideal is precisely the opposite of the pattern [of specialized cropping and market dependence for selling products and buying labor and food] called for in theories of economic development. . . . Instead the Embu, like many other African farmers who must deal with uncertain market and state institutions, find it wiser to reduce the risk of hunger or starvation by producing a number of different crops for home consumption, and to avoid selling food crops (ideally) until they can assess the quality of the next harvest. (Haugerud 1989: 70)

Given the unpredictability of rainfall and destructive storms, of livestock disease and insect plagues, the smallholder strategy must be multiyear and minimax. It must be geared to the "law of the minimum" that posits survival on meeting caloric needs at the low points of the environmental cycle rather than at the average.[7] The efforts involved in producing fodder and seed for the next year, processing food supplies, building storage structures, and maintaining farm equipment all form what Eric Wolf (1966: 6) calls a replacement fund. Whereas shifting cultivators may put away only enough to get part way through the "hungry season" before the next harvest (A. I. Richards 1939; Kay 1964), intensive farmers may select crops for their storage capabilities rather than their high yields. The Kofyar harvest the tiny seeds of acha in part because its low moisture content allows it to be sealed for years in a mud granary without spoilage. Unlike the daily staple of sorghum grain divided among the adult women of the household, the male head keeps the old acha separately, using it for ceremonial occasions or when there is a dearth of local food. Similarly, a Swiss household puts away threshed rye and baked bread under lock and key in its *Speicher*, and old cheeses may be husbanded in the cellar on special, ladderlike racks for years. Wine, an important source of calories in the diet of the Portuguese and other European peasants, can not only

7. The "law of the minimum," stated by the plant physiologist Justus von Liebig in 1840, is the ecological principle that "the occurrence and success of an organism in a given situation are limited by certain materials essential for growth and reproduction, and that those materials most closely approaching the critical minimum (that is, in shortest supply) will tend to be the limiting ones" (Ellen 1982: 34).

be carried over from year to year but may improve with age. Large livestock are a self-reproducing, mobile store of wealth "on the hoof," and the expected value of Swiss dairy cows for milk and calf production is only realized if they are kept for eight years or more.

Household labor produces and preserves such stores, and family wealth and success may be measured in their abundance. Widening the network of people beyond the household entitled to share in these carefully hoarded supplies might encourage the slacking and the free-rider problems that the internal mechanisms of the household control. Though households may make gifts from their stores and fulfill obligations to kin and fellow villagers, replacement food and livestock are jealously guarded against those emergencies when they become crucial to survival. They do not go into an interhousehold communal pool.

The household, with its long personal associations and its authority structure built on age, socialization, and experience, is also particularly well adapted to regulating consumption. The stringencies of crop failure or individual incapacity can be shared within the group with smaller and less frequent meals or a general belt-tightening. Along with hard work, thrift (*Sparsamkeit*) was the virtue most honored by the villagers of Törbel. In the old days they ate from a common pot with wooden ladles or spoons, and the amount of cheese in the soup and the frequency with which dried meat (*Trockenfleisch*) or home-cured sausage (*Hauswurst*) appeared on the table were daily measures of household self-sufficiency and future prospects. During the Great Depression, Swiss who lost jobs in the city retreated to the farm, surviving on potatoes and skim milk and enjoying coffee only on Christmas Eve. The same ethic that called on family members to do extra work at very low marginal rates of return could enforce a frugality in consumption to cope with scarcities that were chronic but not crippling. Shared privation and the deferral of individual gratifications could also be justified as benefiting all family members in the long run and likely to improve the corporate position of the household. A Swiss friend, recalling his family's sacrifices to buy farmland, said his father had roused his sons each morning with the admonition that they must work hard because they were 10,000 francs in debt. The implication that security and eventual affluence could only be achieved by the mutual sacrifice of individual interests to the common aims of the household and its estate was inescapable.

The Fit of Farm Size and Household Size: A Smallholder Adaptation

If a smallholding is to produce optimally, it requires certain amounts and types of labor, and if the smallholder household is to survive and prosper, it needs employment opportunities for its members. The dy-

TABLE 2.2

Chinese Agricultural Landholdings and
Household Size, 1929–33

	Mean size of family		
Farm size	North China	South China	China
Small[a]	3.98	3.94	3.96
Medium	4.57	4.48	4.52
Medium large	5.13	4.93	5.02
Large	6.07	5.49	5.76
Very large	7.92	6.80	7.31
TOTAL	5.44	5.01	5.21

SOURCE: Buck 1956: 368, 370.
[a]Smallest one-fifth in locality.

TABLE 2.3

Hungarian Peasant Landholding and Household
Size, 1941

Landholding	Families (N)	Household size (mean)
Well-to-do (over 11.50 ha)	27	5.18
Medium (5.75–11.50 ha)	73	4.52
Small (2.87–5.75 ha)	89	4.46
Indigent (0–2.87 ha)	237	3.92
TOTAL	426	4.11

SOURCE: Fél and Hofer 1969: 403. Used with permission.

TABLE 2.4

Japanese Rural Mean Family Size, by Size of Landholding,
1716–1823

	Holding		
Year	Small	Medium	Large
1716	1.6	2.5	4.1
1738	2.7	3.8	5.1
1780	3.0	4.3	5.6
1802	4.0	4.5	5.8
1823	4.9	5.1	5.4
Average	3.7	4.2	5.3

SOURCE: T. C. Smith 1977: 123.
NOTE: Holding size is estimated in *koku*, a measure of rice yield equal to
approximately 5.1 bushels. Small holders are those with 0–4 *koku*; medium,
4.1–18.0 *koku*; large, over 18 *koku*.

namic process of continual adjustment between the numbers and skills of household members and the agrarian resources to which they have access is particularly important among intensive cultivators. It is apparent that smallholders everywhere strike some kind of economic balance between household numbers and land size. John Lossing Buck's (1956: 358, 360) huge survey of pre–World War II Chinese farmers shows a strong rank ordering of household size, with farm sizes stratified into five levels. Where the smallest holdings had on the average just under four household members, very large farms had mean household populations of 7.92 in North China and 6.80 in South China (Table 2.2). Less intensive but still diversified, permanent, smallholder cultivation in Hungary over the range up to above 11.5 ha gives correlated household sizes of from 3.92 to 5.18 (Table 2.3). The mean size of Japanese farmers' households in the village of Nakahara showed a constant ranking to the three classes of farm size (Table 2.4), even though growing population from 1716 to 1823 forced up the numbers at each level. That farms as small as 0–4 *koku*[8] were finally occupied by families averaging 4.9 members suggests that not only agricultural intensification but a growth in part-time employment and off-farm occupations had been taking place over the 107-year period. The same relative ranking is seen when categories of wealth from tax rolls or numbers of draft horses per farm are correlated with household size (Netting 1982b).

As a rule, it is safe to say that where resources gather in rural societies, so do people. There may be higher levels of nutrition, more births, and better survival of children on larger and richer farms. Kin may be drawn to an affluent household to enjoy its "relative" advantages. They, as well as unrelated servants, apprentices, and retainers, may also increase the productive labor force, which in turn adds to household prosperity. But people who are not members of the primary family or who belong to associated "branch" families (T. C. Smith 1959) may not share equally in household resources. If the farm is too small to absorb more labor productively, children may be sent out as servants to households with a labor shortage, or earnings from nonfarm work may be pooled to acquire more land. Older couples or solitary householders may lease out land, and widows or widowers may move in with adult children, combining farms. With little fallow or opportunity to use land in a suboptimal manner, smallholders are under pressure to match land, labor resources, and consumption needs by regularly fine-tuning the household enterprise.

8. The *koku* is a measure of rice yield approximately equal to 1.8 hl (5.1 bushels). Fields were assessed for tax purposes in terms of rice or equivalent crops produced, and houses were similarly given a value translated into rice (Kinoshita 1989: 48). One koku would barely feed one adult for a year (T. C. Smith 1959: 25).

The Household as a Flexible Source of Farm Labor

If land availability in areas of intensive cultivation acts to constrain household growth, an expanded supply of the limiting resource should both allow and promote household growth. My first awareness that such a relationship might exist came from an analysis of my Kofyar household-census material (Netting 1965). Despite polygyny and occasional multiple-family households, Kofyar homestead farmers had average household memberships of only 5.1, as compared to their near neighbors on the Jos Plateau, the Chokfem Sura, whose households numbered 14.8 individuals (Table 2.5).

On the terraced, manured homestead farms surrounding their residential compounds, the Kofyar could supply the careful, sustained labor needed by their grain, tuber, vegetable, and tree crops and their domestic goats and chickens with just a few people. A married couple, or even a single adult, could keep the farm in high production (Netting 1968: 135), and expanding the labor force on the plots, which averaged 0.4–0.6 ha (about 1–1.5 acres) each, did not bring a corresponding increase in returns. Because homestead farms formed a dense honeycomb of adjoining cells in the most desirable agricultural areas, it was usually not possible to expand onto immediately contiguous land. The general tendency was for young married couples to move onto homesteads made vacant by death and establish households independent of their parents. The Chokfem and the Angas on the grasslands to the north practiced shifting cultivation in areas of lower population density (Table 2.5), and households often included two or more conjugal units and their children:

Both the Kofyar and their Chokfem neighbors interpreted their household compositions in economic terms. Kofyar informants said that it was good for a young man and his wife to develop an independent homestead, that because they provided directly for their own subsistence they worked more conscientiously. [A multiple-family] household, on the other hand, allowed the support of lazy, parasitic members by their industrious relatives. Chokfem people justified their larger households by the advantages of a cooperating group which could be mobilized to complete work swiftly at the optimum time. The nuclear family household can successfully manage an intensive farm, while larger fields require, in the absence of machinery, a more extensive labor group. (Netting 1965: 425)

That households that had accommodated to the requirements of intensive agriculture in what may have been a long evolutionary process could adjust rapidly to new labor demands was demonstrated by the experience of Kofyar who migrated to a farming frontier on the Benue Valley plains about 30 miles south of the plateau escarpment. When land was no longer scarce and slash-and-burn techniques could be used to plant cash crops of yams, sorghum, and millet in temporary fields of 4–6 ha, the only limit

TABLE 2.5
Household Composition of Nigerian Hill Ethnic Groups

Ethnic group	Villages or hamlets (N)	Total population	Households (N)	Population/ household	Pct. multiple-family
Kofyar	16	2,999	588	5.1	8.5%
Chokfem Sura	1	133	9	14.8	89.0
Hill Angas[a] (Ampam area)	?	3,043	294	10.3	70.0
Plain Angas[a] (Kabwir area)	?	2,522	187	13.7	81.0

SOURCE: Netting 1965: 423.
[a]Angas census figures from Fr. J. Donnelly, S.M.A. (MS, n.d.).

on production was the labor that could be mobilized. Though the Kofyar expanded their use of beer-party exchange labor and even began to hire outside workers for the first time, the bulk of their labor came from the household. In the first stage of expansion, homestead farms were maintained and household members commuted to newly opened "bush" fields near Namu. With more work and simultaneous tasks going on in widely separated locations, the larger family household, which could divide its efforts, coordinate activities under central direction, and pool its production, had obvious advantages over the nuclear or elementary family household (Sahlins 1957). Whereas nonmigrant Kofyar homestead farmers retained rather small households of 4.54 in 1961 and 4.17 in 1966, those with bush cash-crop farms had 6.44 members in the 1960s (G. D. Stone et al. 1984). By 1984, the average household size in the Namu area was 8.38 people (Netting et al. 1989). The longer households farmed on the large frontier tracts, the more they expanded in size, adding people of all ages, but especially adult women aged 15–64, who were very productive (Fig. 2.1). The largest households, with a mean membership of 9.26, are now those with 16 to 20 years in the cash-cropping area. Older households are beginning to decline in size as adult married sons leave to form autonomous units and fragmentation begins to decrease farm size.

Household growth was accomplished most rapidly by adding adult women as wives in polygynous marriage. Of 601 traditional homestead farm households censused in 1961, 48 percent were polygynous, while among 106 migrant households cultivating a bush farm, the rate of polygyny was 59 percent (Netting 1965: 426). Men could turn cash-crop profits into bridewealth, giving them immediate adult labor and the expectation of children who would contribute to the work force of the future. Because of cooperation with co-wives in individual agricultural production and in domestic tasks, polygyny was also preferred by Kofyar women (M. Priscilla Stone 1988b). It was also advantageous to keep young married men within the household rather than have them form

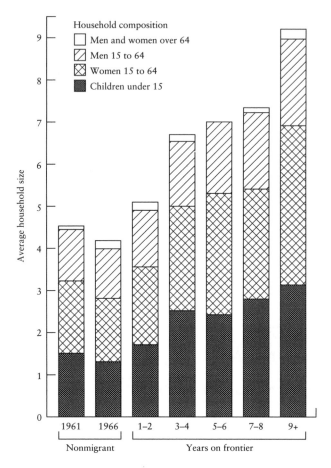

Fig. 2.1. Kofyar household size and composition, by years spent in cash-cropping.

their own separate units. Fathers encouraged sons to stay in their households, providing them with cash for clothing, motorcycles, and especially bridewealth. The proportions of multiple-family households among homestead farmers in 1961 was 4.6 percent, but among 1984 bush farmers, 17.0 percent were multiple-family households. Massed household labor could continue to provide subsistence and insurance while pooling family efforts to bring in the labor-intensive cash crop of yams. As fallow declined and intensification by interplanting, fertilizing, and elaborate crop rotations became more important, large households could better supply the various types of specialized and complementary labor required.

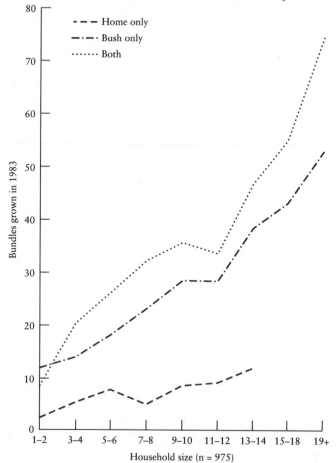

Fig. 2.2. Kofyar household size and millet production (including home production). "Home only" indicates homestead subsistence farm in traditional homeland; "Bush only" indicates cash-crop bush farm in frontier area.

Because the Kofyar were increasing agricultural production without animal traction or other sources of mechanical energy, mobilizing labor through the household used expeditious and socially familiar means to provide economic growth. Household labor applied to household fields constitutes 63 percent of all hours devoted to farming (G. D. Stone et al. 1990), and for the traditional subsistence grain crops of millet and sorghum, 76 percent of the labor comes directly from the consuming household (Netting et al. 1989). Indices of the major crops went up steadily along with household enlargement (Figs. 2.2, 2.3). Higher household

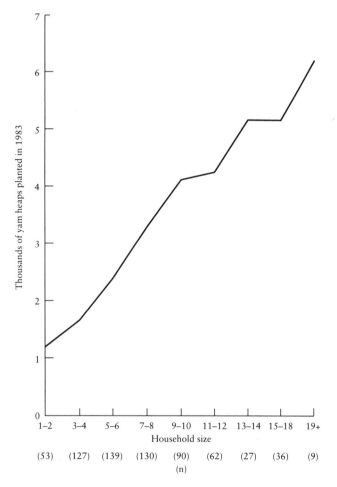

Fig. 2.3. Kofyar household size and yam production in bush cash-crop area.

cash incomes and more evidence of consumer goods, such as galvanized roofing and motor vehicles, also accompany the growth in household size (Figs. 2.4, 2.5). Though per capita returns to household members did not show the same increase, suggesting that the marginal productivity or the total labor input of individuals might be declining, a large household's potential for accumulation is still attractive, particularly from the viewpoint of the household head. As Chihiro Nakajima points out, an increase in the number of dependent members of a farm family will increase the family labor utilized, output, and income, but will reduce the marginal productivity of labor. If farm acreage is fixed, a larger household of

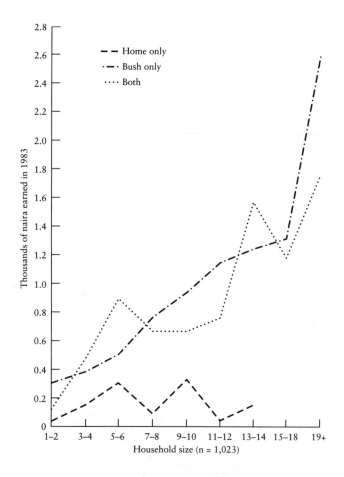

Fig. 2.4. Kofyar household size and cash-crop sales (including home production) in bush cash-crop area.

workers will decrease the income per worker and increase average labor hours (Nakajima 1969: 174).

Chronic problems among intensive cultivators, which may result in household fission even if there is no overt conflict, are the divergent economic interests of the household heads, who enjoy the benefits of accumulation from the efforts of pooled labor, while at the same time members of this labor force may see their productivity decline and their share of household income decline even faster. If land becomes increasingly restricted for newly formed households, we might expect the trend toward larger Kofyar households to be reversed. Cash-cropping house-

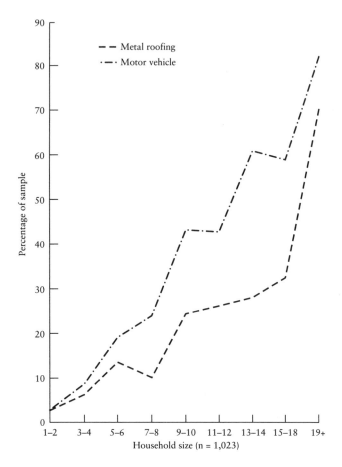

Fig. 2.5. Kofyar household size and capital purchases.

holds established in recent years have tended to be less complex and smaller, averaging only 5.70 members (Netting et al. 1989), but this more limited size may merely reflect an early stage in the developmental cycle.

The close linkage between household size and resources, especially land area, appears to hold for intensive cultivators in general. It reflects an adjustment among the factors of land, productive labor, and consumption that can alter over the life course of individual households as well as in the longer-term trends of rural societies under changing economic and ecological conditions. In making decisions, individual household members respond both to implicit contracts and culturally normative expectations and to specific economic changes in labor, allocation of resources, and property rights that may engender conflict and negotiation.

Although household size in relation to farm size appears to be a relatively simple, directly correlated variable, household type classified by the composition of family kin groups and the presence of non-kin like servants or boarders shows less regularity in its association to farming patterns (Hammel and Laslett 1974, Laslett and Wall 1972). Similar labor requirements do not necessarily result in identical forms of family and work organization, and varying patterns of inheritance and succession could generate very different household structures (Viazzo 1989: 255). The simple, or nuclear, family household, consisting of a conjugal pair and their offspring, may predominate among smallholders, especially among those with very limited land. This form is often accompanied by extended family households consisting of a conjugal family unit with the addition of one or more relatives other than offspring. The presence of kin such as a wife's mother or a husband's unmarried younger brother could indicate the inclusion of dependent kin or the enlargement of the work force. Intensive cultivators such as the alpine Swiss (Netting 1979) and the Philippine Ifugao (Conklin 1980) may show a high frequency of simple and minimally extended household types and emphasize the setting up of new, economically independent households, with their own farms, at marriage. Traditional Kofyar homesteads housed domestic groups that were on the average as small as simple family households, though many included a household head with more than one wife.

Multiple-family households including two or more conjugal family units connected by kinship or by marriage (Laslett 1972: 30) may also practice intensive cultivation, and a subdivision of this type, the stem-family household composed usually of a head couple and their married son, appears frequently in northern Portugal, Austria, and Japan. Stem families are often associated with impartible inheritance and co-residence of a senior couple and the child who is heir to the smallholding and who represents the continuation of a family line on a particular property. Where land is more plentiful or where segments of the population have access to larger-than-average holdings, multiple- or joint-family households can include a head couple and several of their married children. This may be the preferred or elite household form, as in rural China, even when the large majority of families lack the resources to maintain such a group.[9] A group of married Indian brothers may continue to pool income

---

9. In Chinese "larger joint family households, a mechanism to deal with the emergence of a 'free-rider' dilemma is to divide into smaller units, which is why the joint family household tends to be unstable. The marriage of two sons typically results in tensions within the joint family over the relative consumption and contribution of each family to the household economy" (Nee 1986: 190).

and retain common ownership of land and cattle even after they move to separate dwellings and begin to eat separately (Hawkesworth 1981). Fission usually follows the father's death and is complete when there is no longer any property held jointly (Shah 1974). In early modern Europe, multiple or joint families, more than any other kind of household, had to be "united around some common economic project" (Wheaton 1975: 174); the household might pool income from its smallholding, seasonal migrant wage work, and craft production. Unlike situations in which there is little valuable property in developed agricultural land, livestock, and substantial, durable houses, the household bonds of intensive cultivators, founded on pooled labor and heritable means of production, seem to be more enduring and significant than those based on wage income or the immediate exchange of domestic services.[10]

Detailed Japanese records show directional change in household composition as related to type and scale of farming. In the central hinterlands, large, multiple-family households of 16.6 members continued to practice swidden cultivation of widely dispersed fields into this century (Befu 1968a, 1968b). In most of the country, however, the Tokugawa period (1600–1870) saw a decline of shifting cultivation of cereals by large households and the emergence of single-family farms with irrigated rice, double-cropping, selection of seeds, commercial fertilizers like dried fish and oil cakes, and oil used as insecticide (T. C. Smith 1959: 86–98). With smaller, more specialized, higher-yielding farms, labor requirements per unit of land increased, and hereditary servants or persons indentured for life disappeared from households (Cornell 1987). Household structure changed from more to less complex, with joint families declining in one community from 26 to 13 percent, while the smaller stem-family households increased from 30 to 37 percent (Hayami, cited in Cornell 1987: 156). Multiple families whose labor forces had grown too large for their land divided holdings unequally between the main family and collateral branch families. The branch household occupied a subordinate social position, continued to furnish services like rice transplanting to the main family, and intensified their use of the poorer land they had been granted. This process limited the size and responsibilities of the main family, reduced an unwieldy labor force to manageable size, fixed single-heir in-

10. Where there is a mixture of swidden cultivation, fishing, and hunting subsistence strategies, as among the San Blas Cuna of Panama and the Miskito of Nicaragua, a matrifocal core of mothers and daughters or married sisters may share a house and some productive activities, but there is less emphasis on the conjugal family as a work group and corporate property holder. Among the Black Caribs of Belize, migrant wage-worker "visiting" husbands may provide only intermittent support for the households of their wives or their mothers (Helms 1976). Unlike valuable but fixed and less divisible farmland, major livestock, and permanent houses, wage income is fungible, and it is not locked into visible income-producing property like that of the smallholder household.

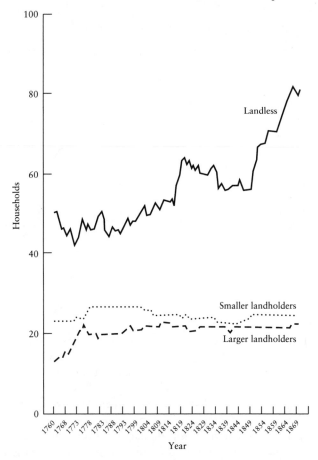

Fig. 2.6. Japanese village households, 1760–1869, by land-holding status. (From Kinoshita 1989: 194)

heritance within the co-resident stem family, and increased the productivity of the remaining household members (ibid.: 19). The growth of commerce and a national market enabled noninheriting offspring to migrate to the cities or take side jobs and perform agricultural wage labor in rural areas (Cornell 1987).

In Yambe, a village of north-central Japan whose population characteristics and trends from 1760 to 1870 have been reconstructed by Futoshi Kinoshita (1989), the number of households of richer farmers (20 koku and above) and of smaller landholders (0–20 koku) remained almost constant, while the households of landless farm servants and tenants increased in number (Fig. 2.6). The landholding households had inherited

TABLE 2.6
Japanese Rural Household Size and Type by Socioeconomic
Group, 1760–1870

	Landless	Smaller landholders	Larger landholders
Household size			
1760–1870	4.55	5.69	6.79
Average			
1760–99	4.08	4.94	7.09
1800–1835	4.47	5.73	6.63
1836–70	5.12	6.44	6.64
Household type (%)			
Solitaries	5.12%	3.43%	1.59%
No family	1.90	2.21	1.90
Simple-family	44.07	28.61	22.50
Extended-			
family	19.20	23.79	20.84
Multiple-family	27.36	43.91	54.76

SOURCE: Kinoshita 1989: 199, 219.

peasant rights (hyakusho kabu) which allowed them privileged access to irrigation water, village common lands (McKean 1982), and eligibility for village offices. The fixed numbers of such rights and the passing down of undivided farm properties to single heirs created a remarkable stability in the number of smallholder households (Kinoshita 1989: 195). The groups with more access to land had households that were both larger in size and more frequently characterized by the stem-family (extended and multiple) form than those of the landless group (Table 2.6).[11] It is clear that the corporate stem-family household was attached to an estate that was perpetuated with little parcelization over long periods of time, and that population growth took place among the landless and non-inheriting branch families who could not support larger and more complex family households (Fig. 2.7). A relatively constant set of resources was matched with smallholder households that perpetuated themselves with relatively stable size in the stem-family form over generations. The continuity of the smallholder households without land fragmentation or expanding population was ensured by inheritance rules and a system of subordinate social statuses that channeled noninheriting children into wage labor or nonfarm occupations.

Smallholder agriculture can be carried on under a variety of household regimes. While average household size is closely adapted to land re-

11. Perhaps the first to point out the regular association of stem- and extended-family household organization with larger and more valuable farms was Lutz Berkner (1972) in his classic paper on an eighteenth-century Austrian manor. In contrast to the more prosperous peasants, poor families of lodgers and landless laborers almost universally had small, nuclear family households with no co-resident relatives.

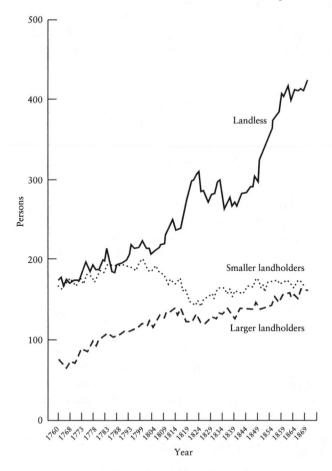

Fig. 2.7. Japanese village population, 1760–1869, by land-holding status. (From Kinoshita 1989: 198)

sources, household type depends on cultural factors such as inheritance rules and marriage age, in conjunction with the specific economic situation of the household. In most societies, however, landless farm laborers show a characteristically high predominance of small, simple conjugal units. With only daily wages and little if any real property to pass on, there was little incentive to delay marriage or to remain within a parental household in adulthood. Members of a rural proletariat might also be individually mobile in response to the demands of a regional labor market. Wage workers could not afford large numbers of unproductive dependents, and they lacked land where family members could be assured of gainful employment. Most households of day laborers in Tuscany were simple, with only 12.7 percent extended and 5 percent multiple, while the proportions for sharecroppers in the same area were 29.3 percent extended and 46.7 percent multiple (Benigno 1989: 182).

A line can be drawn across the Iberian peninsula and Italy separating regions characterized by smallholdings that are owned or rented on a long-term basis from areas in which dry-grain growing on large estates is typical. For considerable historical periods, nuclear families have been typical of day-laborer households on the *herdade* of the Alentejo region of Portugal, Andalusian latifundios, and the estates of Sicily and southern Italy. Farm workers resident in densely populated agrotowns were hired on a daily basis by the managers for absentee proprietors. The system of intense, but episodic, seasonal labor led to serious unemployment (Benigno 1989: 184). With few alternate occupations and estate owners monopolizing farmland, an impoverished rural proletariat was created. The cereal-growing estates of the plains also had a high-pressure demographic system, with high birth and death rates, and with adult male mortality exceeding that of females, who did not generally work in the fields. Peasant smallholdings of fruit and olive trees in the hilly sections of the Campania in southern Italy had more favorable demographic rates than those on the nearby grain-producing estates (Benigno 1989: 177). Low mortality and longer life expectancy allowed delayed marriage and more complex household forms, while the poorer life chances of the wage workers fostered small, nonextended, and relatively short-lived households. Marriage and household patterns thus appear responsive to environmental constraints and the contrasting nature of smallholder and estate-farming enterprises.

Smallholding as a Farm-Household System

If social units broadly designated as households are practically omnipresent in human society, there would seem to be no point in attempting to associate households with the particular type of land use we classify as intensive agriculture. This chapter attempts to make the case that the household as a corporate, co-resident domestic group is a recurrent, solvent form of social organization with a multitude of direct functional relationships to the smallholding. There are practical social and economic reasons why the individual, the descent group, the managerial landlord, the collective, and the business firm may be less well adapted to the needs of intensive farming. The family household characteristically mobilizes and allocates the labor and manages the resources of the smallholding: the household is the key productive unit. Though household members may also carry on individual agricultural production or have nonfarm occupations, they generally contribute to the farm enterprise in material ways and derive a part of their consumption from pooled household subsistence production and income.

The household grouping reproduces both its own work force and the skills and ecological knowledge so important to careful husbandry. The long-term, reciprocal relationships of members are in part constituted by enduring economic relationships, which can be seen as implicit contracts or covenants. The important contributions of children and the superiority of household labor to hired workers reflect the diversity of tasks and the requirement for skilled, responsible, unsupervised task performance in intensive cultivation. The household enjoys advantages in transaction costs, incentives, and security over farm enterprises that are differently structured. Property rights transmitted within the household cement the long-term interests and mutual dependency of household members. The close relationship of social unit and farm is demonstrated by the regular links between the average size and composition of households and the size of the smallholding. Despite great cross-cultural variation in marriage practices, preferred household types, and inheritance rules, there appears to be a remarkable congruence between the social organization of family households and the practice of smallholder farming.

CHAPTER THREE

Labor–Time Allocation

THE PRACTICES and technologies that the smallholder uses to intensify production from limited resources are means of coping with constraints on space and time. The shifting cultivator who lacks land because of encountering a geographical barrier like a range of mountains, the increased competition of a growing population of other farmers, or the removal of land by a warring enemy or plantation owner is faced with the problem of occupying and using the remaining land more continuously. Smaller available space means some combination of producing more crops per unit of area and using the area for a longer period in time. Tommy Carlstein (1982: 151) represents this process as the occupation of a "space-time volume" defined as the maximum volume of a land area under crop multiplied by the time a crop is in the field—for example, hectare-days (Fig. 3.1). The use volume may expand along either the space or the time dimension or both.

The names of some high-altitude meadows in the Swiss village of Törbel indicate that at one time they were forest land. Cutting and burning or removing the timber created open space where natural pasture grasses could grow, thereby enlarging the grazing area and the number of cows that could be supported. The construction of an irrigation system to water that land increased the summer time period when the grass could develop at its maximum rate, giving sufficient biomass for two crops of stored hay rather than merely rain-fed pasture. Meadow replacing forest meant that a larger area was producing more of a desired agricultural good for a longer time. When the Kofyar village of Bong received refugees from neighboring settlements that had been overrun and pillaged in war, the newcomers built homesteads on poorer-quality grassland at the village perimeter, which had previously been fallowed fields (Netting 1974b). By building terraces, manuring the homestead field with stall-fed goats, and planting oil palms, the immigrants patiently transformed a periodically productive resource into a permanently cropped homestead with high, diversified yields. The space of cultivation had not been en-

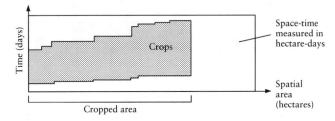

Fig. 3.1. The space-time budget for a small agricultural region. The demand on space-time for crops is indicated by the shaded area. (From Carlstein 1982: 151)

larged, and indeed the new residents had lost their old homesteads in the dangerous no-man's-land of their deserted villages. The use-volume of Bong's territory, however, had been increased as the time under crop and production per unit of land on the old fallows increased. If the number of homesteads on the best lands also increased, their average size would likely diminish, and intercropping and crop rotation could increase the number of months of each year that they were in use. Carlstein illustrates such an "impaction" under the pressure of a shrinking territory as space per occupant declines and time of resource use increases (Fig. 3.2).

Intensification in the use of land where smaller areas are used for longer time periods has a cost. For resources to give larger agricultural returns more constantly, unless they are sites such as alluvial valleys that have their fertility naturally renewed each year, a significant input of energy is required. This may be mechanical energy from manufactured implements and fossil fuels or chemical fertilizer and pumped water made available by the expenditure of various kinds of power. For smallholders, the cost is often human energy, the product of sinew and sweat. Digging and turning earth, irrigating a dry field, foddering livestock, and collecting manure are all tasks that take time and effort, and Ester Boserup's key evolutionary insight was that people do not undertake these tasks unless they have to. Boserup suggests that until quite recently, intensification took *more* work, rather than there having been successive evolutionary stages of technological invention in which the digging stick, animal traction and the plow, iron sickles, and eventually the tractor and the combine each increased production as it "saved labor." Reducing fallow in favor of annual or multiple cropping did not repeal the law of least effort or contradict the assertion that "the best general rule to the behavior of primitive farmers is that they work to get the maximum return for the minimum effort" (Nye and Greenland 1960). We may go further and say that farmers, although they may be seeking a variety of returns and attempting to increase their security over the long term by reducing risk and

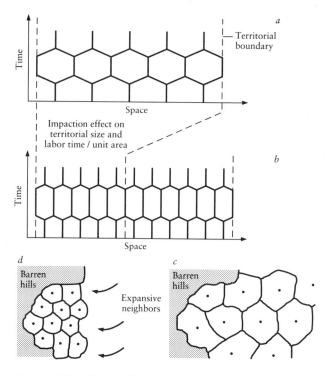

Fig. 3.2. The effects of impaction on settlement and land-use intensity: a hypothetical example of an impaction process and its effects on territorial size and range of labor time per unit area. Cases *a* and *c* show a population and its sections and subunits prior to impaction. As the result of neighbors' expansion on the one side and barren or inhospitable hill country on the other, both segments of the affected population are pushed into a much smaller region, with consequent effects on settlement and level of land-use intensity. Cases *b* and *d* show the situation after impaction. (A similar pattern of development may occur when population growth leads to the gradual segmentation of villages, and territories are subdivided into smaller units, although this is a different kind of impaction). (From Carlstein 1982: 198)

avoiding mindless maximizing, are aware of the costs and benefits of specific practices. Even though farmers may not be in a market economy with money prices on inputs and outputs, they make choices between labor and leisure, and they act so as to economize on the expenditure of time in relation to what the returns on that work effort are.

Boserup had a few sources indicating that irrigated agriculture in India took more labor days per hectare and per crop than dry farming, and that long-fallow systems in Africa, with their minimal tillage and lengthy "dead seasons," made for short hours (Boserup 1965: 40, 44–51), but the case for higher labor inputs both in total and per person as a result of intensification was not demonstrated in detail. The additional implication that labor efficiency or productivity declined, that "in the typical case, an intensification of the pattern of land use reduces output per man-hour, or, in other words, that agricultural employment increases at a higher rate than agricultural output" (ibid.: 43), appeared logically compelling, but was still unproven. Because input-output ratios were seen as days or hours of human time and amounts of crops, neither being very accurately measured, and because a single comparable measure like kilocalories of energy was not applied, it was difficult to generalize to a model of intensification that could be applied to modern conditions, under which labor had become a relatively minor factor of production. Since the ability of the smallholder to mobilize and manage work is one of the most distinctive features of this adaptation, I shall here review some of the evidence with respect to (1) labor-time expenditure; (2) productivity, or returns on labor; (3) diversified, nonagricultural labor; and (4) indigenous methods of intensifying labor use.

How Much Work Is Too Much Work?

Any attempt to describe the daily round of a smallholder's working activities or the labors of a peasant family is apt to provoke somewhat contradictory responses. On the one hand there is an appreciation of the "good life" of honest toil in a cooperative, mutually supportive household, founded on the moral bedrock of a shared "work ethic." But there is also the equally strong conviction that farming is the epitome of grueling, unrelenting labor if it is "primitive"—that is, done manually on a small scale without benefit of modern machines. Women rhythmically bending their backs to transplant rice seedlings in a flooded paddy may be picturesque, but what they are doing is seen as stultifying drudgery. Hoeing a long corn row from dawn to dusk or milking cows by hand are tasks to which not even the most workaholic modern farmer could imagine returning. Isn't the 40-hour week (about 1,920 hrs/yr with a two-week vacation and some holidays off) a triumph of modern civilization? Perhaps, but before we assume a scale of evolutionary progress coinciding with fewer person-hours of work, let us look at the comparative evidence.

Hunter-gatherers who forage for naturally occurring plants and animals seem to get by on a lot less time and effort than food producers. The !Kung San of the Kalahari Desert in Botswana put in only a little over 40 hours a week, including food preparation and tool making. Subsistence work for men averages 21.6 hours of work a week (1,123 hrs/yr) and for women just 12.6 (655 hrs/yr) (Lee 1979: 272–80), and neither teenagers nor adults over 60 contribute much to the food quest. Rather than hunter-gatherers being engaged in an unrelenting search for a meager subsistence, the more recent conventional wisdom of anthropologists speaks of them as "the most leisured peoples in the world" (Service 1966: 13), and, more memorably, as "the original affluent society" (Sahlins 1968: 85–89). One begins to wonder if the myth of starving savages, always anxious about where their next meal was coming from (John C. Frémont, quoted in Steward 1938: 8) was not a cautionary tale used by farmers to keep their wayward children from straying into the wild, where the living was so much easier.

Even when agriculture begins to play a larger role among shifting, slash-and-burn cultivators, time investments may still be quite moderate. Among the Machiguenga, who grow cassava, bananas, and corn in their rain-forest clearings in Peru, men garden for about two and a half hours a day, and women for less than an hour; this is supplemented by fishing, fruit gathering, and hunting, which occupies men for another two hours a day (A. W. Johnson 1975). Machiguenga subsistence agriculture produces 3.5 kilos of food for each hour of work (Baksh n.d.). The total work of food getting and processing is fairly evenly divided between the sexes and comes to about four hours a day (Carlstein 1982: 365), and of this, crop production is only an average of about 1.6 hours a day per adult (or less than 600 hrs/yr).[1]

The Ushi, who practiced the *chitimene* system of pollarding or felling trees in the dry woodlands of Zambia and planted finger millet, cassava, peanuts, and beans in the burned swidden, spent about a third of their agricultural labor time in cutting and piling tree branches and another third in harvesting (Kay 1964: 41). Because of high rates of labor migration, young men (less than 45 years old) were present in the village for only about half the year, and did only an average of 427 hours of work. Older men, on the other hand, expended some 1,144 hours, while the average totals for women below and above 45 were 944 and 1,247 hours respectively (ibid.: 49). Since there were only 17 older people in the sample of 46 adults, the annual average for agricultural labor was 892 hours. This disproportionate contribution of women, 1,053 hours to men's 700 hours, is not unusual for African shifting cultivators. Some

1. A more recent survey of Machiguenga labor time puts work at subsistence and cash-crop agriculture at 1.35 hrs/day for men and 0.73 for women, while the total of hunting, fishing, and collecting time is 2.18 hrs/day for men and 0.83 for women (Baksh n.d.). Agriculture for a male would thus involve a total of only 493 hours a year, but it would result in 1,725 kilos of edible food.

time went into collecting wild produce like mushrooms, caterpillars, leaves, and roots, as well as into fishing, but this was significant only for younger women (343 hrs) and older women (218 hrs). Summarizing the results of 18 studies in various parts of Africa where long-fallow agriculture is practiced, Boserup (1970: 147) reports an average weekly labor input by men and boys of 14 hours (728 hrs/yr) and by women and girls of 15 hours (780 hrs/yr). The surveys reviewed by J. H. Cleave (1974: 189) show with remarkable consistency that adult members of African farm families work in their predominantly swidden fields for only 120 to 160 days a year, with working days of from four to six hours, giving individual totals of 480 to 960 hours annually.

It is obvious that shifting cultivators have fewer agricultural operations to perform and that they devote much less time to such indirect inputs as constructing irrigation networks or collecting and distributing manure. In dry-rice swiddens such as those of the Iban in Sarawak, stumps and unburned tree trunks are not removed from the steeply sloping fields, and the crop seeds are merely dropped dexterously into holes made by the point of a dibble stick, without turning or breaking of the ash-covered soil (Freeman 1955). With a good burn in a newly cut woodland field, there may be little need for weeding, and the Zambian Ushi devote only about 4 percent of their labor time to this task (Kay 1964: 41). Certain crops are relatively undemanding of cultivation time. Cassava, or manioc, can be planted over a wide range of times, be left to grow over 18 months in a swidden reverting to natural vegetation, and be harvested at the convenience of the farmer. Hoeing this staple crop takes the Maku of Cameroun just 16 days' work each by a husband and wife to provide a major part of the family food supply (Burnham 1980). Given the low population density and the abundance of land capable of supporting cassava in the area, the Maku could easily produce a surplus, but the roads, transport system, and market prices that might motivate more agricultural work are not yet present. In the topical forests of Belize, where cash crops such as maize and rice can be produced by shifting cultivation, Kekchi Maya men may increase their agricultural labor inputs to an average 1,672 hrs/yr (range 784–2,524); returns of rice, for example, remained relatively attractive at about 1.5 kg/man-hr (Wilk 1988: tables 6.8, 6.9). Where a mixed system of shifting cultivation of food crops and commercial tree-crop production (cocoa orchards and oil palms) exists, as among the Nigerian Yoruba, males may work 710 hours on the food swiddens, with an additional 561 hours in arboriculture. Women, most of whose time is spent in trading, work only 160 hours on the farm, with another 373 in palm-oil processing (Galetti et al. 1956, cited in Cleave 1974: 42).

Intensifying Labor: The High End of the Spectrum

Intensification is achieved, in the absence of mechanization, by increasing the amount of labor applied per unit of land in the total farming operation. In contrast to shifting cultivation, this generally means inputs of 1,000 to 4,000 hrs/yr per adult household member, either directly in field and garden work or in the associated animal tending, food- and cash-crop processing, and domestic upkeep. Smallholders who lack sufficient resources in land and livestock to raise their labor commitment on their own farms will redirect their time to crafts, trade, and wage-labor employment, at what is often a lower rate of return. One comparison using time-allocation studies suggests that with high population density, intensive farming systems demand some 3½ more hours per day from men and 4 hours more from women than extensive agriculture under conditions of low population pressure does (Minge-Klevana 1980; A. W. Johnson 1980). When man-hours are calculated in terms of adult male–equivalents, the Ukara Island intensive millet-manioc system in Tanzania requires 1,400 hrs/yr. Irrigated wet rice in a Thai case uses 1,252 hours; in an Indian case, 1,182 hours; and in a Taiwanese example, 2,545 hours (Ruthenberg 1976: 147, 200–201).

In Bangladesh villages with irrigated double- or triple-cropped rice, jute, mustard, potatoes, wheat, and orchard gardens, men work between 3,138 and 3,700 hours and women from 3,772 to 4,275 hours (Wallace et al. 1987: 76, 81). Though women may do little work in the fields, they manage all the poultry and livestock feeding, cleaning the cow shed, cutting grass for the animals, milking the goats and cows, and taking the ducks to the pond and collecting snails for them. Women also perform the chores of the kitchen garden, fencing it with cut bamboo, hoeing, staking up climbing plants, weeding, watering, and harvesting fruit. The post-harvest activities of stirring, cleaning, threshing, winnowing, parboiling, and storing the paddy rice are extremely time-consuming, as is the drying and extracting of the jute fiber for sale (Wallace et al. 1987: 74–75). Women from landless households work over 4,000 hours, with 460 hours for paid services or producing goods to be directly sold. Even women in households with large landholdings, who may occasionally hire helpers, work more than 3,800 hours because of the greater volume of crop processing and livestock care (Wallace et al. 1987: 78, 80). As in many societies that use the plow (Ember and Ember 1971), much of the field work is done by men, and Moslem religious customs of female seclusion may further accentuate this gender division of labor. Men driving ox teams may plow, harrow, and level a field for 10 hours a day, and they also break clods with a wooden hammer or iron spikes (Ali 1987: 282).

TABLE 3.1
Japanese Labor by Farm Size
(Pre-1952 figures)

	Japanese average		Farms less than 0.5 ha	
	Male	Female	Male	Female
Persons/farm engaged in farm work[a]	1.41	1.45	0.63	0.98
Hrs/person/yr total work[a]	2,306	1,770	2,810	1,377
Hrs/person/yr nonfarm work	474	148	1,320	238

SOURCE: Clark and Haswell 1967: 138.
[a]Average working day about nine hours.

The process may be repeated three or four times for both the rainy season and the cold-weather crops. Although men participate little in domestic work, their labor hours are not far short of those of women, and they may be three to seven times as long as the average working hours of males in shifting cultivation.

As farm size declines in a specific region or local area, the same phenomenon of higher labor inputs per individual farmer and per unit of land is apparent. The labor time put in by Japanese smallholders with less than 0.5 ha was 22 percent higher than average: 2,810 as against 2,306 hrs/yr for men (Table 3.1). Farmers with more than 0.5 ha had more than double the number of working males per household and spent only about a third as much time in nonfarm occupations as men with less land. This suggests that there are limits to how much the occupant of a very small landholding can raise production by labor and intensive techniques alone. Those with access to very little land (for instance, 0.2 ha of irrigated rice, or *sawah*, in Java is not enough to produce the minimum annual rice requirement for an average family of 4.5 members) must seek other occupations (Benjamin White 1976). Even by double-cropping their land and tending a garden and livestock, they lack the space on their minifundia to grow more or to invest more labor. In systems of rain-fed agriculture without irrigation, the problem of filling slack times in the dry season with income-producing work is even more severe. In such situations, the members of a smallholder household typically intensify their labor *outside* of agriculture with part-time employment as craftsmen, traders, artisans, and wage laborers. Smallholders work more, but a larger proportion of their labors are off-farm.

Population Density and Labor Use: Some African Comparisons

Labor intensification in agriculture appears most clearly when we compare the mean cultivated area (farm size in ha minus fallow area) with total hours spent in work on that farm and hours per cultivated ha. Three

TABLE 3.2
Hausa Labor and Productivity, 1965–73

	Dan Mahawayi	Doka	Hanwa Cattle owners	Hanwa Non–cattle owners
Population /km²	30	147	264	
Farm size (ha)	4.8	4.0	3.7	2.2
Fallow (ha)	1.0	1.1	0.1	0.1
Fadama (ha)[a]	0.4	0.5	0.1	0.3
Cultivated (ha)	3.8	2.9	3.6	2.1
Farm work per worker Hrs /day (including travel)	4.9	5.0	5.4	5.7
Days worked	140	159	127	116
Hrs /yr (including travel)	683	792	682	661
Other work Days worked	123	39	124	86
Hrs worked	604	179	696	479
Total (hrs)	1,287	971	1,378	1,140
Total hrs on farm[b]	1,516	1,634	2,109	2,405
Hrs /cultivated ha	399	563	584	1,124
Returns to labor-hr (naira)	₦0.14	₦0.12	₦0.11	

SOURCE: D. W. Norman 1969; D. W. Norman et al. 1982: 104, 107.
[a]*Fadama* is bottomland with a high water table, allowing flood-recession or irrigation agriculture.
[b]Includes males, females, and hired labor.

Hausa villages near Zaria in northern Nigeria all practiced rain-fed cultivation of millet, sorghum, peanuts, and cowpeas, and they differed in
population density and access to the city (D. W. Norman et al. 1982). The
expected inverse relationships between local population density and average farm size, as well as proportion fallowed, appear (Table 3.2). Cultivated area also declines regularly, except that the richer cattle owners in
Hanwa have somewhat larger farms than would be predicted on the basis
of population density. As farm size and cultivated area go down, hours
on the farm, and especially hours per cultivated ha, go up as farmers
compensate for less land and shorter fallows by more intercropping,
manuring, and tending of irrigated *fadama* (land suitable for flood recession or irrigation agriculture). Hrs /ha almost triple between Dan Mahawayi and Hanwa. This produces some small lengthening of the working
day (from an average of 4.9 hrs to 5.7 hrs), but the number of days
worked in agriculture declines, and this is not fully compensated for by
days spent in other, nonagricultural work. What we see is in part the result of choices related to market opportunities. Dan Mahawayi is 32 km
from Zaria along poor roads, and its farmers must supplement agriculture with traditional dry-season crafts and trading, which have comparatively low cash returns. Doka is 40 km from Zaria, but on a paved road,
allowing more farm produce to be sold with low transport costs, and

TABLE 3.3

Ibo Labor, Productivity, and Income, by Population Density, 1974–75

	Okwe (low density)	Umuokile (medium density)	Owerre-Ebieri (high density)
Population/km²	250	500	1,200
Farm size (ha)	2.40	1.00	0.40
Fallow (ha)	2.00	0.73	0.17
Cultivated (ha)	0.40	0.27	0.23
Length of fallow (yrs)	5.30	3.87	1.38
Labor			
Man-equivalents available for work	3.9	2.4	2.5
Man-hrs family labor/ household	734	458	577
Man-hrs family labor/ man-equivalent	188	191	231
Man-hrs non-family labor/ household	470	111	200
Total man-hours	1,204	569	777
Man-hrs/cultivated ha	3,010	2,107	3,378
Man-hrs/compound ha	–	2,772	3,353
Man-hrs/outer-field ha	3,010	2,023	3,388
Total production (naira)	480	321	272
Returns/hr	0.40	0.56	0.35
Returns/hr (compounds)	–	1.12	0.53
Returns/hr (outer fields)	0.40	0.14	0.11
Income (naira)			
Net farm income	421	308	225
Nonfarm income	300	347	721
Total household income	721	655	946

SOURCE: Lagemann 1977: 22, 23, 42, 93, 95, 101, 109, 111, 114.

fewer hours of "other work" are needed to raise cash. Doka has more high-quality fadama land than the other two villages, permitting agricultural work to continue through the dry season (ibid.: 8).

For commercial purposes, Hanwa is best situated, being on the outskirts of Zaria City, and its cattle owners can expend much time profitably on their livestock, while those who own no cattle take modern service-sector jobs, such as those of driver, commission agent, messenger, and bicycle repairer, which are "almost certainly more remunerative than those in the traditional sector" (D. W. Norman 1969: 12–13). Both hire a good bit of labor for their own farms, thereby raising the total farm-labor input. Returns to labor per hour of farm work do decline with higher population density and smaller farm size, but only minimally (Table 3.2), suggesting that trade-offs with other occupations and possibly more efficient agricultural practices may keep productivity from falling even while intensification is taking place.

A similar three-village comparison has been done by Johannes Lage-

Umuokile (medium density)

Owerre-Ebeiri (high density)

Okwe (low density)

Houses

Distant fields (D)

Compounds

2d year cassava & fallow (2)

Near fields (N)

Fallow

Forest

Oil palm trees

Fig. 3.3. Ibo village population and land use. (From Lagemann 1977: 28, 29)

mann (1977) in the Ibo area of southeastern Nigeria. With rainfall of some 2,200 mm per year, over twice that of the Hausa area near Zaria, the natural forest vegetation is that of the lowland humid tropics, and important food crops include yams, cassava, maize, coco yams, palm oil, sweet potatoes, and a wide variety of fruits. These are grown in both outer fields, with two years of cropping followed by a two- to seven-year fallow period, and in compound gardens, which are fertilized, highly diversified, and maintained in continuous production. Population density in the sample villages takes up where the Hausa examples leave off, at approximately 250/km² and goes to some of the highest rural densities known in sub-Saharan Africa, represented by Owerre-Ebieri, with 1,200/km² (Table 3.3). Along this scale from least to most dense villages, farm size falls by 83 percent and fallow area shrinks from 83 percent of the farm to 43 percent (Fig. 3.3). Such minuscule cultivated areas as 0.23 ha cannot accommodate unlimited labor inputs, so total man-hours per farm go down, even while hours per man-equivalent, representing individual work inputs, are rising.

The more intensively tilled compounds, with their wealth of tree crops, are somewhat less demanding of labor (2,107 hrs/ha) in Umuokile, a village of moderate density, than are the fallowed outer fields of Okwe (Fig. 3.3), which has sufficient land not to need compound gardens (Lagemann 1977: 94). But in crowded Owerre-Ebieri, both outer fields and compounds absorb over 3,350 hrs/ha (Table 3.3). With declining fertility owing to shortened fallow in the outer fields, returns per hour to work there falls by about 75 percent, and even the higher productivity of the compound drops from Umuokile's 1.12 ₦/hr to Owerre-Ebieri's 0.53 ₦/hr (Table 3.3) (Lagemann 1977: 91–94). The drop in total production and in returns on labor in the compound cultivation with the highest labor investment per unit area suggests that land degradation is taking place, reducing yields, labor productivity, and farm income per capita. Intensification, an efficient strategy at lower levels of density like Umuokile, can no longer compensate for the too-intensive use of increasingly scarce resources, as in heavily populated Owerre-Ebieri.

Intensifying Labor Off the Farm: Why Smallholders May Have It Better

The paradox of high population density and its correlates of land-use intensification and possibly declining returns on labor (involution) (Lagemann 1977: 6) is that these conditions do not mean automatic impoverishment for the mass of rural people. If there exist gainful employment activities outside of agriculture, time and effort are directed there, though the smallholding is not necessarily abandoned. All 74 of the Ibo farmers

TABLE 3.4
Ibo Landholdings, Nonfarm Income, and Education, by Population Density, 1974–75

	Okwe (low density)	Umuokile (medium density)	Owerre-Ebeiri (high density)
Population/km²	250	500	1,200
Cultivated area/ household	0.4 ha	0.27 ha	0.23 ha
Farm income (naira)	₦421	₦308	₦225
Nonfarm income	₦300	₦347	₦721
Total household income	₦721	₦655	₦946
Education/school fees	₦43	₦45	₦139
Off-farm income of educated[a]	₦323	₦561	₦968
Off-farm income of uneducated	₦287	₦264	₦494

SOURCE: Lagemann 1977: 22, 101, 109, 110, 231.
[a]A farmer who has attended primary school for more than two years is classified as "educated."

in the three villages Johannes Lagemann (ibid.: 109) surveyed had income from other sources than farming, and off-farm income was inversely related to farm income (Table 3.4). The Ibo are famous in Nigeria for their success in a variety of occupations, from the construction trades and trucking to bar ownership and watch repairing. The nonfarm-derived percentage of total family income went up from 42 percent in the low-density Okwe to 76 percent in high-density Owerre-Ebieri, and the proportion of this income generated by trading went from 24 percent to 55 percent (ibid.: 109–10). Total family income and income per household member is highest in the most impacted village where, without off-farm employment, the households could not provide their daily food requirements (ibid.: 116). An effective way of preparing for nonagricultural jobs is through education, and increasing cash resources, from ₦43 to ₦139, are devoted to school fees (Table 3.4) along the density continuum (Lagemann 1977: 231). Off-farm income is always higher on the average for an "educated" farmer who has attended at least two years of primary school than for the uneducated, and the advantage of the educated group is highest in the high-density village (Table 3.4). A correlation of population pressure with educational level and off-farm employment suggests that smallholders are making an economically reasonable allocation of time and resources to the production of income, both within and outside of agriculture.

Decisions on allocating labor among various activities, dividing tasks by gender and age within the household, and securing the best returns on labor according to available land and other resources emerge with special clarity from studies of Javanese wet-rice cultivators. Under conditions of very high population density (730/km²) in a village in central Java, Benjamin White (1976: 285) found that adult men of 15 years old and above were working 3,173 hrs/yr, of which 956 went into irrigated rice and garden cultivation and 482 into livestock care and feeding (Table 3.5). In the characteristic plow-cultivation division of labor, men applied some 45 percent of their working time to agriculture, while women contributed only 6 percent of their total of 4,056 hours' work to it. Women made large labor investments, on the other hand, in crafts, trading, the preparation and sale of food, and wage labor, both in and out of agriculture (1,862 hrs, or 46 percent of labor time). About a third of men's work went into similar cash-producing activities. The domestic activities of childcare, cooking, firewood gathering, and shopping absorbed another 1,885 hours of female time, bringing an average working day to 11.1 hours, as contrasted to males' 8.7. Younger women (aged 12–14) contributed more than three-quarters of an adult labor input, concentrated in the domestic and craft-trading spheres (Table 3.5). Boys of the same age did little field work but more than doubled their fathers' work times on livestock.

A powerful determinant of how labor is allocated is access to the multicropped, terraced wet-rice lands, or sawah. In another Javanese village, investigated by Gillian Hart (1986: 124), those with holdings of rice land or a fishpond above 0.5 ha (Hart's Class I) could expend 1,048 hours, or more than 55 percent of their 1,893 hours of work time (average for males 10 and older) on agriculture and/or aquaculture (Table 3.6). Those with less than 0.2 ha, including the landless (Class III), could devote only 4.3 percent of their somewhat greater total labor time to their own farming and had to make up for it with 1,289 hours of wage labor. They also did more fishing and gathering than the groups with more adequate landholdings.

Comparing returns on labor in the various occupations of rural Java (Table 3.7) makes explicit the rationale for working in agriculture if one has or can gain access to land (Benjamin White 1976: 279). Owner-cultivators of 0.5 ha and above could get returns of 50 Rp/hr, garden cultivation brought 25 Rp/hr, and even sharecroppers made 12.5 Rp/hr, putting their earnings substantially above those of unskilled construction labor, jobs in a weaving factory, food preparation for sale, and such crafts as making hats or bamboo mats. Even artisans with their tools, like carpenters, and male traders on foot with a capital of Rp 1,000 could command only 15 Rp/hr for their work.

The extent of a household's participation in wage labor or other activities where the returns to labor are much lower than in rice cultivation

TABLE 3.5
Average Annual Work Inputs by Age, Sex, and Type of Activity in a Sample of 20 Households, Java, November 1972–October 1973
(Hrs / person / yr)

	Age group							
	Female				Male			
	6–8 (n = 7)	9–11 (n = 4)	12–14 (n = 6)	15+ (n = 33)	6–8 (n = 6)	9–11 (n = 7)	12–14 (n = 10)	15+ (n = 31)
1. Childcare	606	180	596	376	425	173	99	133
2. Housework	97	390	400	386	51	22	75	28
3. Food preparation	14	140	413	994	11	14	23	37
4. Firewood collection	95	100	28	32	218	285	332	75
5. Shopping	4	28	20	100	2	–	19	16
6. Handicrafts	35	657	958	847	–	–	37	163
7. Preparation of food for sale	–	–	38	151	11	2	1	125
8. Animal care/feeding	326	249	29	53	620	534	899	482
9. Trading	–	5	279	523	–	15	7	264
10. Garden (own)	6	2	8	39	4	38	45	267
11. Sawah[a]	18	78	134	151	7	13	84	689
12. Unpaid work[b]	25	18	42	55	14	12	45	279
13. Agricultural wage or exchange labor[c]	23	110	206	279	–	2	33	107
14. Nonagricultural wage labor	–	–	–	62	–	29	17	406
15. Other	3	–	4	11	5	53	33	102
Total directly productive work (nos. 6–15)	436	1,119	1,698	2,171	661	698	1,221	2,884
All work (nos. 1–15)	1,252	1,957	3,155	4,056	1,368	1,192	1,769	3,173

SOURCE: Benjamin White 1976: 275–76, 285.

[a]*Sawah* cultivation includes agricultural work on owned, rented, and sharecropped land.

[b]Unpaid work includes work on village projects and reciprocal labor for other households, (e.g., house-building and repair).

[c]Wage or exchange labor includes all agricultural work done on land farmed for another household, whether for a cash wage or a share in kind (as in harvesting) or on an unpaid labor-exchange basis.

"depends primarily on its access to sawah" (ibid.: 278). Because 20 percent of the local population do not have the use of land as owners, sharecroppers, or renters, and a further 43 percent had less than 0.2 ha, they were forced to allocate substantial portions of working time to lower-paying nonagricultural occupations. Even households with rice land had slack periods in rice cultivation or had female or younger members who did little field work. That enjoined both high total labor inputs and "occupational multiplicity and a highly flexible division of labor among household members" (ibid.: 280). Smallholders who intensify the use of their own household labor can seldom be full-time farmers, though if they had sufficient land, they would prefer the higher returns from agriculture. Regardless of the proportion of time devoted to agriculture,

TABLE 3.6
Household Patterns of Labor Allocation in a Javanese Village, 1975–76

| | Average annual labor/"worker unit"[a] | | | | | |
| | Class I (> 0.5 ha rice or fishpond) | | Class II (0.2–0.5 ha rice) | | Class III (< 0.2 ha rice) | |
	Hrs.	Pct.	Hrs.	Pct.	Hrs.	Pct.
Own production[b]	1,048.4	55.4%	431.9	21.4%	92.9	4.3%
Wage labor	151.9	8.0	770.9	38.2	1289.0	59.7
Trading	169.6	9.0	123.1	6.1	73.7	3.4
Fishing and gathering	26.0	1.4	252.7	12.5	314.6	14.6
Housework[c]	497.5	26.3	437.4	21.7	389.7	18.0
Total work	1,893.4	100%	2,016.0	100%	2,159.9	100%

SOURCE: Gillian Hart 1986: 124. Used with permission.
[a]Persons aged 10 and over.　　[b]Rice, fish ponds, gardens, livestock.　　[c]Excluding childcare.

TABLE 3.7
Returns per Hour in Various Occupations, Rural Java, 1972–73

Occupation	Rupiahs/ hr	Occupation	Rupiahs/ hr
Rice cultivation		Trade	
Owner-cultivator (0.5 ha)	50	Men on foot (Rp 1,000	
Owner-cultivator (0.2 ha)	25	capital)	15
Sharecropper (0.2 ha)	12.5	Men on bicycles (Rp 15,000	
Garden cultivation	25	capital)	20
Agricultural wage labor		Preparation of food for sale	
Plowing (own draft		Coconut sugar (own trees)	5–6
animals)	70–90	Coconut sugar	
Hoeing	9–11	(sharecropped)	2.5–3
Transplanting	6–7	Fried cassava	3.5
Weeding	9–11	Fermented soybeans	5
Harvesting	16–20	Animal husbandry	
Nonagricultural wage labor		Ducks	5–12
Carrying/construction	10	Goats	1–2
Crafts (carpentry)	15	Cattle	4–6
Weaving	7	Goats/cattle sharecropped	2–3
Trade		Handicrafts	
Women on foot (Rp 1,000		Pandanus-leaf hats	1.5
capital)	5–10	Bamboo mats	3

SOURCE: Benjamin White 1976: 279.

members of smallholder households work a great deal, and their labor inputs, which are already substantial, rise as their access to land declines. Intensifying labor by increasing the number of hrs/person/yr and the number of hrs/unit of land may entail some decline in agricultural productivity, but differences in return/hr from primary field work may not be great. Much more marked are the lower returns per hour of domestic animal tending, crop processing, and subsidiary crafts, trading, and unskilled wage-labor occupations. It is the necessity to intensify time use outside of agriculture, because of lack of access to land and other resources, which brings with it acceptance of decisively lower returns to labor and marginal productivity.

Putting in More Time on the Farm: Strategies of Scheduling

Because agriculture is tied to seasonal factors such as rainfall, day length, and temperature and to the cycles of reproduction, growth, and maturity in plants and animals, labor is not normally applied in fixed, standard amounts, and neither can work be increased by the simple expedient of lengthening every working day. Even or perhaps especially in systems of shifting cultivation, there are bottleneck periods when every hand is needed in the fields and when production is restricted by the lack of labor. In tropical forests, there may be only limited and somewhat unpredictable periods when rains let up enough to allow the cut timber and brush to dry so that a good burn and ash layer on the field can be created (Freeman 1955). If the jungle growth is slashed too early, new wild plants will have the opportunity to grow, and if too late, an inadequate firing may leave sprouting vegetation and pests to compete with the crops. Effective cutting must occupy a narrow window of opportunity, and everyone's swiddens need it at about the same time. Once the crop of dry rice is up, the most important task may be weeding, and when this is a largely female task, as it is among the Iban, whose young men migrate in that season for wage work, the ability of women to weed only about 1.5 acres each limits the extent of fields (Freeman 1955). The highest labor inputs into the highland New Guinea yam/taro and sweet potato swiddens of the Tsembaga are the planting and almost continuous weeding, which require 32 percent of the effort put into gardening and cannot be delayed without reducing yields (Rappaport 1971). Seasonal bottlenecks in shifting-cultivation labor alternate with periods in which little or no agricultural work is required, and individuals may have plentiful time to spare for social, religious, and recreational activities.

In African savanna environments, where the rhythm of farm work is dictated by the rains that follow a months-long dry season, labor intensification presents particular problems. The Kofyar who have moved onto forested land in the Benue Valley have a seven-month period in which practically all of the 1,200 mm of precipitation falls. Their major grain crops, millet and sorghum, do best when planted as early as possible after the first soaking rain, and this means that men, women, and children must work long hours as soon as the rain comes, perhaps repeating the planting if there is an early drought. From daily labor records of all workers of age 15 and above (G. D. Stone et al. 1990), we can see that the late March–early April planting of 1984 required peak labor of about 7½ hrs per person per day (Figs. 3.4, 3.5). Another labor bottleneck

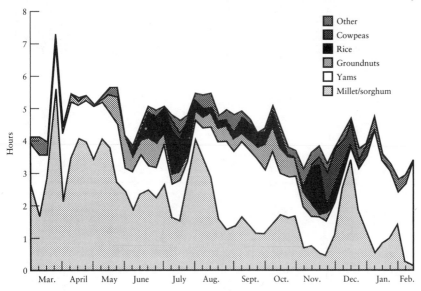

Fig. 3.4. Kofyar agricultural labor, 1984–85: Kofyar mean daily labor inputs per individual (scale in hours), broken down by crop. (From G. D. Stone et al. 1990: 11)

comes in July / August, when the early millet must be harvested, bundled, and stored under a thatched roof on a drying rack during the short drier period between the early and late rains. Pennisetum millet is a desert ephemeral that matures rapidly and becomes ripe in a short period of time, unlike the sorghum that grows in the same field and can be harvested in a more leisurely manner during the dry months of December and January (Fig. 3.5).

While meeting the labor demands of the millet / sorghum / cowpeas farming calendar, which continues to supply the bulk of their subsistence food, the Kofyar have added the cultivation of yams and rice to sell in the internal Nigerian market. Farmers attempted to increase agricultural production for cash at the same time that they went from shifting cultivation of an open frontier to permanent occupation of intensively tilled and fertilized fields. Since they continued to rely on hoe cultivation, higher production along with shorter fallows could only be achieved by additional labor inputs. Average annual farm labor came to 1,549 hours, 1,729 for men and 1,473 for women. Hausa cultivators living in a somewhat drier (1,105 mm of rainfall) area near Zaria, but growing the same subsistence crops by similar methods, worked only 661 to 792 hrs/adult male, including travel time; Hausa women follow Moslem practices of seclusion and contribute the negligible amount of 7 hrs/yr (D. W. Norman 1969:

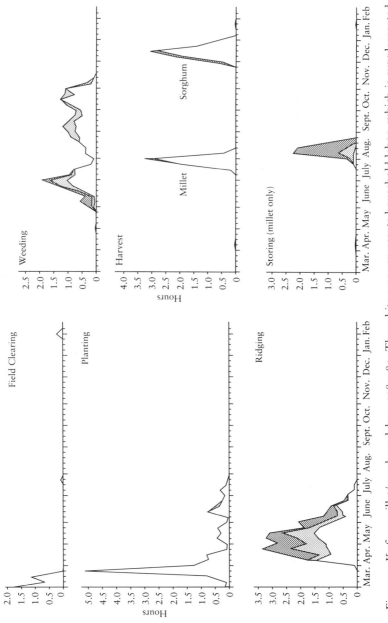

Fig. 3-5. Kofyar millet/sorghum labor, 1984–85. The white area represents household labor, which is supplemented for certain tasks by exchange labor (light shading) and beer-party labor (dark shading).

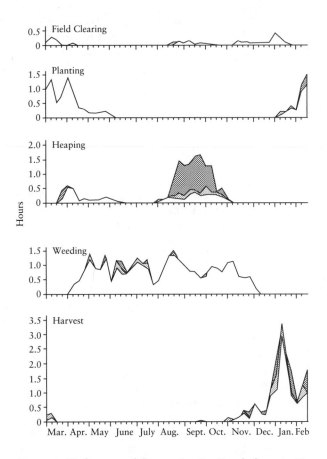

Fig. 3.6. Kofyar yam labor, 1984–85. Symbols as in Fig.
3.5. (From G. D. Stone et al. 1990: 18)

7, 11). Female farm labor is recognized by the Kofyar as a major component of their ability to raise total production, and together with heavy
male work inputs, women's labor allows the average Kofyar couple to
expend 3,202 hrs/yr on agriculture, almost 4½ times the Hausa average
of 716 for a male and female. Unlike both shifting cultivators and plow
users, the Kofyar have a relatively equal division of labor, with few tasks
or crops rigidly assigned to one gender or the other. Men do 59 percent
of the weeding and transplanting work, as well as 52 percent of harvesting, storage, and processing. Women do 42 percent of the heavy ridging
and mounding, but they perform the complementary role of grinding
and brewing the millet beer that is given to work parties (G. D. Stone et
al. 1990).

Since the time and energy available for the strenuous manual labor of agriculture are limited, even when the tasks are shared by male and female members of the household, the Kofyar have extended their working day and year by two strategies. They have filled in slack periods during the growing season with additional activities, and they have extended their labor into the dry season, which was formerly free of farm work. The new cash crops, yams and rice, along with a marketed surplus of millet, absorb up to 50 percent of Kofyar labor, suggesting that they have consciously and voluntarily increased their work time (Netting et al. 1989). The extra work has been in part scheduled between the peaks of the traditional millet/sorghum labor cycle. Yam weeding must be done fairly regularly throughout the rains from May through November, so this task is superimposed on the millet/sorghum ridging of April through June and the weeding of June through November (Figs. 3.4, 3.5, 3.6). The major effort of making two-foot-high conical yam heaps comes in the period from mid-August to late November, between the millet and the sorghum harvests, when the work load was formerly quite light (Figs. 3.5, 3.6). The yam heaps are made in the same field where the early millet has been reaped and where sorghum and cowpeas are still growing. The work cultivates the existing crops, hills up the sorghum stalks, and mounds the still wet and easily worked soil for the yam tubers that will form the following year's rotation. At the same time, a catch crop of sesame may be planted on the new yam heap.

The current crop of yams may be harvested in the formerly slack dry-season months of January and February (Fig. 3.6), and seed tubers, protected from the sun by a cap of straw and dirt, may be planted so as to be ready to sprout at the first rains. Though labor falls to an average of 2.8 hours per day during the latter part of the dry season, the need for it never disappears, as is usually the case with farm work in the savanna. By farming rice in low-lying, swampy areas unsuitable for grains or yams, the Kofyar also utilize somewhat less busy time periods just before the millet harvest and preceding the picking of cowpeas at the end of November (Figs. 3.4, 3.5). The next step in cash-crop introduction may well be the growing of bananas in streambeds, where they can receive some dry-season irrigation, and the planting of cassava, which continues to develop during the dry months but must be fenced as protection from free-ranging livestock. All of these activities are scheduled so as not to exacerbate existing labor bottlenecks, to fill in slack periods in the agricultural calendar, and to extend productive activities into the dry season. Their effect is to smooth labor inputs over the year, intermeshing various agricultural operations on an increasing diversity of crops, and thereby increasing the total productive application of labor.

CHAPTER FOUR

Energy Inputs, Outputs, and Sustainable Systems

WITH THE CLOSING of the agricultural frontier and increasing limitations on the supply of arable land, a growing rural population and /or an increased demand for farm products results in intensified cropping systems. The basic smallholder strategy is to rely on practices like complex tillage, manuring, terracing, and irrigation, which maintain the soil under more permanent use with sustained yields. Labor substitutes for land, and people work longer, either directly on the farm or in crop processing, agricultural support actvities, or other employment to supplement their incomes from their own fields and livestock.

But isn't this the old-fashioned solution, the primitive and now outmoded route that is appropriate, though regrettably so, only for the teeming masses of earthbound Asia? The "modern," "scientifically and economically proven" approach is that of replacing manual labor with machines, breeding high-yielding strains of plants and animals, raising fertility with chemical fertilizers, and controlling risks by pesticides and herbicides. The fantastic growth of American agriculture and dramatic shrinkage of the U.S. farm population in this century testify to the fact that labor productivity can increase in almost unlimited ways with the help of technology. Moreover, the need for large acreages, expensive equipment, specialist skills, and especially capital to make this system operate effectively must inevitably doom the smallholder (Gladwin and Zulauf 1989).

No one, even the most optimistic of technological determinists, ever claims that there is such a thing as a free lunch in food production. You cannot get higher and more predictable production from the land, you cannot intensify its use, without cost. The principal input may be labor, involving physical effort, the allocation of time, and the management of people, or it may be energy of other kinds. Boserup's model of preindustrial agricultural intensification understandably emphasizes human labor, both its quantitative increase and the potential for declining returns as more work is applied. But the systemic relationships she analyzes are not invalidated when other sources of energy like fossil fuels are brought into the equation. By measuring energy in a common calculus of kilocalories or megajoules, students of energetics like David and Marcia Pimentel (1979), Gerald Leach (1976), and Tim Bayliss-Smith (1982) have compared farming systems from Iban shifting rice cultivation to American corn and soybean industrial agriculture. Their findings bear out the general contention that intensification requires higher energy inputs, but the comparisons of inputs and outputs provided the revolutionary and quite unexpected evidence that ratios of energy expended and returns of edible material *declined* under modern methods to the point where they were strongly negative.

Swidden cultivators with low labor requirements and few other energy inputs (discounting for the moment the heat lost in burning off forest growth) achieve a ratio of about 7.08 calories of output to 1 calorie of input in Iban dry rice (Pimentel and Pimentel 1979: 75), 36.2 to 1 in African millet, and 61.0 to 1 in African cassava (Gerald Leach 1976: 119). Commercially grown U.S. rice, though it absorbed an almost unbelievably low 17 man-hours per hectare, and produced yields of three times the Iban 2,016 kg/ha, required so much nitrogen fertilizer, diesel fuel, pumped water, drying, seeds, and herbicides that the energy ratio was 1.55 to 1 (Pimentel and Pimentel 1979: 75, 77). If edible grains are converted to animal proteins, returns are lower, with 19 kcal expended for each kcal of chicken, 20 for 1 in eggs, 30 for 1 in milk, and 65 to 1 in pork (Pimentel and Pimentel 1979: 55–58). Energy ratios for entire farm operations range from 5 to 50 calories produced for each calorie expended in preindustrial systems, to 1 unit of output for 10 units of input in the United States, with specialized greenhouse vegetable production reaching 1 for each 600 calories invested (Evenari 1988). Neither approach to agricultural production is economically irrational, given the cost and availability of inputs and the relative prices of outputs, but they have different implications for the presence or absence of smallholder production and for the ecological sustainability of the system.

Energy and Evolution

The observation that there are two paths that lead to increased agricultural production appears to be obvious, even banal, but the labeling of these trajectories as traditional and modern, preindustrial and industrial, Western and non-Western, or even extensive and intensive, obscures the significant differences and imposes an evolutionary straitjacket on our thinking. Technological and scientific "progress" is an unquestioned good in manufacturing and distributing commodities, so it *must* be the key to "getting agriculture moving," to relieving human want and removing drudgery. The "truths" of Western scientific and engineering knowledge are deemed universal, and only isolation, "peasant conservatism," illiteracy, and poverty impede their transmission and implementation. Each stage of technological advancement from Stone Age to Iron Age, from human muscle power to horsepower, from the steam engine of the Industrial Revolution to the electricity generated by atomic fission, represents an increased capture of energy.

Cultural evolutionists from Lewis Henry Morgan, Sir Edward Tylor, Marx, and Engels to Leslie White (1943) never doubted that the discoveries and inventions that tapped larger sources of energy were the prime engines of change, providing not only more material goods but a higher standard of living, if only their fruits could be distributed equitably throughout society. The corollary view was that supplies of mechanical energy were practically limitless, and that the efficiency of transforming one form of energy to another inevitably increased.[1] Some disillusionment with the side effects of power-hungry civilizations, the degraded soils, the polluted air and water, may now have set in, but the conviction that food production has a fundamental call on energy supplies, and that only a bit of technological rejiggering is needed to spread the Western pattern successfully to a waiting Third World of peasant farms, dies hard.[2]

1. Leslie White's "law of cultural evolution: culture develops when the amount of energy harnessed by man per capita per year is increased; or as the efficiency of the technological means of putting this energy to work is increased; or, as both factors are simultaneously increased" (L. A. White 1943: 338) explicitly focuses on variable nonhuman energy in tools and practices such as agriculture, while the human energy factor, along with particular skills, is treated as a constant. More "need-serving goods" come, not from more person-days of work with equal or even declining returns to labor, but only from the technological capture of energy that increases "the productivity of human labor" (ibid.: 346). "Efficiency" is ambiguously defined as "the efficiency with which human energy is expended mechanically, . . . the efficiency of tools only" (ibid.: 337), but no attempt is made to measure human or other energy inputs quantitatively or to address the inverse relationship between increasing returns on human work and potentially declining returns on mechanical energy. (Analogies between low-cost electricity and the energy of a human slave [ibid.: 345] are merely anecdotal.) When evolution is modeled in this reductionist manner, technological change raising the amount of energy used per capita precedes and produces population growth, improves human well-being and comfort, grants "independence of nature," and raises output per unit of labor (ibid.: 342–43). To the degree that the smallholder adaptation is a low-energy alternative with less mechanical and more human energy expended, it would presumably be judged evolutionarily retrograde or reflecting a barrier to cultural development.

2. The evolutionary assumption that manual labor in agriculture is backward, extremely time-consuming, onerous, and coerced, and that replacement of such labor by technological energy is therefore the only route to abundance and freedom, is still very much with us. "An old saying has it, 'slavery will persist until the loom weaves itself.'" All ancient civilizations, no matter how enlightened or creative, rested on slavery and on grinding human labor, because human and animal muscle power were the principal forms of energy available

All energy is not, however, created equal, or equally procreative. Of the fundamental physical sources of energy, sunlight, water, land, and labor are all renewable over time, but finite in any given period. The technically useful energy of fossil fuels is both finite and nonrenewable. Food production, always a major user of land and solar power, is differentially dependent on human labor and on fuel energy in developing and industrialized countries (Gerald Leach 1976: 3). Which factors of production will be used most freely and which will be conserved depends on their relative costs and benefits. Where land is plentiful, readily appropriated, and cheap, and where population is sparse, as on a settlement frontier, or where aridity or mountainous terrain make ordinary farming techniques marginally productive, the first choice is to economize on labor with extensive techniques like slash-and-burn cultivation or open-range herding. This is true regardless of whether we refer to the expansion of Neolithic farmers into Europe or the establishment of cattle ranches in Brazilian rain forests (National Research Council 1992: 67–75). If there are few people present and they have a variety of ways to make a living with relatively little effort, the cost of labor will be high. For intensification to take place under these circumstances, less expensive sources of energy will be sought, and there will be a heavy emphasis on increasing labor productivity, usually by mechanical means (ibid.: 15). With a market that prices the inputs of labor and fuel energy and the outputs of food, practical economic decisions can be clearly specified. The economically appropriate level of energy use is the point at which the marginal monetary value equals the cost of the increment of energy (Lockeretz 1984). Where fossil fuel energy and the machines to utilize it are relatively cheap, as in North America and Europe, and labor is costly, technology will replace manpower. For example, in 1978,

a gallon (3.79 litres) of gasoline sold for about $0.65 in the United States. . . . Based on a minimum wage of $2.65 per hour, this gallon can be purchased with only 15 minutes of work. However, that one gallon of gasoline in an engine will produce the equivalent of 97 hours of manpower. Thus, one hour of labor at $2.65 will purchase the equivalent of 395 hours of manpower in the form of fossil fuel. . . . If energy is cheap relative to the price of food, then obviously fossil energy use in food production is an excellent investment. This is true today in the United States. One thousand kilocalories of sweet corn in a can sells for about $0.93 whereas 1000 kcal of gasoline has a value of $0.02. Hence, 1 kcal of sweet

for mechanical work. The discovery of ways to use less expensive sources of energy than human muscles made it possible for men to be free. The men and women of rural India are tied to poverty and misery because they use too little energy and use it inefficiently, and nearly all they use is secured by their own physical efforts. A transformation of rural Indian society could be brought about by increasing the quantity and improving the technology of energy use" (Revelle 1976: 974).

corn is worth 47 times more than 1 kcal of gasoline energy. (Pimentel and Pimentel 1979: 27)

Because energy has its price and because nonrenewable fuel sources can be expected to rise in price as supply declines over the long term, even the rich nations that are most prodigal with energy may in time reduce its use. David Pimentel and his colleagues speculated that if U.S. fuel costs increased two to five times from their 1973 level, an average return of 2.8 kcal of corn per 1 kcal of fuel input might become uneconomical, and minimal tillage, substituting manure for chemical fertilizer, and reinstituting rotations might result in economically justified energy savings (Pimentel et al. 1973).

Whether the energy demands of agricultural intensification are met by human or mechanical/chemical sources of energy is not solely a function of their relative costs. Even in countries with access to large stocks of capital and to scientific agronomy, differing amounts of land per capita may result in contrasting trajectories of historical growth in productivity. Whereas the United States remained consistently ahead of Denmark in agricultural output per male worker from 1880 to 1970, Danish farms have far exceeded U.S. averages in output per ha (Prologue, Fig. P1). With much larger farms, U.S. farmers could afford a more land-extensive strategy and increase labor productivity with fuel energy. Although the Danes also used chemical fertilizers and modern techniques of livestock rearing, the marked limitations on arable land in Denmark meant that there had to be great stress on higher production from each land unit, achieved presumably on much smaller holdings with higher labor inputs. Japan's curve in Figure P1, defining a path still higher and to the left of Denmark's, shows even lower work-force productivity, which does not rise as fast as yields that are four to eight times the output of equivalent U.S. land.

This analysis says nothing about either total labor per farmer or total use of nonhuman energy. It does point up the divergence of strategies appropriate to land-short and land-abundant environments (Fig. 4.1). Irrigated, high-yielding, labor-intensive systems in Asia and Egypt are paralleled by European systems with considerably lower returns to land and higher output per worker. Both Asia and Western Europe contrast decisively with the "frontier" model of the United States and other relatively recently settled countries, such as Canada and New Zealand, where growth has been carried almost completely by increases in labor productivity powered by technical energy (Ruttan 1984: 108). Although Green Revolution technologies with high-yielding crop varieties and chemical fertilizers have been applied to the wet-rice systems of the Orient, they are largely supplementary to the preexisting "conservation model," in

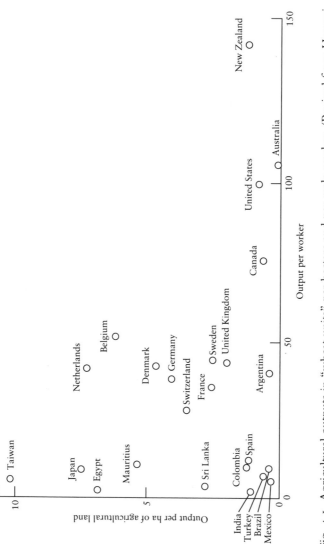

Fig. 4.1. Agricultural outputs in "wheat units" per hectare and per male worker. (Revised from Hayami and Ruttan 1971: 71)

which inputs of plant nutrients, animal power, land improvements, physical capital, and labor force are largely produced or supplied by the agricultural sector itself (Ruttan 1984). To the extent that smallholder intensification tends to increase production per unit of land at the expense of output per worker, and does so with more reliance on renewable as opposed to fuel or nonrenewable energy sources, it is an agricultural system appropriate to relatively high rural population densities and relatively expensive nonhuman energy inputs. Conversely, a relatively low man/land ratio, expensive labor, and cheap fuel just as surely conduce to big, land-extensive farms with few workers and lavish energy use.

Is Farm Labor an Anachronism?

To avoid the sterile dichotomies of evolutionism, and to dissect the amounts and efficiencies of energy flows in real systems, we need to look at comparative examples. At the outset, we should realize the approximate nature of our energy measures. Estimates of agricultural work (Gross 1984) may be based on random visits over time (A. W. Johnson 1975), on reports from individuals gathered two or three times a week by enumerators (D. W. Norman 1969), or by daily schedules filled out by farmers or local residents (Stone, Netting, and Stone 1990). They may be based on timing a task in a measured area and then generalizing it to the entire plot (Rappaport 1971) in a small sample of farms. Hours vary in different tasks of the agricultural cycle, and they must then be translated into some rough common measure of energy, as Gerald Leach (1976: 119) assumes 0.8 MJ/hr for African agriculture, or given different ratings according to tests of respiration on subjects engaged in hoeing, weeding, and so on (Rappaport 1971).[3] Inputs of labor for the major activities of New Guinea Tsembaga shifting cultivation can be diagramed in this manner, and labor energy in kilocalories compared to the energy of the resulting yam and sweet potato crops, showing a favorable 16 to 1 return on the human-energy investment. Because much of this vegetable food is fed to pigs, the final return to human consumption from pork may be only 2 or 1 for each calorie expended (that is, energy output/input ratios of 2:1 or 1:1), but calories have been traded for more valuable high-quality proteins. Further elaborations of data gathering could distinguish

---

3. Studies of unmechanized agriculture in Hungary, Russia, Italy, Germany, and Gambia measured average male energy expenditure at 6.0 kcal/min in various agricultural tasks, while samples of women in Russia, Italy, Gambia, and Nigeria averaged 4.7 kcal/min (Passmore and Durwin 1955, cited in Revelle 1976). Reducing these totals by one-third to exclude the portion of working time spent at rest would give expenditures for adult males of 240 kcal/hr and for females of 182 kcal/hr. For rural India, Revelle (1976) uses estimates of male cultivators using 250 kcal/hr over 180 ten-hour days, while females expend 200 kcal/hr in 125 eight-hour days.

TABLE 4.1
Energy Inputs and Crop Outputs in Three Countries

	Mexico	Guatemala	Nigeria
Input			
Labor time (hrs/ha)	1,144	1,415	620
Seeds (kg/ha)	10.4	10.4	10.4
Chemical fertilizer (kg/ha)	–	–	–
Energy (Kcal/ha)	642,338	781,903	555,778
Output			
Corn yield (kg/ha)	1,944	1,066	1,004
Corn yield (Kcal/ha)	6,901,200	3,784,300	3,564,200
Efficiency (ratio of Kcal			
output to Kcal input)	10.74	4.84	6.41

SOURCE: Pimentel and Pimentel 1979: 63–65.

NOTE: The Mexico figures are ca. 1951; the Guatemala figures ca. 1940; and the Nigeria figures ca. 1971. The source table breaks down Kcal expenditures for several different stages in the agricultural cycle (clearing, seeding, cultivating with axe and hoe, etc.), but I have opted to give only the total figure under the label "Energy."

the different work effort and caloric expenditure of adult males, females, and children, standardizing them as a man-equivalent (Ruthenberg 1976: 51). Conventions for calculating average energy input for draft animals and number of work days or hours per crop or per season may be equally arbitrary and imprecise. But there remains a basis for gross comparison between farming systems and, using the more accurately measured energy inputs of fossil fuels and manufactured items, comparing energy ratios and costs among various types of industrialized agriculture.

A simplified schematic energy budget for slash-and-burn maize cultivation in Mexico, Guatemala, and Nigeria suggests energy output/input ratios of 5:1 to 11:1 (Table 4.1). Small inputs for making the axe and hoe and producing the seed are estimated, and in the Nigerian case inputs include some fertilizer. Corn yields/ha in Mexico are almost double those for Guatemala and Nigeria, and there is no mention of either average field area per household or the supplementary legume, cucurbit, tuber, and tree crops so frequently interplanted in maize swiddens. Nevertheless, the caloric production of corn alone from a single ha would in every case surpass the 3,504,000 kcal that would provide 2,400 kcal/day for four man-equivalents over an entire year (Gerald Leach 1976: 9).[4] With this level of consumption and using the not unreasonable energy ratio of 25:1 for swidden cultivator labor alone, the caloric needs of a family could be provided with 1.6 MJ, about two hours of physical work

4. Maize is an especially productive swidden crop because of its biologically efficient energy-conversion system. Utilizing the carbon-4 photosynthetic pathway, maize produces two to three times more biomass per unit of land than crops of the carbon-3 type, such as wheat, barley, rye, and oats. Its yields ratios to seed are 100–300:1 as compared with the 4–6:1 of European grains, and the returns to hoe labor of maize are correspondingly high (Truman 1989: 169).

TABLE 4.2
*Comparative Net Human Energy Inputs for Rain-fed
and Irrigated Upland Agriculture Using Draft
Animals, India, 1956–57*
(MJ/ha)

Production task	Rain-fed crops	Irrigated crops
Land preparation	53	77
Sowing	25	36
Manuring	1	6
Weeding	13	64
Harvesting	72	98
Threshing	38	74
Irrigation	–	140
TOTAL	202	495

SOURCE: M. J. T. Norman 1978: 362 (adapted from Boserup 1965: 40). Used with permission.

per day (for comparison, see the Machiguenga case, Chapter 3) (Gerald Leach 1976: 9). A total of 720 hours a year in agriculture falls within the range given for Hausa male farmers in the preceding chapter. Yearly work input in extensive agriculture remains low, though adequate for subsistence, even with the quite modest returns of 0.75 to 1.70 kg corn/hr. Dry-rice swiddens in Sarawak, the Philippines, and Tanzania (Freeman 1955; Conklin 1957; and Ruthenberg 1971, cited by Bayliss-Smith 1981: 34–35) show a similar range of 700 to 1,081 hrs/ha with returns of 1.59 to 1.95 kg/hr. Nutritional energy requirements for a household of five persons could be met by 1,671 kg of unhusked rice, including 60 kg of seed, with 1,078 hours of work on 1.156 ha of land (Bayliss-Smith 1981: 35). When scarce land prevents the shifting of fields and laborsaving practices of swidden cultivation, energy inputs/ha necessarily rise, but smallholders at the same time may tap increasing amounts of nonhuman energy.

Under the severe land constrictions of Ukara Island in Tanzania, with total household landholdings of 2.83 ha, of which 1.88 ha are intensively tilled, with manuring, terracing, and sometimes irrigation, labor inputs are estimated at 6,300 hours for a household of 10.9 members (Ruthenberg 1976: 147). The returns in kcal/man-hour under this system may be less than half of those achieved under shifting cultivation of upland rice (Ruthenberg 1976: 52, 147). Intensification involves both increasing time devoted to agricultural tasks and the addition of new labor demands such as those of irrigation. A comparison of rain-fed and irrigated crops in the same Indian villages (Boserup 1965: 39–40) suggests that every operation, but especially weeding, harvesting, and threshing, consumes more time as production per ha climbs. Irrigation alone is equal to 69 percent of the entire expenditure for the rain-fed crops, and total human-energy inputs/ha go up almost 2½ times in the irrigated regime (Table 4.2).

Draft Animals and the Plow—An Energy Revolution?

Where draft animals are used for traction or for turning a wheel that raises irrigation water from a well, there is a major increment in effective energy. A 550-kg ox working a six-hour day can expend six times the energy of a human worker, and a man plowing with a pair of draft animals harnesses 13 times the energy of his unaided efforts (M. J. T. Norman 1978: 357, 360). Human labor achieves much higher returns when it is combined with animal power, but total human labor must increase as well. A system of more frequent plowing and irrigation of wet rice using an animal-powered water-lifting wheel in Madras, India, demands four times the human labor of rain-fed rice in Thailand. The efficiency of labor declines from 40:1 to 13:1. If animal energy is counted as an input, the ratio of caloric output to input sinks from 12.5:1 to a meager 1.7:1 for Madras (ibid.: 363). Moreover, these figures do not take into account the human effort devoted to maintaining draft animals. Free-ranging pasturage on natural grass would require little work, but many densely populated peasant communities have little forage and must grow fodder, lead animals to water, and build fenced paddocks and sheds for their beasts. One estimate of the costs of keeping two male draft animals and a replacement herd of 2.2 adult females and 2.4 immature animals without free grazing sets their consumption at 400 MJ/day, which is ten times that of a farm family of six (Odend'hal 1972, cited in M. J. T. Norman 1978: 358).

Because animal power allows the initial plowing and harrowing of a field to be done so much faster than any hand method, bigger fields can be tilled, and more land can be prepared during the most propitious periods of the cropping season.[5] If land is plentiful and if rainfall or annual flooding in a river valley provides moisture without artificial watering, a technological change to plow and draft animals can rapidly raise production. In the 14 Asian examples of farming technology surveyed by L. M. Hanks (1972: 56–64), shifting dry-rice cultivation had an average field size of 1.54 ha, while the systems that employed the plow, animal traction, and the broadcast sowing of rice had fields of 4.25 ha. Irrigation and transplanting systems used fields averaging 2.23 ha. Labor in man-days per single rice crop fell from 255 in shifting to 179 in broadcasting, and, although yields/ha also declined somewhat, the efficiency of direct labor in rice output per man-day went up 3.6 times. In both broadcasting and

5. It was just this speed in springtime plowing that influenced medieval European peasants to replace ox teams with horses, even though the literate agricultural "experts" of the day worried that growing oats for horse fodder made these animals much more expensive to keep than the grazing oxen (Lynn White, Jr., 1962).

irrigation systems, however, there were indirect costs of equipment plus substantial investment in buying, training, and maintaining draft animals (ibid.: 62). Irrigation also required an original investment in leveling and diking fields, digging the irrigation network, and then maintaining the canals and ditches. Plow technology has a clear advantage in labor efficiency only when naturally watered land is still plentiful and the additional labor inputs of transplanting and irrigation are not necessary.

Although agricultural intensification by definition refers to the process of increasing the long-term energy output of land resources, it does not follow that technological evolution represents a smoothly declining curve of human labor input and an increasing efficiency of that labor. Animal traction and the plow may additionally contribute to an upward spike in total production and an improvement in labor efficiency, but these advantages may diminish under population pressure. Cross-cultural comparative evidence (see Chapter 9) refutes the notion that where large domestic animals and an amenable environment are present, along with knowledge of the plow, it will always be adopted, because it universally increases labor productivity (Pryor 1985). In fact, where low population density and shifting cultivation coincide with staple crops that do not particularly benefit from the plow (root crops, tree crops, millet, sorghum, dry rice, and maize), the plow is almost totally absent, whereas intensive cultivation and "plow-positive" staples (wheat, barley, rye, buckwheat, teff, and wet rice with lower yields and/or time constraints in land preparation) strongly predict plow use (Pryor 1985: 738). Labor productivity may indeed be higher in plowless long-fallow cultivation than in cultivation with the plow. For areas of dense population where intensification is achieved with irrigation, terraces, intercropping, and hoe methods, such as among the the Philippine Ifugao, the Tanzanian Wakara, and the Nigerian Kofyar, the plow, while its use is possible, is unnecessary and inconvenient. The returns on nonhuman energy are not worth the cost.

African farming systems where there has been a transition from hoe to plow methods show common economic characteristics (Pingali et al. 1987: 33–36). As agricultural intensity measured by frequency of land use increases, labor inputs for land preparation, weeding, and maintaining soil fertility rise rapidly (Fig. 4.2). Since bigger plowed fields would require more hand weeding and a major labor investment in destumping, fallow systems demand less work. Acquiring, training, and maintaining draft animals increases the labor overhead, as does the accompanying practice of manuring. The switch point at which hand-cultivation labor costs per unit of output have climbed to the point where the shift to animal power becomes economically justified is schematically represented in Figure 4.2. Heavier soils that are difficult to work by hand or a good market for animal by-products such as meat, milk, and hides may shift

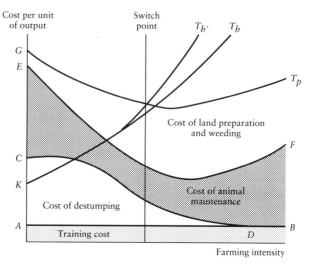

Cost per unit
of output

Switch
point $T_{h'}$ T_h

G

E

T_p

Cost of land preparation
and weeding

F

C

K

Cost of animal
maintenance

Cost of destumping

A   B

Training cost   D

Farming intensity

T_p = Total labor costs for land preparation, early-season weeding,
and manuring, using animal traction

T_h = Labor costs for land preparation and early-season weeding,
using the hand hoe

$T_{h'}$ = T_h plus labor costs for maintaining soil fertility without
manure from draft animals

Switch point = Farming intensity at which animal traction becomes
the dominant technology

Fig. 4.2. Comparison of labor costs of hand cultivation and
animal-powered cultivation. (From Pingali et al. 1987: 34)

the switch point to the left (Pingali et al. 1987: 35), but shifting cultivators or nomadic pastoralists are unlikely to accept the costs of plowing.

Altering the relative costs of the factors of production—that is, making labor more expensive, cutting the cost of fuel and tools, and raising the price for which agricultural goods are sold—will lead to increased reliance on nonhuman energy sources, from animal-drawn plows to tractors, chemical fertilizers, and large amounts of fossil fuel. While labor becomes more productive, energy efficiency as measured in the ratio of inputs to outputs almost inevitably declines. Table 4.3 compares crops produced in several areas as draft-animal and then mechanical energy is substituted for the predominantly manual methods of subsistence shifting cultivation. Local environmental conditions and differences in average field size and fallowing cycle may lead to a range of efficiency values—for example, Mexico's 10.74 value for slash-and-burn corn and Guatemala's 4.84. For some normally land-extensive crops with lower

TABLE 4.3
Energy Input-Output Ratios by Crop and Type of Technology, Eleven Countries

Crop	Shifting cultivation		Animal traction		Mechanized	
Corn	Mexico	10.74	Philippines	5.06	United States	2.93
	Guatemala	4.84	Guatemala	3.11	United Kingdom	2.34
Wheat	–		India	0.96	United States	2.41
					Britain	3.51
Rice	Sarawak (Iban)	7.08	Philippines	3.29	United States	1.55
					Japan	2.45
Sorghum	Sudan	14.43	–		United States	1.96
Peanuts	–		Thailand	6.46	United States	1.40
Cassava	Tanzania	22.93	–		–	
Potatoes	–		–		United States	1.23

SOURCE: Pimentel and Pimentel 1979: 63–94.
NOTE: All figures are for various years in the 1960s and 1970s with the exception of Mexico (ca. 1951), Guatemala (ca. 1940), and Sarawak (ca. 1955). Readers interested in the exact dates for any given country should consult Pimental and Pimental 1979: 63–94.

yields, like wheat, Indian bullock plows have an efficiency of 0.96, below that achieved by U.S. (2.41) and British (3.51) mechanization, though India continues to utilize largely renewable energy sources (Pimentel and Pimentel 1979: 72–73). But it is fair to conclude that there is a trade-off between labor and other sources of energy, a bargain that depends on their relative costs as well as on the supply of energy in the ecosystem, whether it comes from on-farm sources, and whether or not it is renewable.[6]

Renewable Energy and Sustainable Agriculture

Behind the comparative calculations of labor and nonhuman sources of energy lies the fundamental question of how to value the factors involved. The monetary costs and returns as measured by agricultural economics may suggest little reason for drastic change in energy use, but market forces may be far from perfect when it comes to handling depletable or nonrenewable energy (Lockeretz 1984). Concern about a growing population of food consumers, dwindling fuel supplies, the dangers of industrial energy generation, international conflict over energy resources, and

6. For a given intensity of farming (measured by the R value of years of cultivation × 100 years of cultivation and fallow), animal traction, where feasible, reduces labor use by 64 percent, and tractor use leads to an additional saving of 35.5 percent of the remaining labor. The switch from hand to animal methods increases yield per man-hour by 71 percent, and tractor use gives an additional 61 percent. Although an intensification of the agricultural system leads to longer and more regular working hours, a switch from hand to animal and finally to tractor cultivation increases labor productivity. It is important to note, however, that at any given level of intensity, changing tools alone does *not* increase yields even when it saves labor. See Pingali and Binswanger 1983: 9–11.

disproportionate energy use in some parts of the globe have brought questions of sustainable agriculture to the fore. The technological emphasis in agricultural development on raising production has encountered a slowdown in the growth of cereal yields worldwide since the beginning of the 1970s, and there is now evidence from Japan, Holland, and the United States of definite limits on the ability to boost output with more fertilizer (Douglass 1984). Pests that flourish in monocropped fields are freed from their natural enemies and at the same time become resistant to pesticides, demanding the development of more toxic chemicals. The same industrial systems threaten soils with topsoil erosion, waterlogging and salinization through irrigation, and the buildup of toxic substances. As larger proportions of petrochemically derived energy are used on the farm, the costs of these inputs rise.

Sustainable agroecosystems (Conway 1985, 1987; Marten and Saltman 1986; Douglass 1984; Chapin n.d.) can be in part defined in energy terms over time. The attributes of such systems include:

1. Relatively stable production per unit of land. Yields do not decline, because soil fertility and water supplies are maintained within acceptable levels. Weeds, animal pests, and diseases do not progressively reduce the energy needed for plant and domestic animal growth. Land productivity can be at various levels (an "integral" system of shifting cultivation achieves stable though infrequent crop yields from a given unit of land without environmental degradation), but increases in production should not be temporary, and the system should be resilient in the face of short-term or seasonal perturbations. Soil erosion or "mining," falling ground-water supplies, or declining resistance of crop plants to drought or insects, suggest that production will decline, even if it has not already begun to do so.

2. Predictable and relatively stable inputs of energy. Rapid increases in demand for labor, mechanical power, fertilizer, or water suggest ecosystem imbalance and the difficulty of maintaining the higher level of inputs.

3. Economically favorable rates of return between inputs and outputs, both in energy and in monetary terms. Consistently exceeding the energy budget, even if unrenewable energy sources are cheap, risks decreases in necessary energy inputs when their costs increase or when farm prices decline. When a significant portion of energy inputs come from the farm itself, when energy is conserved and recycled, stability in the face of uncontrollable changes in outside energy costs and prices is easier to preserve. Diversity limits risk and strengthens stability (Norgaard 1989).

4. Returns to labor and other energy inputs sufficient to provide an acceptable livelihood to the producers. This includes not only subsistence, with nutrition that sustains strength, health, and normal growth,

but also sufficient saving and accumulation to meet contingencies and to make the investments necessary for long-term productivity (Brokensha 1989).

The combination of stable and diverse production with high yields, internally generated and maintainable inputs, favorable energy input/ output ratios, and articulation with both subsistence and market needs is effectively achieved with a smallholder strategy of intensification.

Certainly the most enduring and best-documented systems of highly intensive agriculture are the irrigated rice smallholdings of China. For the Jiaxing region with its rich soils, ample 1,300 mm rainfall, irrigation water, and favorable temperatures, *Shen's Agricultural Book* (Zhang Lu-ziang 1956) from the seventeenth century allows a reconstruction of en-ergy flows in farming. Dazhong Wen and David Pimentel (1986) have evidence that heavy production of rice, wheat, mulberry leaves for silk-worms, and livestock has been sustained over centuries by human labor, using intensive practices of composting and green manuring, crop rota-tion, irrigation, and animal husbandry. Draft animals or other sources of imported mechanical energy were not used, and the soil was tilled with a rakelike iron tool that reached more than twice the 10 cm depth of a plow drawn by water buffalo. A major labor investment amounting to one-third of the 2,330 hrs/ha involved in rice cultivation went into processing and distributing 10,000 kg/ha of animal and human manure composted with rice straw. Summer rice could be alternated with winter crops of wheat, barley, rape, or beans, producing, for example, 3,900 kg/ha of rice and 1,300 kg/ha of wheat from the same field. Another type of crop rotation that was somewhat less labor-intensive involved growing one-third of a hectare of Chinese milk vetch, which provided 7,500 kg of green manure for a hectare of the following rice crop. Both the compost-ing and the green-manuring methods required heavy labor inputs in til-lage, transplanting, and harvesting. Both were also irrigated by foot-powered waterwheel pumps that required three men to operate them.

Although the production of irrigated wet rice and the energy inputs in an intensive system will always be relatively high, the factors of produc-tion are all subject to adjustment. A Philippine system (Pimentel and Pi-mentel 1979: 76) lowered human labor to less than a third of the seven-teenth-century Chinese figures, partially compensating with energy from water buffalo traction and small amounts of nitrogen fertilizer (Table 4.4). Philippine rice production/ha was only 42 percent of Chinese totals, and energy efficiency dropped to 3.29. Among contrasting con-temporary cases, Dawa, China shows the potential for increasing even the remarkable traditional yields from pond fields by adding still more labor, using animal power, and introducing large amounts of fossil-fuel energy (Table 4.4). The energy ratio falls to 3.16, but it remains higher

TABLE 4.4
Comparison of the Energy Inputs and Crop Outputs for Philippine, Chinese, and U.S. Rice-Production Systems

	Philippines (1962–63)	Dawa, China (1979–81)	Louisiana, United States (1977)
Input			
Labor (hrs/ha)	576	3,045	25
Animal power (hrs/ha)	272	332	–
Seeds (kg/ha)	108	164	n.a.
Energy (Kcal/ha)	1,825,432	7,577,139	11,460,694
Output			
Rice yield (kg/ha)	1,654	8,094	4,114
Rice yield (Kcal/ha)	6,004,020	29,382,672	14,933,820
Efficiency (ratio of Kcal output to Kcal input)	3.29	3.16	1.30

SOURCE: Wen and Pimental 1986: 4.

NOTE: As in Table 4.1, I have given only the total Kcal expenditures under the heading "Energy." The variations in the forms that energy expenditure takes are very great: for example, human energy accounted for 1,568,381 Kcal/ha in China, but only 12,772 in the United States; yet fossil fuel (including chemical fertilizers) accounted for only 128,568 Kcal expended in the Philippines but 11,447,992 Kcal in the United States.

than in a mechanized system in Louisiana, where human labor has almost disappeared and the energy represented by diesel, gasoline, and natural-gas fuels; nitrogen, phosphorous, and limestone fertilizers;[7] herbicides; drying; electricity; and insecticides reaches almost 11.5 million kcal/ha. A phenomenal labor productivity of over 600,000 output calories for each calorie of direct human work effort has been gained at the expense of an overall energy ratio only 11 percent of that achieved by the Chinese green-manuring system. Where the energy subsidy reaches its epitome in the Central Valley of California, a technology of laser-leveling of fields, aerial seeding and fertilization, and combines that harvest and thresh 8 ha/hr, the enormous energy inputs are almost equally divided between irrigation, equipment, and biochemical uses (Pudup and Watts 1987: 367), a dependence that has emerged in only 70 years and that raises serious questions of sustainability.

7. Manufacturing fertilizer is energy-intensive. Nitrogen is especially costly, requiring 14,700 kcal of fossil fuel for 1 kg of nitrate fertilizer, compared with 3,000 kcal/kg for phosphorus and 1,600 kcal/kg for potassium. Livestock manure contains less concentrated nutrients (0.56 percent nitrogen), but it aids in reducing soil erosion and improving soil structure (Pimentel 1984). Although considerable fuel is required to transport and spread manure with a tractor and spreader, the energy expenditure for commercial fertilizer is almost three times that of manure if it is transported no more than 1.5 km (Pimentel 1984: 128). The issue with the manufacture of chemical fertilizer, as well as with pesticides and herbicides, and with the release of CO_2 and other greenhouse gases as products of motor vehicles and factories, may be less the exhaustion of fossil-fuel energy resources than the environmental damage they cause. Industrial agriculture may very directly threaten its basic resources of soil and water.

Rather than importing fuel and mechanical energy for large-field monocrops, the Chinese farmers of the seventeenth century practiced a diversified, organic strategy that recycled internal and renewable energy resources. The persistence of smallholders in an area that supported 7.8 people/ha of farmland in the seventeenth century and provides for 15 people/ha today testifies to the maintenance of high yields without serious environmental degradation. Sustainability refers not only to the stability and favorable ratios of energy inputs and outputs but also to the source and costs of inputs and the range of economic needs met by outputs. To the extent that inputs are produced on the farm and by means available to the household, the farm family is less dependent on outside forces and less vulnerable to rapid changes in the market or failures in external economic and political systems. Labor energy, though substantial, could not exceed the potential of the household and occasional hired help, and double-cropping, tree cultivation, animal husbandry, and cottage industry provided the diversity of tasks that smoothed seasonal swings in work demand (see Chapter 3) and accommodated gender and age differences in strength and skills through a division of labor.

The stress on collecting or producing, processing, and returning organic material to the soil meant that diverse "waste" materials were never wasted. Of the 930 kg/ha manure (dry weight) applied to the crops, 34 percent came from the human residents, 47 percent from the 3.6 pigs/ha or a combination of pigs and sheep, and 19 percent from silkworm excrement, which could also be fed to fish. Rice straw was either composted or burned as household fuel, with the ashes being composted. The residue from processing rape-seed oil was applied as top dressing to cropland, and the fertile bottom muck from ponds used for irrigation and fish culture was dredged up to fertilize mulberry plantations. Vetch grown solely for green manure fixed substantial amounts of nitrogen for the system (Wen and Pimentel 1986: 8).

It is significant that the farming system included integrated subsystems that were market- rather than subsistence-oriented. Ten percent of the farmland was devoted to mulberry trees for silkworm cocoon production. Labor, at the rate of 1,450 hr/ha/yr, produced 13,800 kg/ha of mulberry leaves, which were in turn fed to sheep and to silkworms. Mulberry-leaf yields had an energy ratio of 0.57. Women reared the silkworms, putting in some 300 hours to feed 1,000 kg of mulberry leaves, producing 63 kg of cocoons and 5 kg of silk (Wen and Pimentel 1986: 20–21). Clearly, large amounts of energy were here being transformed into small but very valuable quantities of commodities. Harvested grain was not merely consumed or sold in raw form. Perhaps one-third of the crop was converted into alcohol, and the residue from the distilleries was fed to pigs (ibid.: 22). Meat for consumption or for sale also came from sheep, from barnyard fowls tended by women, and from the fish raised in ponds, and the droppings of all these animals entered the fertilization cycle. Moving up the food chain, as in feeding 180 kg of rape cake or 250 kg of barley to produce 50 kg of pork in six months, entailed a ratio of feed to meat of 5:1, but the addition of high-quality protein to the diet evidently justified the cost in energy.

With negligible fossil-fuel energy embodied in hand tools, the major source of power for the system was human labor, with an input of 3,756 hr/ha, of which 70 percent was for cropping, 4 percent for the mulberry plantation, 8 percent for rearing silkworms, and 18 percent for tending pigs and sheep. "The other energy used in the agroecosystem was renewable bioenergy produced within the agroecosystem" (Wen and Pimentel 1986: 26). Perhaps the system would have been more effective if wood from trees had replaced crop residues for household fuel, and if both fewer pigs and a somewhat smaller human population had been supported. But there seems little doubt that the system demonstrated ecological and energetic sustainability. Its overall production could have been further increased (although at the cost of some decline in efficiency) with a small input of industrial energy, but this might have risked the balance of energy flow that has been sustained for centuries.

It is striking that intensification of agriculture in the Chinese heartland of the Zhu Jiang (Pearl River) Delta did not approach some climax state of maximum energetic efficiency and then remain static or become involuted. Since the low-lying land was reclaimed by digging ponds and constructing dikes in the ninth century A.D., it has gone from wet rice, fish, and fruit trees to fish culture, mulberry trees, and silkworms for the nineteenth-century silk industry (Zhong 1982). Photosynthetic energy flows from the mulberry leaves to the silkworms. Silkworm excreta in turn feed fish, whose droppings, along with aquatic organisms, create fertile pond mud, which is used as manure for mulberry trees (Fig. 4.3). Vegetables, bananas, and grass (for fish food) can also be grown in rotation with the mulberry trees, and silk cocoons, fresh fish, sugarcane, and vegetables are all marketed. A similar integrated system of nutrient recycling is used for shrimp propagation by Vietnamese smallholders (Fig. 4.4). The Chinese mulberry dike–fish pond complex absorbs twice as much labor as the alternate land uses of rice or sugarcane growing, and it provides a considerably higher income per unit area (Zhong 1982: 201). This agroecosystem has gone through stages of increasing intensification for "thousands of years without apparent deterioration of soil fertility" (Zhong 1982: 192).

My global comparisons between labor-intensive smallholders and large farmers using major amounts of technological energy have tended to contrast different world regions or highly distinctive farming systems

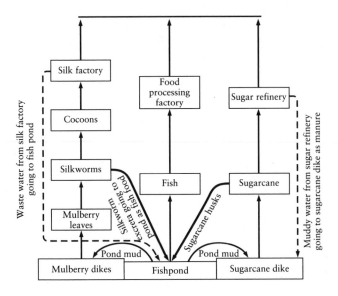

Fig. 4.3. A Chinese mulberry dike / fish pond ecosystem.
(From Zhong 1982: 197)

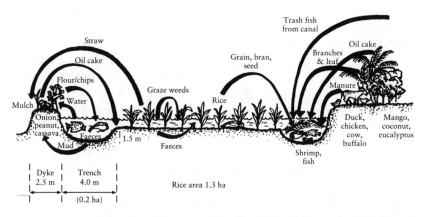

Fig. 4.4. Material flows in a Vietnamese integrated farming system. The diagram
is based on sketches by Mekong Delta farmers of their own enterprises with tree
crops, vegetables, and livestock on dikes; trenches for fish, shrimp, and ducks;
and irrigated rice. The commercially valuable shrimp are protected by sub-
merged mango and eucalyptus branches, and they are fed germinated rice, cassava flour, rice bran, coconut, and peanut oil cake. Vegetable matter and animal
droppings reduce but do not eliminate the requirement for chemical fertilizers
(Lightfoot and Tuan 1990: 18–19).

TABLE 4.5
Energy Ratios in Amish Farming
(1970)

	Size (ha)	Cows	Input (Mcal)	Output (Mcal)	Yield (Mcal/ha)	Energy ratio
Pennsylvania						
Old-Order						
Amish (dairy)[a]	32.6	31.0	113,367	134,527	3,151	1.009
"English" (dairy)[a]	73.4	47.3	444,453	245,715	3,071	0.553
Wisconsin						
Amish (mixed)[a]	60.8	14.5	31,379	50,631	1,305	1.614
"English" (small dairy)[b]	71.6	24.5	–	–	1,668	0.274
"English (large dairy)[b]	107.6	40.9	–	–	2,069	0.395
Eastern Illinois						
Amish (mixed)[a]	38.9	–	177,821	173,134	3,165	0.974
"English" (hogs)[a]	200.6	–	1,230,769	2,466,156	11,444	2.003

SOURCE: W. A. Johnson et al. 1977: 376. Used with permission.
[a]Interviewed by W. A. Johnson et al. Three groups of Amish were studied.
[b]Wisconsin Farm Business Survey, 1975, with supplemental information from *Selected U.S. Crop Budgets: Yields, Inputs and Variable Costs*, U.S. Department of Agriculture Bulletin, 1965.

for single staple crops. Are contrasting yields and energy ratios also present between big farms and smallholdings within single agroclimatic regions and economies? The Amish in the United States selectively employ some modern inputs (Hostetler 1980) while rejecting the tractor and motor vehicles. Horse-traction farming and continued use of labor-intensive intermediate technology has limited the Amish to family farms of ca. 40 ha (about 100 acres), usually less than half the size of those of their "English" (non-Amish) neighbors (Table 4.5). For milk production, which is expensive of energy, "English" farmers in Pennsylvania use 83 percent more energy (not including labor) than do Amish, and their energy ratio is only half as good. The Amish strategy is also better fitted to the long, narrow valleys, steep, wooded hills, and marginal pastures of those remaining Pennsylvania areas that are not yet intensively tilled. Though Wisconsin Amish farms have fewer cows than those of the "English," and somewhat lower yields, their input/output energy ratio is four to six times as high. Large mechanized farms typically show economies of scale and operating efficiency over small ones, and in certain activities, such as Illinois hog raising, big "English" farms are more productive than those of the Amish. But with a religious ethic of self-sufficiency, frugality, and the conservation of energy resources for future generations, the Amish have both provided for their physical needs and accumulated cash to purchase land for their offspring. Technical energy procured from external sources in the form of fertilizer and the means to preserve homegrown food is used efficiently and without the need for large capital investments.

Where a single group of people practice different farming systems at the same time, the energy balance may also be strikingly disparate. Russian collective farms of several thousand hectares devote only about 7 percent of their land to private plots (averaging 0.3 ha), but these intensively tilled areas of potatoes, cabbage, orchards, and domestic animals have provided 25 percent of total national agricultural output (Bayliss-Smith 1982). A household may expend 135 days/yr on the collective and 145 days/yr on the private plot. Energy subsidies for the collective in fuel, machines, fertilizer, and electricity are high, but returns are only 6,680 MJ/ha, compared with 41,590 MJ/ha on the private plots, and energy ratios are less than 1.3:1 for the collective (Bayliss-Smith 1982: 94–96). Returns on the private plot, which is near the house, receives constant care and attention, especially from the housewife, and is provided with abundant manure and compost, may be comparable to those of a British allotment garden at 11.2:1 (Bayliss-Smith 1982: 86–90). The amount and quality of work on the private land devoted to household consumption and sale for profit is obviously superior to that mobilized on the mechanized commune.

Sustainability: In the Eyes of Beholders and Smallholders

Sustainability is a term that has buzzed rapidly into the popular consciousness trailing clouds of positive affect, which is also evoked by *ecology, conservation,* and *environmental protection. Sustainability* is a prime candidate to be the watchword of the 1990s, and it is increasingly attached to the agroecology of the smallholder. I have especially emphasized the existence of favorable energy input/output balances on household-operated smallholdings and the dangers of environmental degradation, but the concept of sustainability in common usage covers a multitude of values and goals (Lockeretz 1990; Barbier 1987). Terry Gips (cited in Francis and Youngberg 1990: 4) maintains that "a sustainable agriculture is ecologically sound, economically viable, socially just, and humane." In an Agency for International Development concept paper, sustainability is "the ability of an agricultural system to meet evolving human needs without destroying and, if possible, by improving the natural resource base on which it depends" (cited in Francis and Youngberg 1990: 5). Sustainable production is an "average level of output over an indefinitely long period which can be sustained without depleting renewable resources on which it depends" (Douglass 1985: 10). These definitions combine environmental parameters with economic and social characteristics in the context of changing interactions.

Several dimensions of sustainability, the physical, chemical, biological, and socioeconomic, are identified in the literature (Schelhas n.d.), with the degree of emphasis and analytic detail often depending on the scientific specialization of the investigator.[8] There is also a prevailing assumption that traditional cultivators, because of their low-energy technology, diversified production, small-scale operations, subsistence rather than market orientation, settlement stability, and lack of manufactured inputs, will occupy the sustainable end of the continuum, as opposed to commercial and industrial agriculture. In fact, the presence of these characteristics and their presumed interaction through time must be demonstrated, especially in the case of intensive cultivators, who modify the natural environment more profoundly and permanently than certain other types of land users. Unfortunately, measurements of the following relevant factors through time are seldom available in the case of either smallholder systems or large industrial farms:

1. Physical: soil degradation through erosion, weathering, compaction; diminished water supply, flooding, salinization; depletion of nonrenewable energy sources. Smallholders' techniques of terracing, contour mounding, drainage, irrigation, and diking may in fact be highly developed, and their use of fossil fuels minimal, but environmental deterioration owing to climatic perturbations or gradually increasing overuse may become apparent.

2. Chemical: decline in soil-nutrient status; decreasing responses to chemical applications, necessitating higher dosages; buildup of local or regional toxicity from the residues of fertilizers, pesticides, and herbicides. Rapid population increases among intensive farmers with no other economic options or the drive to raise production rapidly for the market may put pressure on resources so great that yields decline. There are unresolved questions as to whether the high-yielding seeds, chemical inputs, and mechanization of the Green Revolution as adopted by many smallholders will compromise their agricultural sustainability (Mellor 1988).

3. Biological: loss of biodiversity; declining ecosystem stability and resilience. Only groups of low-density foragers or shifting cultivators in large natural ecosystems may pose no threat to biological diversity (Schelhas n.d.). Intensive cultivation can replace natural ecosystems, prevent their regeneration, and cause absolute declines in natural biodiver-

---

8. Gordon Conway and Edward Barbier point to a source of confusion in the different definitions that various disciplinary groups attach to the term *sustainable agriculture* (1990: 9). Four interpretations are: (1) agriculturalists: food sufficiency by any means; (2) environmentalists: responsible uses of the environment, stewardship of natural resources; (3) economists: efficiency, the use of scarce resources to benefit present and future populations; and (4) sociologists: production consonant with traditional cultures, values, and institutions. Clearly, the productivity, stability, and equitability that are the goals of sustainable development projects may be in conflict, and there are necessary trade-offs among them (Conway and Barbier 1990: 39–43).

sity. The substitution of an artificially diversified system of polycultures or interplanting, integrated crop/livestock regimes, and crop rotation can, however, increase total yields, while reducing yield variability, insect predation, and weed competition (Altieri 1987; Gliessman 1984). Such systems appear to be biologically more stable and more energy-efficient than the monocultures characteristic of largeholders.

4. Socioeconomic: providing sufficient sustained economic returns over the long run on existing cultivated lands so that people can achieve a continuing adequate livelihood (Schelhas n.d.). Since the goals are social and economic, variable cross-culturally, and potentially changing through time, such sustainability is particularly difficult to measure objectively (Barbier 1987). Stable production may not be consonant with rising subsistence needs, greater market participation, lower agricultural prices, or higher input costs.

My emphasis on the process of intensification suggests that smallholders do indeed adapt to changing population and market forces, and that households have a variety of off-farm production strategies. This book is, in fact, more directly concerned with the dynamics of smallholder social and economic systems as they encounter the challenge of long-term biological sustainability than it is with the physical stability of such ecosystems. The management choices that the smallholder makes in the light of intimate knowledge of the land are unlikely to involve short-range maximization of production. Farmers who survive must hedge against the uncontrollable fluctuations of the climate *and* the market. The very long time-horizon of the family's intergenerational security and its valuable, heritable property give the smallholder household a unique perspective on sustainability. There is room to question the doctrinaire position of many "deep ecologists" that sustainable production and economic growth are incompatible goals (Hildyard 1989: 62), or that a market economy, population increase, and the new technologies of capitalism are inevitably at odds with sustainable systems (Weiskel 1989). But the suggestion that smallholder systems that can be shown to be sustainably productive, biologically regenerative, and energy-efficient tend also to be equity-enhancing, participative, and socially just (Barbier 1987: 104) is stimulating and provocative. Chapter 6 on land tenure and Chapter 7 on inequality will examine these postulated associations. Indigenous smallholder systems that show a favorable energy input/output balance, achieved by the application of labor and management rather than large amounts of unrenewable energy, exhibit a feasible solution to the problems of resource exhaustion, pollution, and environmental degradation that so often accompany large-scale, energy-intensive agriculture.

Detail from *Landscape with the Fall of Icarus*, oil painting by Pieter Brueghel the Elder, ca. 1555. Reprinted by permission of the Musées royaux des beaux-arts de Belgique. This detail of Brueghel's painting depicts a man plowing a terraced field with a wheeled plow, while a second man tends sheep grazing on the stubble of a lower terrace, suggesting a three-field rotation system. The nearby city and the ships in the background indicate that the people represented here are not of an isolated peasant community, but are connected in some way with trade and transportation.

The two major elements of the northern European agricultural revolution of the early Middle Ages (see Lynn White, Jr., 1962: 57) are represented in this image. The first element is a small version of the wheeled heavy plow, which replaced the scratch plow. The heavy plow had three functional parts: the coulter, or knife, which cut the sod; the flat plowshare; and the moldboard, which turned the slice of turf. This plow was advantageous as it did not require cross plowing, unlike earlier techniques, and less labor was thus needed to plow a larger plot of land. It also allowed plowing of dense, heavy, fertile alluvial bottomlands and improved drainage. Long, narrow fields became typical elements of the agriculture of this era (ibid.: 48). The wheeled heavy plow dates at least to the sixth century in some parts of northern Europe and was most often used in areas of higher population density because it was an expensive tool to purchase and operate (ibid.: 53).

The second set of innovations involved the nailed horseshoe, from at least the tenth century (ibid.: 58), and the horse collar, from at least the ninth century (ibid.: 60), which together enabled farmers to replace oxen with horses. This change was significant because horses were quicker and had more endurance, enabling people to plow larger areas farther from their homes in shorter amounts of time. However, the feeding of draft horses required more intensive grain-fodder production. The plow horse was common in northern Europe by the end of the eleventh century (ibid.: 63). See Chapters 1 and 4.

Bush cash-crop field: previously harvested millet stalks on ground with standing sorghum, interplanted peanuts, and yam heaps prepared for the following year's crops. On the right is a yam vine, and the man points to recently planted sesame on the yam heap. See Chapter 1.

Kofyar intensively planted homestead field with interplanted millet, sorghum, and cowpeas near the huts, conical heaps for yams, and ridges for sweet potatoes in the foreground, and a scatter of economically important oil palms, giginya palms, and mango trees. See Chapter 1.

Stall-fed goats in a Kofyar homestead in the cash-cropping area. The woman is bringing weeds and thinned sorghum plants to goats that remain tethered during the entire cropping season. The composted manure is applied to the homestead field. Millet growing in the background will be harvested and stored on the rack above the goats. Author's photo. See Chapter 1.

Kofyar manuring: removal of composted goat manure, grass, and leaves from circular stone corral. See Chapter 1.

An exchange-labor team of Kofyar young women and men weeds a millet-sorghum field in the cash-cropping area. Author's photo. See Chapter 3.

A Kofyar household of a couple and their six children at their homestead in the cash-cropping area south of Namu. They came originally from Bong village on the Jos Plateau. Author's photo. See Chapter 2.

Kofyar manuring: distributing compost on homestead field to maintain its fertility for annual cropping. See Chapter 1.

Left: A Swiss household enjoying a midday meal as they work on their field near Törbel. Author's photo. See Chapter 2.

Below: The village of Törbel, Canton Valais, Switzerland, 1,500 m above sea level. The old log houses cluster above the modern concrete church. Gardens lie just below the village, and irrigated hay meadows with scattered fruit trees occupy the surrounding steep slopes. See Chapter 1.

Above: Irrigated hay meadows with barns and temporary dwellings above the village of Törbel, Switzerland. Note the small size of the unfenced, privately owned fields. The grassy, forested area along the crest of the hills is part of the communal alp used for summer grazing and strictly regulated woodcutting. See Chapters 1 and 6.

Left: Watering a vegetable garden in alpine Switzerland. Note the small size of the garden beds of onions, lettuce, and cabbages and the fruit tree growing in the patch. See Chapter 1.

Irrigating a hay meadow above Törbel. A metal plate is used to block the flow of water in an irrigation ditch, causing it to overflow and cascade down the slope. Barns and temporary houses cluster in the background. Author's photo. See Chapter 1.

Swiss farmers filling pack baskets with dirt to carry it back up the slope to the top of the garden. See Chapter 1.

Left: Swiss farmer sharpening a scythe in a hay meadow. The agricultural technology of steel and wood hand tools in the Alps had remained largely unchanged until recent years, when garden tractors with cutter bars were introduced. See Chapter 1.

Below: Swiss farmer pitching cow manure composted with straw and pine needles into mule panniers for transport to the meadow. See Chapter 1.

Right: Swiss farmer carrying a load of hay from the meadow to the loft of a log barn. The door to the cow stall is at the bottom of the picture. Several families may own shares in a single barn, each using it in turn for a few weeks or months. The cutting, raking, bundling, transport, and stamping down of the hay in the loft requires the work of the whole household. See Chapter 1.

A Swiss family harvesting rye. The grain is being reaped with a scythe, bundled into sheaves with the help of a sickle, tied with straw, and laid in the field to dry. Rye was formerly the major bread grain of the mountain communities, and it was grown, milled, baked, and stored within the village territory. See Chapter 2.

The summer grazing alp has been traditionally held as common property by the citizens of a Swiss community. There is an annual celebration when individual families bring their cows to the alp in the spring. Herding, milking, and cheese-making are carried on by a village communal enterprise, and the cheese is divided among the cattle owners in proportion to the milk their animals have given. See Chapter 6.

Communal labor on the common-property alp. Men repair a stone wall damaged by an avalanche. Author's photo. See Chapter 6.

Communal labor on the common-property alp. Women rake up twigs before cows come to the pastures in the spring. Author's photo. See Chapter 6.

Northern Portuguese farmers planting intercropped potatoes and kale. Wilted kale seedlings, lying in two widely spaced rows, will revive and flourish. In the background, piles of manure wait to be spread on the grass before the sod is turned by the plow. Posts for a grape arbor stand around the plot, and behind sits a two-storied grain barn. (Bentley 1992: 106–7). See Chapter 1.

Northern Portuguese farm family loading hay. The draft power is provided by dairy cows. Grape arbors suround the field. See Chapter 1.

The Spinner, by Adriaen van Ostade (Dutch, 1610–84), 1652. Reprinted courtesy of the Art Institute of Chicago. A Dutch mother spins, while the father and a child relax in front of their homestead. The house and barn of this scene are typical of large mid-seventeenth-century Dutch buildings (Stone-Ferrier 1983: 182). The buildings depicted here are divided into a separate area (to left, under thatched roof) for stall-fed farm animals like the two pigs shown here, living quarters (to the right), and a storage area under the stoop. Such homes were constructed by the farmer even if he leased land from another individual. Tenant farmers were protected against loss because landowners were legally obliged to purchase tenants' buildings if they were evicted (ibid.: 184). The house in this print is constructed of brick, a new material used by the stable and increasingly prosperous farmers of this era, whose affluence resulted from agricultural specialization, intensification, and market participation (De Vries 1974: ch. 4). Farmers of this period also began to obtain some goods from the growing number of non-farmers who produced clothing, foodstuffs, glass, candles, and so on (ibid.: 120, 236). See Chapters 2 and 7.

Scooping up mud. Reprinted, by permission, from M. D. Coe, "The Chinampas of Mexico" (1964: 95). *Chinamperos* dredge muck from the bottom of the canal with long-handled scoops. The muck is taken in canoes to plots where it is formed into seedbeds. See Chapter 1.

Fresh muck chinampa seedbed. Reprinted, by permission, from Gene C. Wilken, *Good Farmers* (1987: 87). Thick, rich canal bottom muck is spread out to form seedbeds. When the muck solidifies, it is cut into squares and seeds are planted in each cube. See Chapter 1.

Overleaf: Ancient Aztec map. Reprinted, by permission, from M. D. Coe, "The Chinampas of Mexico" (1964: 93). As Coe notes, this map of a portion of Tenochtitlán–Tlatelolco shows that it was a *chinampa* city. Six to eight plots are associated with each house, and a profile of each smallholding householder and his name in hieroglyphs and Spanish script appear above each house; footprints indicate a path between plots or beside canals, which served as thoroughfares for farmers in flat-bottomed canoes. See Chapter 1.

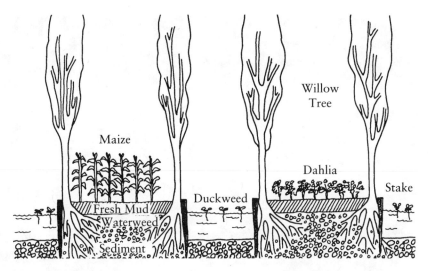

Diagram of chinampa agriculture. Redrawn from Mac Chapin, "The Seduction of Models: Chinampa Agriculture in Mexico" (1988: 10). This highly productive and intensive agricultural system was established in marshy areas or along lakeshores in the pre-Columbian Valley of Mexico, and is still used today. When properly maintained, chinampas retain their fertility for long periods without fallowing. Chinampa beds are constructed with alternating layers of aquatic weeds, bottom muck, and earth, packed in rectangular cane frames firmly attached to the lake or marsh bottom. *Ahuejote*, a type of willow, is planted along chinampa borders to provide shade and help secure beds to marsh or lake bottom. See Chapter 1.

Seed nursery. Reprinted, by permission, from M. D. Coe, "The Chinampas of Mexico" (1964: 95). The seed nursery, an essential element of chinampa farming, is constructed at one end of a chinampa, near the canal. When the seedlings sprout in these squares of manured muck, they are transplanted to the chinampa field. See Chapter 1.

View of a rice-field system at the beginning of the rice cycle, in the southern part of Kalinga-Apayao Province, Philippines. Photo by William Longacre, courtesy of Kalinga Ethnoarchaeological Project (KEP) and Arizona State Museum, University of Arizona. There are two rice crops per year in this region. Note the banana trees planted between terraces. The field in the immediate foreground is in the preparation stages. See Chapter 1.

Opposite: A farmer walks along Ifugao stone-walled terraces supporting multipurpose irrigated pond fields. On the left and above are seedbeds from which rice will be transplanted. Taro may be intercropped in the pond field, and vegetables such as beans, garlic, onions, and sweet potatoes are often cultivated on the dikes and mulch mounds (right center). Mud fish may be stocked in the ponds and collected in a pit when the field is drained. See Chapters 1 and 6. Photo courtesy of Harold C. Conklin.

Two women transplant rice from nurseries to inundated rice fields, Kalinga-Apayao Province, Philippines. Photo by William Longacre, courtesy of Kalinga Ethnoarchaeological Project (KEP) and Arizona State Museum, University of Arizona. See Chapter 1.

Opposite: A property-line dispute in Ifugao. Conklin (1980: 22) describes how, when dikes and terraces are completed and rice is being transplanted to fields, there is a period of visiting and comparing notes between districts to determine whether any walls deviate from their original alignment by even a fraction of a meter. Lack of agreement can lead to duels, a traditional method of settling such disputes. "Such incidents underscore the importance assigned to maintaining precise boundaries for all parcels of permanent pond-field property" (ibid.). See Chapter 6.

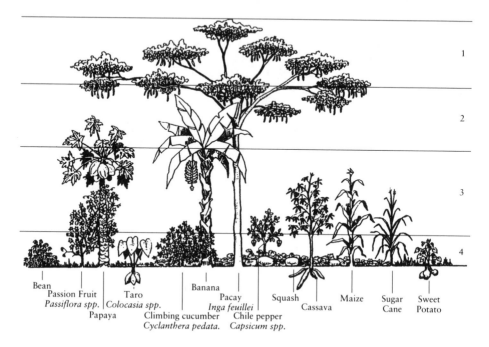

														1
														2
														3
														4

Bean
   Passion Fruit   Taro   Banana
   Passiflora spp. |*Colocasia spp.*   Pacay   Squash   Maize   Sugar   Sweet
       Papaya   Climbing cucumber Chile pepper   Cassava   Cane   Potato
             Cyclanthera pedata. *Capsicum spp.*

Vertical stratification of a tropical garden, with fruit trees, grain and tuber crops, vegetables, and vines in a four-tiered structure. Redrawn from Vera Niñez, "Household Gardens: Theoretical and Policy Considerations" (1987: 172), with permission. See Chapter 1.

Opposite: Ifugao swidden vegetation showing taro and sweet potato foliage. Clearings are made in steeply sloping forests or canelands unsuited for irrigated terracing. They are cropped for about three years and fallowed for two or three times that period. Households provide the labor, but they exercise only impermanent usufruct rights in the swidden plots. Photo: Conklin 1980, p. 25. See Chapter 6.

Thai family transplanting rice into a flooded field. Photo, Rhonda Gillett. See Chapter 1.

Kokihashi: Stripping rice with chopsticks. Reprinted, by permission, from Francesca Bray, *Agriculture* (1984: 354, fig. 170). The harvesting of wet rice, like the transplanting, required the rapid, coordinated labor of units larger than the household. In this Japanese print, the grain is first reaped, bundled, and transported back to the farmstead, then the sheaves are hung to dry, and finally the rice is separated from the stalk using chopsticks. Labor may be exchanged, or branch households may be obligated to help with the harvest of the main households to which they are related. See Chapter 8.

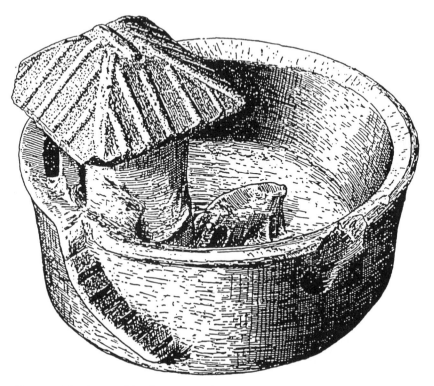

Han model of a combined pigsty and privy. Reprinted, by permission, from Francesca Bray, *Agriculture* (1984: 291, fig. 116). The human feces and pig manure were channeled to a lower area, where they were allowed to "ripen" or "cook," the heat of this composting process killing harmful microorganisms. Fresh manure was known to burn plants as well. See Chapter 1.

Opposite: The dike-pond system of the Zhujiang (Pearl River) Delta. Reprinted, by permission, from K. Ruddle and Gongfu Zhong, *Integrated Agriculture-Aquaculture in South China* (1988: 76, fig. 26). Through natural processes, soil and organic materials settle in ponds. Two or three times per year, the rich muck is scooped from the pond bottom and used to build up dike walls and fertilize mulberry trees and sugarcane planted on the dike. See Chapter 8.

CHAPTER FIVE

Farm Size and Productivity

Do SMALLER FARMS produce more per unit area than larger farms, and if so, why? Do patterns of land use that regularly combine more labor with fewer units of land result in higher yields as well as greater energy efficiency, or do the inputs of mechanical energy associated with bigger farm areas maintain equivalent or greater land productivity? Arguments about the nature of smallholder agriculture are not just exercises in theory, and when it comes to the relationship between size and productivity, questions of personal and national economic interest, development policy, and land reform may hinge on the understanding of this issue.

There is a good deal of ambivalence on the subject. The *New York Times*, which applauded the breakup of big Vietnamese collective farms into private fields (Erlanger 1989), subsequently criticized the "notorious inefficiency of Japan's farming system," in which average farm size is about two and a half acres, and 85 percent of the country's four million farmers work only part-time on their family fields. As is so often the case, cultural values are said to be responsible for economic irrationality. "The Government of Japan has tried for 30 years to encourage farmers to consolidate their holdings—or at least to encourage sharing, leasing, or running farms on consignment—to improve efficiency by expanding the size of Japan's farms, reducing the need for price supports. But the efforts have largely failed in the face of tradition and the almost spiritual love of the land among the farmers" (Weisman 1989: 4).

The obvious question is, efficiency of what? The newspaper photograph of a bending woman pulling weeds while her husband pauses by his garden tractor to wipe his brow suggests part of the answer. Manual labor is hard and has low returns, but clearly there is enough incentive in subsidized rice prices to elicit weekend work. Or is the atavistic call of the soil just too strong for otherwise hardheaded Japanese smallholders to resist? The efficiency of land use in agriculture (as opposed to its value for house or business construction) cannot be at issue, since with improved varieties and scientific fertilization, combined with heavy inputs of skilled labor, Japanese yields are among the world's highest. Given Japan's climate and topography, the assumption that larger farms with more machinery, operated by full-time agribusinesses, would produce cheaper rice is probably fallacious. Would it be more "efficient" for many farmers to stop cultivating rice and work more in industry or merely enjoy their leisure? The subtext, that Japan should buy and become dependent on rice from California and Louisiana, concerns world energy prices and international trade surpluses and deficits, but it obscures smallholder farm productivity.

The logic of agricultural intensification, with labor and techniques increasing land use and raising production in order to cope with scarce land resources, suggests that small farms might be expected to produce more per ha than much bigger cultivated areas in the same region. When labor is substituted for land, both yields and energy efficiency rise, as shown by the farm-system comparisons in Chapter 4. At the most basic level, empirical information from surveys confirms the inverse relationship of farm size and land productivity. A contentious literature, much of it based on Indian cases (Sen 1962; Khusro 1964; Rao 1967; Bardhan 1973; all summarized in Berry and Cline 1979), comes to the same conclusion. But it is only when we disentangle the quite different costs of the factors of production, land, labor, and capital facing small as opposed to large farmers that we can understand the economically rational choices that distinguish size classes of farms (Ellis 1988: 191–209).

Farm Size and the Choice of Land-Use Methods

Where, as in India, there are greater variations in farm size than are found in the most intensively tilled parts of China and Java, the value of production per acre falls as average farm size increases. Because the average farm size in each class is markedly different, total production, measured by income, does go up appreciably. But the rate of growth in returns is much smaller than the increase in farm size, so that when the mean size goes up 3.2 times from 2.9 to 9.3 acres, total income increases only 2.6 times. In comparison to the largest average size of 42.6 acres, 14.4 times bigger than the smallest, average income has increased only 6.8 times (Table 5.1).

It might be claimed that this is merely a reflection of the fact that small farms are concentrated in areas of more fertile soils, while larger, less fragmented operations occur in more marginal environments. However, although this may indeed characterize very different types of land use, such as cropping and extensive cattle ranching, it cannot be demonstrated as a general rule (Ellis 1988: 198). Even when big landowners possess

TABLE 5.1
Farm Size and Productivity, India, 1970–71

Size group (acres)	Average farm size (acres)	Income per acre (rupees)	Total income (rupees)
0–5	2.9	737	2,174
5–15	9.3	607	5,645
15–25	19.5	482	9,399
> 25	42.6	346	14,740

SOURCE: Berry and Cline 1979: 149.

comparable or better land than their poorer neighbors, their tendency is to use it less intensively. They can enjoy the luxury of fallowing land longer, depending on natural regeneration of soil nutrients rather than enhancing fertility by more expensive manuring or tillage practices (Berry and Cline 1979: 13). Yields on portions of the property can also be raised by selecting only the fields with the highest natural fertility for cultivation.[1] The smallholder, on the other hand, must keep his land producing regardless of its relative quality, often interplanting, multicropping, and irrigating in order to raise yields (Ellis 1988: 198–99). The underutilization of total land area that results in lower production per ha on large farms may be increased by a choice of land-extensive enterprises like livestock grazing or lower-value crops. Such land use—for instance, sheep pasturage on land that would support annual wheat/turnip rotations with interspersed gardens and orchards—might take advantage of a market for wool while lowering labor costs. "Larger farms consistently select products using relatively more land and less labor than those chosen by small farms" (Berry and Cline 1979: 15).

Even on the extremely small plots of Bangladesh, where a "large" farm may have only five acres, there is an inverse relationship between farm size and the factors of intensification that result in high land productivity (Boyce 1987: 245). In 1976–77, the farm-size classes ranging from 0.5–1.0 acre to > 7.5 acres exhibited regularly declining cropping frequency, percentage of area irrigated, fertilizer use, and area under high-yielding

1. Where fallowed land is excluded from the comparison and where there is no difference in the proportion of land irrigated among different size groups of holdings, output per acre in some Indian samples tended to remain constant (Rao 1967). By excluding evidence for intensification and differential land use, this comparison downplays those elements that account for the inverse relationship of size and productivity elsewhere. Barbier (1984: 194) points out the methodological problems of using large samples and disaggregated data to support conclusively the size/productivity relationship, and he refers to the low yields of the tiniest farms. His own data from West Bengal, however, do show an inverse association between farm size and multiple cropping.

varieties (Boyce 1987: 207). Less land was also devoted to high-value garden crops such as potatoes, vegetables, chilies, and tobacco. Smaller farms achieved higher levels of yield per unit of cropped area and higher rates of agricultural growth in the 1965–80 period (Boyce 1987: 246). / More family labor *and* more hired labor per acre was used on small as compared to large farms (Boyce 1987: 204, 245).

So-called economies of scale are often used to justify a trend to replacement of smallholdings by large farms. The scale of farm enterprise measures differences in the overall economic size of farms, and *increasing* returns to scale refers to a simultaneous and identical increase in all resources (e.g., land, labor, equipment), resulting in a *larger* percentage increase in output (Ellis 1988: 193). The received wisdom of contemporary agricultural economists is that returns to scale in farming tend to be constant. "In other words if 100 days of labor, one bullock pair, and one hectare of land produce 3 metric tons of paddy, then 200 days of labor, two bullock pairs, and two hectares of land would produce 6 metric tons of paddy" (Ellis 1988: 194). In fact, such across-the-board increases in resource use are difficult to achieve. Doubling farm size and fertilizer use may be possible, though expensive, but increasing the number of farm buildings or expanding family labor by an equal amount might be out of the question. As Frank Ellis (1988: 194) points out, farm mechanization and the use of a machine like a tractor at its optimum level is often inaccurately termed an economy of scale. Because a tractor is an indivisible resource and cannot be added in continuous small increments, merely combining a tractor with a fixed resource like a small, irregularly shaped field that is already fully farmed by hoe cultivation will not necessarily achieve higher output or lower expenses of production. Much larger farmed areas must be available for such a technology to achieve minimum average costs. Tractor plowing may necessitate large increases in hand weeding (Baldwin 1957) or the replacement of labor by even more costly herbicides. Rather than looking at increases in scale as identical with economic efficiency, we might better seek the optimum farm size that minimizes the average cost of production for a given technology, which obviously varies with different crops, technologies, and factor prices for land, labor, and capital. Intensive cultivation generally refers to small farm size with relatively large inputs of other resources, while large farms most often practice much less productive extensive land use, with low inputs of other resources (Ellis 1988: 196).

Economists have no difficulty in explaining the proclivity of smallholders to increase their labor per unit of land. "Farmers maximize returns to their relatively scarce factor of production (land) by intensifying inputs of their relatively abundant factors (labor and managerial time)"

(Herring 1983: 242). But they find inefficiencies in extremely small farms based on low marginal returns to labor and the inability to use capital inputs effectively. Among the smallest Indian peasant farmers, "the costs of family labor are to a large extent fixed; the family (and bullocks) must be maintained whether they contribute five, ten, or zero hours of labor per day" (ibid.: 243). There may not be enough land for the available human hands and animal traction to be fully employed. Such farmers cannot achieve a more adequate holding size because of a low investible surplus and inadequate credit and capital (ibid.: 262). But it is a matter for empirical determination just how small too small is, and we must also consider the role of part-time cultivators with other sources of income.

Access to Land and Contrasting Farming Strategies

In a single geographic area with one state administration and a common market system at a specific point in time, we can see not only different farming systems side by side but contrasting movements in the direction of intensification and extensification of agriculture toward both small and large holdings. In Paso, a community of 77 households in a mountainous region of Costa Rica, some farmers carry on the shifting cultivation of maize and beans on rented plots, others practice multicropping of tobacco and corn and beans on chemically fertilized terraces, and a few pasture beef cattle on range grasses. Peggy Barlett (1976, 1977, 1982) has shown how these systems developed from a traditional maize / beans swidden with one year's planting followed by a five-year fallow. Land was the factor of production that changed first, both in terms of supply, as population grew to 239 per square mile, and in fertility, as yields declined and purchased chemical fertilizer became necessary. There was a marked difference in returns to labor between good land with a kcal input/output of 1:101 and the 75 percent of the land of poorer quality with 1:63. Farmers who began to construct annual contour terraces, transplant tobacco seedlings, apply fertilizer, and rotate their tobacco with corn and beans grown in the same year were able to pull their energy ratio back to 1:74 and at the same time produce a high-value market crop (Barlett 1976: 134–37). Agricultural intensification raised land productivity and countered a fall in labor efficiency.

But access to land and to the capital needed for investment in fertilizer and tobacco seed was not equally distributed in the community. People with little land or cash had to rent small plots for maize /beans subsistence crops. The marginal land that they could use had to be fallowed and was not suitable for tobacco. Wealthier landowners could use their more extensive holdings for cattle. Though beef animals require an initial investment for stock and fencing and time to mature, they need little labor, especially as compared to the high inputs in tobacco cultivation. Though return on land was less than it might have been with either tobacco terracing or maize /beans swiddens, labor time and costs were minimized, and risk was reduced to a point below that of grains, coffee, or rentals (Barlett 1982). With the development of a market for Central American beef in the United States (the so-called "hamburger connection"), total cash return also became attractive. The middle class of real smallholders who took the route of intensification "began tobacco when they had to, either from land scarcity or from soil depletion" (Barlett 1976: 137). About half of them have bought enough land with profits from tobacco and grain to return to less intensive fallowed agriculture, but the "law of least effort" does not mean that people who have developed needs that can only be met through the market will return to subsistence standards (Barlett 1982). Access to land and capital differentiated the landless, middle-sized smallholders, and wealthy farmers, accounting in rational economic terms for their decisions to continue swiddening, to adopt terraced tobacco, or to disintensify land use by converting to cattle pasture (Barlett 1977). The productivity of land declined sharply both on large farms and where land could only be used in small amounts on a temporary basis. At both ends of the size spectrum, the risks of deforestation, erosion, and possibly permanent environmental degradation increased. Only the option of smallholder intensification increased output per unit of land, while conserving natural resources.

Large-Scale Crops and Capital: No Business for the Little Guy

The kinds of agricultural goods produced, their prices on the world market, and the capital required to produce and process these goods efficiently also have a pronounced effect on farm size. Densely planted, high-yielding tropical crops like bananas, tea, and sugarcane may be so dependent on an infrastructure of costly warehouses, shipping networks, driers and sorters, or mills that a small producer may suffer sizable disadvantages. Where initial processing can be done in the household with little costly equipment, and where tight schedules of tasks demanding technical and financial knowledge are not necessary, as in coffee, cocoa, coconut, or rubber growing, the smallholder may compete with plantations (Bachman and Christensen 1967). Economies of scale may be much more significant in marketing, the provision of inputs, and other support institutions than in production. Although national survey data in Kenya show a common pattern of inverse correlation between farm size and output per ha, Angelique Haugerud (1989: 64) points out that the magnitude and direction of the relationship are likely to vary by crop and to alter as the agrarian sector becomes more capitalized and proletarianized. It is probably true that monocropped tea or sugarcane may give high, continuous yields under plantation cultivation, but peasant farms with a subsistence component almost never attempt such specialization, creating "apples and oranges" difficulties in any such comparison.

Lower-yielding crops grown in pure stands with frequent fallowing, like dry wheat, may also correlate with larger farms. A monocrop, be it sugarcane or wheat, with concentrated planting and harvest tasks and little labor in between does not provide the diversified and nearly continuous occupation of a small intensive farm. Indeed, a temporary force of wage workers like the laborers in an Andalusian agrotown hired by the day on a latifundio may meet the labor demands more cheaply than resident family households. Philippine sugarcane plantation workers with only six months' employment a year serve a vertically integrated industry but are consigned to poverty much greater than that of smallholders or rural workers in a diversified economy in a neighboring region (Yengoyan 1974).

A small-farm model may not fit the constraints of certain physical environments, their associated crops, and the markets in which these agricultural commodities move. The laws of the United States and Canada in the 1860s that gave settlers free 160-acre homesteads obviously envisioned tracts of optimum size for midwestern areas of mixed maize and livestock production (Feeny 1988). Wheat or other small-grain monocropping or extensive cattle ranching, which are better adapted to the more arid open grasslands of the Great Plains, demanded more space and more capital. In an effort to open up its Northwest Territories around the turn of the century, the Canadian government encouraged railroad building, surveyed the land on a one-mile section grid, and provided homesteads of 160 acres, with an adjoining additional 160 acres at a modest price (Bennett 1969: 103–4).

These [320-acre] amounts were set by Eastern planners who based their decisions on the amount of land one man and a team of horses could farm without extra labor. This was a valid criterion for the humid East, with its diversity of crops. In the Plains, however, the farmer was limited to grain and a few livestock, and if drought hit, there was no grain. Even at the lower price levels of the time, the land simply was not productive enough to permit a decent living for the homesteading family and also to supply needed capital for further development of the farm—a vital consideration as the North American economy developed and prices increased. While in the 1910–15 period a homesteader could break his 320 acres of land, build his modest buildings, and finance his simple horse-drawn machines for about $1,400, a low figure and within the capacity of most serious homesteaders, it required considerably more to increase the productivity of the land in order to bring it to a level that would provide enough income in good years to cushion the bad. This extra capital was precisely what the majority of homesteaders did not have. (Bennett 1969: 104–5)

With droughts and economic depressions leading to abandonment of some homesteads and expansion of the remaining farming enterprises, one 36-mile-square township went from 76 homesteads in 1912 to 72 in 1920, 41 in 1940, and 24 in 1962 (Bennett 1969: 44). The average size of farm holdings in Saskatchewan had risen from 432 acres in 1940 to 686 acres in 1960 and was going up by about 2 percent a year (ibid.: 45). Without large capital investments by farmers in tractors, combines, and livestock, and by government in irrigated pasture projects, grain-price subsidies, cash loans, and the development of new, hardier wheat varieties, even the enlarged operations would have been unable to survive (ibid.: 109–19).

In dry climates with highly variable precipitation and few opportunities for small-scale, relatively low cost irrigation systems, smallholders can seldom accumulate the reserves in food, saleable assets, and cash that will see them through one or a succession of bad years. In the *sertão* of arid northeastern Brazil, even those who own small parcels of land may choose to leave them and work as sharecroppers on large estates. Their considerable agricultural knowledge and reservoirs of family labor are useless in the face of serious drought, and the ability to obtain credit at an estate store or enlist the help of the estate owner when illness or other misfortune strikes appears to be a benefit worth paying for in lower returns on labor and a dependent status (A. W. Johnson 1971). Though an estate or latifundio or enormous Plains wheat farm may have relatively low average production, the surpluses accumulated in good years, when concentrated in the hands of the big landowner, may be sufficient to reduce the risk of periodic crop failures and drastic price variations in the major product.

Cheap Land and Capital—The Big Landowners' Edge

The contrary tendencies to intensify and to disintensify land use that accompany differing farm size reflect in part an imperfect market that puts different costs on the factors of production. "Specifically, small farmers confront a low price for labor combined with high prices for land and for capital; while large farmers confront a higher price for labor combined with low prices for land and for capital" (Ellis 1988: 201). The existence of widely differing farm sizes in the same environments and with similar products may signal market imperfections that impede economic tendencies toward an optimum allocation of resources.

The utopian condition of perfect competition, with a neutral price mechanism acting as the arbiter of all economic decisions, is, of course, never attained in any society (Ellis 1988: 10), but smallholders may function in particularly inefficient market systems. Land scarcity combined with rights of tenure that are varied, insecure, and without dependable legal enforcement may cut down on opportunities for new farmers to acquire land and for unsuccessful farmers to exit from the market (ibid.: 11). In such highly inefficient systems, information on prices tends to be poor, favoring merchants and officials, and capital markets for credit are fragmentary or nonexistent, with interest rates that are not standardized (ibid.). Labor-market failures also inhibit large farms from gaining access to low-opportunity-cost labor on smallholdings (Carter 1989: 41). In a perfect neoclassical labor market, land-poor families would allocate labor to other occupations once on-farm productivity had slipped below the market wage. Land-abundant farms would hire labor until marginal on-farm labor productivity fell to the level of the market wage. Productivity would thus be identical between labor-surplus and labor-poor farm units. What Michael Carter calls "the persistence of hyperproductive small-holdings" (1989: 39) reflects multiple market failures in land, capital, and labor.

Effective land prices per ha are higher on small plots as opposed to larger ones of the same quality. There are more potential buyers for small parcels of land because of their low total cost, so competition drives the price up. For a large owner, the cost of selling a big tract is less than that of dividing it into small fragments, so he can accept a lower price/ha (Berry and Cline 1979: 10; Feder 1985). Because land purchase often involves borrowing, and large operators have better credit, they can also buy land at lower rates of interest than are available to the smallholder (Berry and Cline 1979: 10). Where government agencies subsidize loans for land and equipment, give guaranteed prices for products, or make available low-cost supplies of fertilizer or seed, these benefits may go disproportionately to the wealthier landowners, who tend to have better political connections. In similar fashion, large, mechanically operated collective farms may be profitable less because of labor-saving techniques than because of government-sponsored low-cost credit not available to smallholders (Bachman and Christensen 1967).

The capital that is relatively cheap for large owners is exorbitantly expensive for smallholders. Beginning with inherited land that is below the optimum size for farm operation, they may have to reduce family consumption in order to save or borrow at high rates from moneylenders. Small farmers often reject opportunities to borrow in which their land is required to be given as security. The risk of losing their only major asset is just too great. Nevertheless, accumulated debt, food shortages, or family emergencies may force distress sales and allow concentration of land by large owners paying less than the market price (Ellis 1988: 203).[2] Ill-

2. The assumption that high rates of tenancy and the concentration of ownership of land in south China were owing to debt sales by impoverished smallholders has been challenged

advised land reforms may not be able to create a group of self-sufficient smallholders where none existed in the past. When poor-quality common lands were divided among landless laborers in the dry-wheat area of the Alentejo region of southeastern Portugal, the new owners could neither produce reliable grain crops from their land nor pasture sufficient livestock to support themselves. Their parcels were often sold for back taxes or foreclosed mortgage loans, reverting to the big landowners who had the capital to buy them (Cutileiro 1971). Intensification of production from arid plains may only be possible with high levels of investment for wells, pumps, and center-pivot irrigation systems, and these capital-intensive improvements are beyond the reach of smallholders. "The gap between the price of capital to large and small farmers frequently encourages the substitution of capital equipment for labor on the large farms" (Berry and Cline 1979: 10), even when there is abundant labor and a shortage of capital in the general economy. Because land has asset value quite apart from its immediate productive value in agriculture, large landholders may also retain considerable tracts for speculative gain. In countries where capital markets are poorly developed, especially those with chronic inflation, non-farmers may buy land as a store of value that does not depreciate (ibid.: 11–12). Under such circumstances and with protection from government courts and bureaucracy, big owners may prevent both renting and sharecropping of the land, keep it off the market for smallholders, and depress its agricultural productivity to zero.

Labor in an Imperfect Market

Labor markets may also exhibit a dualism in which a highly capitalized modern sector uses labor only to the point where its marginal product (the increase in total product caused by the addition of another unit of labor, holding constant the effects of all other factors of production) equals the wage rate in the modern sector, while the traditional sector has a large supply of labor with a very low marginal product, which may even be less than the income received (Berry and Cline 1979: 8).[3] On small farms where family labor is dominant and income is shared, an individual, before hiring out, would have to have a wage equal to the average return on all labor in the household, even if his own marginal productivity were quite low. Household pooling provides insurance that

---

by John Shepherd (1988). In fact, tenants in late imperial and republican China had many of the rights of owners, and the renting of small plots increased during periods of prosperity.

3. Market imperfections are obviously interdependent. If capital and land-rental markets were perfect, each household could rent in or lease out as much land as required to maintain an optimal operational holding proportionate to the size of the family. Labor inputs would then be identical across farms, and yields unaffected by farm size (Feder 1985: 311).

single members may not be willing to exchange for the going wage in the external economy. Food costs in labor are probably less for household production than on food purchased with wage labor. The worker on the family farm would also have to receive a wage on the outside high enough to cover the costs of transport to that job and the risks of not acquiring full-time work (ibid.).

Large farms use relatively less labor because they can only afford to pay employees up to the point where their wages equal the return on their labor. Because household members on small farms get no wage and require only subsistence, they can continue to work even when their marginal products are below the going wage. They are therefore available to do necessary but very low-return work like scaring birds away from the crops or herding goats. Because of their age, gender, or other domestic activities, household members cannot readily find substitute employment at the existing wage rate. Many of the tasks of intensive agriculture have such low marginal returns that an employer cannot profitably hire someone to do them. "The effective price of labor is lower on the small farm which, as a result, can exploit more marginal land, bring a larger share of its given land under cultivation, and achieve a larger output per unit of available land resource. . . . labor-market dualism leads to the phenomenon of higher utilization of the available land resource on small farms than on large" (ibid.: 9–10). In the next chapter we shall explore some additional reasons for the high production per unit of land of the small-farm household. It is apparent, however, that land, labor, and capital are quite different factors for big and small farmers in imperfect markets, and that the demonstrated differences in the productivity of the land farmed by smallholders and big landowners arise from rational economic decisions in these contrasting arenas.

CHAPTER SIX

Smallholder Property and Tenure

Sᴍᴀʟʟʜᴏʟᴅᴇʀꜱ ᴏᴡɴ private property. They have rights in resources—farmland, livestock, fruit trees, firewood, irrigation water—as well as in the technology that renders these resources productive—granaries, barns, plows, wagons, and hand tools. Some continuing, socially recognized rights in the land and animals belong to the cultivator. These rights of tenure are what smallholders *hold.* Such rights may be embodied in deeds, tax valuations, and wills, they may be enforceable by courts under the jurisdiction of the state, or they may depend on customs relating to the acquisition, use, and social transmission of certain resources and technologies within the little community. Indeed, de facto landholding rights that are recognized and respected by one's neighbors may be more functionally important than de jure legal rights established by national governments (Seymore 1985).[1]

Though rights may be formally assigned to certain individuals and nested within more inclusive systems of rights belonging to lineages, villages, landlords, nobles, estate owners, and the political administration of the county, district, or state, they are associated in a fundamental manner with the farm household and reflect its ongoing relationship with productive property.[2] Household members work on a particular farm,

---

1. Though we shall be discussing principally smallholder rights in such tangible factors of production as land, water, livestock, tools, and buildings, the concept of property is obviously much more inclusive. "By property is meant a set of abstract rules governing access and control, use, transfer, and transmission of any and every social reality which can be the object of a dispute" (Godelier 1986: 75).

2. Following Parker Shipton (1987: 52 n. 9), we might classify rights as: (1) rights of use, including hunting, grazing, cultivation, water, wood, mineral, passage, building, and residence rights (the accompanying rights of disuse may include those of fallow, or of the holding of reserves for future family expansion); (2) rights of transfer, including those of inheritance, gift giving, lending, swapping, mortgages, rentals, sales, and other contracts; and (3) rights of administration, including allocation or withdrawal of use, dispute settlement, regulation of transfer, management of land for public uses, and "reversionary" or "ultimate" rights (e.g., to collect royalties, tributes, or taxes). These rights may vary along scales of time (for what period the right can be exercised), exclusivity (degree to which rights may

they derive appreciable benefits from it, and their investment of labor and capital over time establishes and sustains valuable property rights that may pass to close kin by inheritance. Under circumstances of population pressure and market demands, intensive agriculture emphasizes well-defined, defensible, and enduring private property rights in a qualitatively different manner from hunting and gathering, fishing, or shifting cultivation. Though we are accustomed to thinking of land tenure as a set of jural concepts or legal rules externally formulated and enforced by political bodies, I shall examine property rights here as part of a local agro-ecosystem, testing the hypothesis that, other things being equal, land use by and large determines land tenure. Private individual or household property, frequently accompanied by corporate group rights in common property, is regularly and systematically associated with smallholder intensive agriculture.

Property Rights in a Stateless, Nonmarket Society

I was not prepared for the fact that the Kofyar terrace farmers on the Jos Plateau of Nigeria considered their land, both the intensively tilled homesteads and the fallowed bush fields, to be *owned* by specific individuals and inherited from fathers or brothers. A creek in a ravine, or a large stone partially buried in the earth, might mark a socially recognized boundary line between fields. When I asked about a 6m² triangle of tall grassland between two bush millet fields near Bong, I was told that a dispute over ownership could not be resolved (the witnesses were contradictory), and the village chief had decided that neither party should have the use of the contested land (Netting 1968: 172). Evidence of socially explicit ownership and inheritance of land, fixed boundaries, and litigation over property ran diametrically counter to the conventional wisdom on African land tenure. Because the Kofyar had not been incorporated in a Hausa-Fulani state with Moslem law and courts, and because they were only peripherally involved in the market economy, I had expected that they would practice only temporary use of land with communal territorial rights held by clans, villages, or tribes. Instead, the Kofyar vehemently insisted that all cultivable land had a proprietor, and that claims could be publicly affirmed, disputed, and enforced.

Where land is a scarce good that can be made to yield continuously and reliably over the long term by intensive methods, rights approximating those of private ownership will develop. Kofyar institutions specifically recognized private property and made possible its retention and transfer in ways that lessened uncertainty and conflict. Some homestead residents

---

be shared), and agent (the right exercised by an individual, a collectivity, or a corporate group).

said that they had occupied a compound and used its surrounding manured field for three generations, but that the land did not belong to them, and that they recognized the true owner by giving him the palm oil made from the trees on the plot every year and presenting him with a jar of beer when they brewed. The tenant might have secured a lifetime lease with the single payment of a goat or some cash, or there might be an annual rent in kind. It was difficult to remove a long-term tenant from a homestead, but the tenant's failure to attend and contribute to a funeral feast of the owner's family was grounds for dispossession. Sales of land for cash or valuables, which had to be approved by the patrilineage of the seller, were rare, but not unknown in the past.

Fields were usually inherited from a father or kinsman. An heir was expected to conduct a costly second funeral commemoration sacrifice of a cow or horse to verify his claim, and the prominent stone-cairn graves at the entrance to the compound recalled named ancestors who had once occupied the homestead (Netting 1968: 168–72). When there were no heirs, a non-relative might donate the sacrificial beast, receiving the land as permanent property unless the clan of the dead man returned an equivalent animal to him within seven years (ibid.: 166). The head of the deceased's patrilineage had first claim on his land, but it passed to him as an individual rather than into any sort of communal tenure of the descent group. The household head was normally succeeded by his son, perhaps a younger boy who had remained with his father when his elder siblings married and moved out to vacant compounds. Where land was in shorter supply in the plains villages at the foot of the plateau escarpment, a more rigid succession to the oldest son was followed.

Though Kofyar women do not own land in their own right, a widow is allowed to use her husband's homestead for as long as she wishes. Women also request usufruct of particular fields from husbands, kin, or friends, and the produce of such land belongs solely to the woman who farms it. At divorce, a woman removes her personal crops, livestock, and cash from her ex-husband's homestead. When the Kofyar moved to the frontier cash-cropping area, some farmers used some swidden plots until their productivity declined after six or eight years and then abandoned them to take up new bush land. Other settlers, however, intensified their production on land for which they had originally paid a nominal tribute to the chief of a nearby plains village. With no more remaining free land in the area, rights to the developed bush farms are now held de facto by the occupants, who control assignment to sons, loan or gift to others, and sale (for increasingly substantial money payments). Although there is no provision for individual farm ownership under current Nigerian law (Mortimore et al. 1987), the Kofyar are treating their new permanent homesteads as valuable property, with the same rights of long-term occupancy, inheritance, temporary transfer, and, more recently, sale, that characterized their traditional lands.

Few authorities on tenure may be inclined to accept the claim that the individualization of rights to land and the weakening of communal or collective regulation of use and transfer form a continuum rather than an abrupt and wrenching transformation with the penetration of a market economy. Though lineages or corporate communities might allow some limited forms of private property inheritance and exchange, they draw the line at permanent alienation of land. In Africa the "most commonly cited expression of the community's right in land is that individual land-holders cannot sell their holdings under most indigenous tenure systems" (Bruce 1988: 25). This prohibition tends to break down as new crops and markets give land enhanced value, and as an impersonal market begins to develop (ibid.: 26). Certainly a money economy and an accelerating demand for commodities that can only be purchased with the proceeds from cash crops contributes to this transition, but, among traditional intensive cultivators, an external market furthers existing tendencies rather than initiating them. Traditional Kofyar concepts of property were consonant with land scarcity and household control. Their recent voluntary migration in order to increase their market participation did not require a breaking of cultural rules or an abrogation of descent-group controls.

The Kofyar land-tenure system illustrates an adaptation to demographically induced shortages of arable land and intensive land use, although the private property practices occurred in a formerly stateless society that was not dependent on the market. Smallholders without easy access to new land must invest time and effort in improving the productivity of what they have. Major permanent construction like terrace walls and a stone corral for the stall-fed goats, the annual labor input of composting and distributing manure, and the tending of oil palms and other economic trees on the homestead farm all represent investments in deferred rewards. Even today a household that occupies a deserted homestead site undergoes some privation for the years it takes to restore soil fertility, prevent erosion, and bring the orchard trees to bearing. If returns on the work of intensification and buildup of productive capital in the farmstead are threatened by insecure tenure, reallocation of the land to others, or denial of household members' right to future benefits, the incentive to intensify will decline. Yet because of unpredictable changes in household demography and economic conditions, permanent individual rights to land must be sufficiently flexible to allow both temporary and long-term transfers. Loaning and leasing permit new households to acquire the use of land without removing it permanently from the control of the owners. Specific arrangements for inheritance, appropriating land in the absence of an heir, and securing rights of occupancy mean that land is not kept out of use when it is needed, and that individuals have options in securing a subsistence base. Land does not have to be a market commodity for it to function as definable property in a system of intensive agriculture.

Shifting Cultivation and Usufruct Rights

Where the conditions of land scarcity and continuous production to support a dense local population are relaxed, rights to resources become less strict and explicit. As noted in Chapter 3, the Ushi of Zambia pollarded or felled dry forest trees, burning their branches and planting a four-year sequence of cassava, millet, maize, cucurbits, and groundnuts in the ashes (Kay 1964: 41). Gardens were frequently changed in this region of low population density, and most were within six miles of the village. "With no apparent shortage of either land or trees, and no vested agricultural interests in any parcel of land for an indefinite period, principles of land tenure have not been clearly defined. Such principles as do exist are, in fact, elementary and simple, but they refer to the rights of an individual to live and earn a living in the village rather than to property rights; they are concerned with the use of land rather than with land ownership" (ibid.: 29).[3] One could farm or gather wild produce anywhere, and there was no defined village territory exclusively reserved for occupants of any settlement. Cultivation conferred absolute and free rights over use of the area as long as it was used. A garden that was left fallow for longer than the customary period was regarded as abandoned and might be reoccupied by anyone. The size of a garden was controlled only by the availability of labor at the critical times of the agricultural year (ibid.: 31).

Where land is plentiful and exploited by extensive methods, it will have little value for exchange, and there will be no grounds for dispute or litigation (Biebuyck 1963). In shifting cultivation, all planted crops are privately owned as long as they are productive, but "when commonly recognized harvest procedures are terminated, the tenure has ended, and the land is considered as returned to the regenerative cycle. Such land has then reverted to the common pool of land owned by the group . . . and becomes public domain open to all forms of appropriation" (Spencer 1966: 90). As Harold Conklin (1957) succinctly observes, land for shifting cultivation is a free good: tenure is by usufruct only.

3. Kay's wording indicates that by "property rights" he means some form of privately held rights. Usufruct is also in the general sense a property right, defining who has the right to use a resource in a certain way (Louis De Alessi, pers. comm.).

Though slash-and-burn farmers may return to old swiddens to gather tree crops or cassava tubers that compete with invading bush vegetation, usufruct rights may atrophy during the long period of fallow when natural regeneration of forest vegetation is taking place on the plot. Most shifting cultivators are not on an open-access, largely unpopulated frontier, and they must return periodically to fallow fields, sometimes cultivating large tracts surrounding their village on a rotational basis. An eight- or ten-year optimal fallow period is, however, sufficiently long that individual household needs and labor power might have changed, necessitating a larger or a smaller tract. Since bush fields are used by and for specific households, even when clearing is a cooperative task and the fields are grouped in one place to better protect against predators, old fallow land must often be reassigned or allocated. A village or a localized descent group, as among Ibo shifting cultivators, may meet to portion out parcels for the usufruct of households, or an individual may ask an elder or chief to be shown a place where he may farm. With land resources that are adequate for fallowing, the major role of the community may be to protect the territory from seizure by other groups (Jones 1949).

Shifting cultivators want to retain optimal levels of arable but presently unused land and perhaps to expand into the fallow of their neighbors. With a need to control in regenerating fallow perhaps ten to twenty times the amount of land presently under cultivation, swidden farmers are frequently involved in conflict, and, like the Maori of New Zealand, their wars may promote the acquisition of forest fallow (Vayda 1961). The segmentary lineages of Tiv shifting cultivators in the Nigerian savanna had endemic border arguments in which individuals extended their farms, then called out their kin groups for acrimonious debates on where the boundary should be. Since the outward push was characteristically directed against the most distantly related lineage or against a foreign ethnic group, a direction of expansionist movement was established, with each lineage losing land in the rear and gaining ground in front (Bohannan 1954; Sahlins 1961). Even among more sedentary, less mobile swiddeners, corporate groups function to restrict access to land, defending it against outsiders and regulating its use by group members (Johnson and Earle 1987: 158, 181). A peasant community council of elders or household heads may perform the same task, preventing trespass by nonmembers and allocating plots from the fallow reserve to newly created households or to existing families whose membership has grown. The Russian repartitional commune that seems to have been the tacit model for Chayanov's village collectivity of households could in theory organize such periodic distributions.

As population density increases, with a corresponding decline in fallow time and average field size, householders can no longer depend on the village or the lineage to provide them with well-rested land, and they are reluctant to give up any land that they have brought into permanent productive use. Areas of intensified annual cultivation are in effect subject to usufruct that is not terminated by the need to fallow, so this land does not revert to a common pool. Ibo villagers cultivated shifting plots of yams, cassava, and coco yams in sectors of tropical forest surrounding their settlement, but they simultaneously also tilled kitchen gardens adjoining their houses (Netting 1969a, 1977):

> In the case of houseland, the householder wants to occupy for an indefinite period the land he clears for his compound and his gardens. Gardens, being enriched with household rubbish, can be farmed at more frequent intervals and they also contain permanent crops in the form of oil-palms, kola, and other economic trees. Thus the ownership of houseland is vested in the individual householder and passes to his direct male descendants, while [fallowed] farmland is owned by the community, that is, by the lineage. (Jones 1949: 313)

Where population pressure and land shortage restricted possibilities for fallowing (W. B. Morgan 1953; Lagemann 1977), large multiple-family households in nucleated settlements broke up into smaller household groups that dispersed into contiguous smallholdings on the formerly fallowed fields (Udo 1965). What had been a communal territory of periodic usufruct rights became houselands permanently occupied and individually owned (Jones 1949: 314). On a regional scale, areas of the most dense Ibo rural population coincide with dispersed settlement and intensive tuber and arboriculture in gardens surrounding small residential compounds (Udo 1965). Physically siting the household in the midst of its compound farm not only lessened travel time, allowed diverse gardening tasks to be done promptly, and guarded the crops, it also asserted a continuing and visible claim of occupancy to the land.

Arrangements for renting land, pledging it to a temporary user in return for payment, or selling it outright proliferate in just those areas where there is the greatest competition for access. The reorientation from communal to individual household tenure was not a revolutionary change in the customary rules applying to land. Houseland and bush-fallow farmland had always been treated differently. Rather intensification, under conditions of scarcity and competition for resources, changed the *use* to which land was put and the proportions of available land under contrasting farming systems (Netting 1969a). Smallholder households emerged, asserting continuing claims to property in scarce, improved, permanently occupied farm lands. "In the main, it may be said that availability of land determines the type of tenure and it is where the pressure is greatest that the few remaining areas of common land are most rapidly decreasing" (Chubb 1961: 14).

Descent-Group Territories

Where natural resources, especially in land, are not the critically limiting factors for agriculture, labor and rights in people's productivity are more important than property. With temporary usufruct, it made no sense to develop a system of private rights to a particular parcel of land that could not be protected and had little utility until its fertility was restored. "The crucial element for the continued control and use of land was to have enough people, be they relatives or slaves, to work the land," and a corporate group could do this better than individuals (Feder and Noronha 1987: 47). As shifting cultivation is accompanied by growing population density and settlement stability, tendencies toward the regulated transmission of collective rights may give rise to unilineal kin groups (Forde 1947: 70). Among the horticultural Mae-Enga of highland New Guinea, Mervyn Meggitt (1965: 279) found that areas of land shortage were positively correlated with patrilineal organization or patrilocality. It appears that the descent groups were excluding affines (relatives by marriage) and non-kinsmen from access to land to ensure the continued adequacy of their own subsistence resources. Roy Rappaport (1968: 27–28) points out that a single New Guinea kin group with low population density might grant land rights in its abundant resources to a wide variety of relatives, but that as the supply of open land declined and conflicts over farms and marauding pigs increased, a tendency to confine use and inheritance to the more rigidly defined patrilineage would become apparent. A unilineal descent group can both reduce conflict over land among its members and secure cooperation beyond the nuclear family for the defense of scarce resources (Harner 1970; Netting 1982a: 467–68). There appears to be a continuum, from relatively open local groups or villages with extensive territories of long-fallowed land, to somewhat more restricted shifting cultivation with unilineal descent groups claiming and defending corporate estates, and finally to households with rights to heritable smallholdings.

Running from the eastern to western regions of highland Papua New Guinea, there is a gradient of increasing population density; more intensive cultivation of sweet potatoes, with mounding, composting, and drainage; larger numbers and more careful husbandry of pigs; and dispersed instead of nucleated settlement patterns (Feil 1987). Relatively small linguistic and political groups in the eastern highlands have populations of $30/km^2$ or fewer, while larger western ethnic groups such as the Chimbu (Brookfield and Brown 1963), the Hagen (Gorecki 1979), and the Enga (Waddell 1972) reach densities of 100–150/km² (Feil 1987: 41). These more crowded and sedentary peoples, including the Kapauku (Pospisil 1963) and Dugum Dani of Irian Jaya, Indonesian New Guinea, show a high ranking of agricultural intensity, in most cases practicing complete tillage, with grids, trenches, or mounds; fencing; erosion control on slopes; drainage ditching; fertilizing by mulching or composting; and short fallows of up to six years (Brown and Podolefsky 1976: 215–17). Much of their time, perhaps 41 percent in the case of the Laiapo Enga, is devoted to herding and cultivating tubers for pigs, which are important in exchange and prestige distributions (Feil 1987: 48).

The predominantly shifting cultivators in the east rely substantially on gathering and hunting (Morren 1977), and they keep fewer pigs than the 1.4 to 4 per person characteristic of the groups practicing intensive agriculture. Land tenure—classified as individual, heritable rights to specific marked plots as opposed to predominantly usufruct rights to land in group territory when it is taken out of fallow, or no specific rights in fallow or forest—also varies regularly along this continuum (Brown and Podolefsky 1976: 216). "Individual ownership of land . . . is strongly correlated with high population density . . . , as well as with high agricultural intensity" (ibid.: 221). Individual plots are held and inherited mainly where the fallow period is short and where trees or shrubs are planted by the owner, both for the economic value of their fruits over time and as identifying marks of possession. Where, on the other hand, exogamous clan groups are dominant, their territories for the most part cover between 1.55 and 8.55 km² (ibid.: 232), suggesting that relatively small descent groups would have abundant land for long-fallow agriculture (Rappaport 1968) and would not need to establish more individualized private tenure rights.[4]

Comparing Maya-speaking populations of southern Mexico in Chan Kom, Zinacantan, and Chamula, G. A. Collier found that both sparse and very dense populations had little emphasis on living near and sharing land rights with patrilineal relatives, but the localized household cluster of kin was emphasized by those with an intermediate level of competition for resources. "Generally where land is abundant and a free good, swid-

4. Where a long-term study of land tenure, transfer, and succession was conducted among the Chimbu of the Papua New Guinea highlands (Brown et al. 1990), increasingly intensive and semi-permanent cultivation coincided with a progression from fluidity of land rights in the clan or subclan to individual and family rights. "It seems clear that it is not until a tract of land comes to be cultivated most of the time that tenurial boundaries become fixed. . . . Where it is scarce and valuable, individuals guard access to a small personal range; where it is relatively abundant, they are able to use it entrepreneurially, and both draw on and nourish a wider set of relationships" (ibid.: 46). Such change is not solely a product of either colonialism or the cash-crop coffee economy of the Chimbu.

den farmers do not have descent organization, but where land is scarce and a valued commodity, descent emerges to systematize right to land. When, however, landholdings are overly fractioned by inheritance, farming tends to give way to other occupations and land ceases to motivate descent-based kinship" (G. A. Collier 1975: 206). Among the Chamula intensive agriculturalists, with their tightly packed, privately owned smallholdings and their employment in crafts, trade, and wage labor, corporate clans or lineages have disintegrated. As with most sedentary peasant groups, bilateral kinship and household property are dominant (Goldschmidt and Kunkel 1971).

Parker Shipton (1984) points up the contrast among traditional cereal-cultivating and cattle-herding societies of East Africa between areas of low population density (below 40 /mi²) where chiefs claimed administration over extensive territories and more crowded regions with descent-based lineage systems. The subjects of chiefdoms were shifting cultivators who moved frequently and did not ordinarily establish heritable property rights. Under conditions of land shortage, unilineal descent groups in acephalous societies protected small tracts and regulated land allocation within them. If population pressure increased beyond a high-density threshold, lineage control would yield to a "hardening of individual rights of ownership and transfer":

A pattern of localized lineages is most likely to break down when holdings have been subdivided to such an extent that they can no longer provide (together with non-agricultural activities) a reasonable living to patrilineal heirs and their families, and when these heirs can no longer co-exist peacefully. The densities at which this threshold occurs will be highest where double cropping is possible, where crops of high yields per hectare (e.g., cassava or other bulky tubers, or bananas) or cash crops of high value are grown, and where non-agricultural income (from seasonal or longer-term labour migrations, trade, cottage industries, etc.) is most accessible. (Shipton 1984: 628)

The transition from general land rights, with individual usufruct and reversion of the fallowed plot to lineage administration, to individualized, specific control may be a regular course:

With the development of specific land rights the cultivator can begin to assert certain rights over plots, beginning with the right to resume cultivation of the specific plot after a period of fallow. At a later stage the cultivator asserts—and receives—the right to assign the plot to an heir or a tenant. Thus, the use right to the plot does not revert to the lineage anymore. With increasing population density, the rights assignable by the individual cultivator become more extensive. Eventually they include the right to refuse stubble grazing and, most important, become completely alienable. Thus, a cultivator can lease and sell plots to individuals from outside the lineage. (Binswanger and McIntire 1987: 86)

Households and the Individualization of Tenure

Under conditions of positive land shortage, households become the important institutions for administering land and transmitting the holding to the next generation. Rules may develop that further restrict the inheritance to one or a few eligible heirs. Sons may divide the estate, with daughters given a dowry or a smaller share of land. Primogeniture, ultimogeniture, or some other rule of impartibility may pass on the farmstead intact to a single descendant, while noninheriting siblings may have rights only to stay in the household (usually as celibates), contribute labor, and receive subsistence, without ever sharing in the capital of the enterprise. Such restricted inheritance among Austrian peasants might preserve the farm as an economic unit while non-heirs were forced to migrate, to seek lower-status wage labor, or to work in cottage industry (Khera 1972a, 1972b). In large stem- or multiple-family households with internal differentiation, "fission and morphology may be more a product of property management than of efforts at efficient production. . . . There is always tension between the need to manage [and inherit] property restrictively and the need to procure as much labor as possible" (Wilk 1991: 227).

Specific inheritance rules, depending as they do on long cultural traditions and the historic legal systems of states, do not bear a one-to-one relationship with population/land ratios or with smallholder agriculture. Rights to land and detailed inheritance regulations are, however, unfailingly individualized and privatized when they refer to permanently used, high-yielding, intensively worked household resources. Shifting cultivators, lacking significant units of property, organize their households around labor exchange and the complementarity of male and female production, providing a sustaining shared flow of food, cash, and attention. Richard Wilk (1991: 205–6, 227) suggests that the joint use and ownership of land, the investment of labor and capital, and the expectation of inheritance *do* transform the relationships of spouses, parents and children, and siblings. Where enduring and valuable property in the smallholding contributes to lasting economic relationships within the household, the nature of sentiment, security, and even kinship itself may be palpably different.

The observation that land tenure is not an exogenous variable, an artifact solely of legal and governmental systems, but is fundamentally linked to patterns of land use, was made most forcibly by Ester Boserup (1965: 77). The British classical economists had assumed that private property in land emerges when agricultural land becomes scarce under the pressure of growing population (Boserup 1965: 78), but Boserup described how the process of intensification leads to individualization of rights. "The attachment of individual families to particular plots becomes more and more important with the gradual shortening of the period of fallow and the reduction of the part of the territory which is not used in the rotation" (Boserup 1965: 81). Rights to land did not change overnight, and neither did "private property" become completely dominant over communal or territorial rights in any society. There was no "Tenure Revolution," as Lewis Henry Morgan, Sir Henry Maine, and other nineteenth-century evolutionists seemed to believe.[5]

Linkage of political centralization to the emergence of private tenurial rights remains a persistent, unquestioned premise of both capitalist and neo-Marxist evolutionary scenarios (cf. Netting 1990). Analytically and historically separable economic and social institutions are bundled together in categorically opposed structures, following an implied early/late, simple/complex dichotomy. State formation is seen to mark the transition from "communal cultivators"—a mode of production based

5. It is sobering to consider how strongly the ideology of the nineteenth-century classical evolutionists continues to influence our unexamined concepts of property (Netting 1982a). For Lewis Henry Morgan, the Rochester lawyer, railroad investor, and New York state legislator, cultural development represented both technological invention and intellectual progress: "The idea of property was slowly formed in the human mind, remaining nascent and feeble through immense periods of time. Springing into life in savagery, it required all the experience of this period and of the subsequent period of barbarism to develop the germ, and to prepare the human brain for the acceptance of its controlling influence. Its dominance as a passion over all other passions marks the commencement of civilization. It not only led mankind to overcome the obstacles which delayed civilization, but to establish political society on the basis of territory and of property. A critical knowledge of the evolution of the idea of property would embody, in some respects, the most remarkable portion of the mental history of mankind" (Morgan 1963 [1877]: 5–6). Morgan's latter-day followers insisted that individual property in simple societies was purely personal, while land, the basic source of subsistence, was always collectively held (Leacock 1963: pt. 1, xvi). Critics of the evolutionary view found forms of private ownership along with well-defined communal rights in every society, and they decried the "dogma of a universal primitive communism" built up by Engels and Marx on a foundation of Morgan's work (Lowie 1920: 235). In fact, Engels's (1972 [1884]) understanding of the presumed transition of prehistoric German land tenure from collective clan ownership of land through communistic household communities to individual family holdings had an ecological rationale. With the increasing pressure of population on land resources and the lack of sufficient territory to sustain shifting cultivation, disputes over land could interfere with the common economy and encourage some form of private ownership. "The arable and meadowlands which had hitherto been common were divided in the manner familiar to us, first temporarily and then permanently among the single households which were now coming into being, while forest, pasture land, and water remained common" (Engels 1972: 202). The supposed lessons of cultural evolution reinforced the political stance of historical materialism, dividing the human career into B.P. (before property) and A.P. periods, and emphasizing the lost utopia of cooperatively shared resources.

on communal land ownership, division of labor based on kinship, absent or peripheral markets, and cultural homogeneity—to a "peasant condition" with individual ownership of land, a social division of labor and political hierarchy separate from kinship, the market principle, and the opposition of great and little cultural traditions (Post 1972, cited in Swindell 1985: 59). This transformation is allegedly still going on. Scholars such as Eric Wolf and James Scott, whom Samuel Popkin (1979: 4) refers to as "moral economists," argue that precapitalist structures provided peasants with security, welfare, and a level of insurance against subsistence failure that have been lost under capitalism:

> They assume that peasants are antimarket, prefer common property to private, and dislike buying and selling. They also assume that peasant welfare depends on the closed corporate village so common in precapitalist society and/or on multistranded feudal ties to those who control the land. The transition to open villages with private property and open land sales, and the transition to contractual, single stranded ties with landlords, they argue, force peasants into the market where their welfare invariably suffers. (Popkin 1979: 5–6)

If land use is an important key to functional systems of tenure, we would expect to find household property rights in intensively utilized plots, "while land under more extensive systems of land use is still at free disposal for cultivation by any family with general cultivation rights in the village" (Boserup 1965: 85). Truly diversified land use means that some tracts of low fertility or at a considerable distance from the village can most efficiently be cultivated by shifting techniques or allowed to remain as rough grazing or forest, and rights to such resources are not necessarily individualized. On the other hand, to raise production and reduce fallow on more productive lands, farmers "must adopt fertility-restoring technologies, which require investment of capital and effort—and thus also require incentives for farmers to change their practices. One such incentive is the right to cultivate land continuously and to bequeath or sell it" (Feder and Noronha 1987: 143). Ethnographic cases exhibit just this predicted contrast between extensive / communal or general land rights and intensive / individual or specific land rights (Rosenzweig 1984).

The Bontoc Igorot of mountainous northern Luzon grow sweet potatoes, black beans, millet, corn, bananas, and other vegetables on dry slopes, with the length of fallow directly dependent on how far the field is from the village (Drucker 1977; Prill-Brett 1986). These poor but plentiful slope lands passed from a single original claimant, who first utilized them, to all of his descendants, so that, in time, a large bilateral descent group, including virtually all village members, have inherited common rights of usufruct. Households can open a temporary sweet potato swidden anywhere on land to which they have descent-group rights. Forest lands are communal property, with their timber, firewood, basketry materials, medicinal plants, honey, pasture, mushrooms, and game animals open to all village members (Prill-Brett 1986: 59). Wet rice and dry-season sweet potatoes, which together constitute three-quarters of the annual diet (ibid.), come from the irrigated, intensively worked terraces that can only be built where water is accessible. A restricted inheritance system directs that a father's rice fields and heirloom wealth objects go to his eldest son, and a mother's to her senior daughter. Other rules govern the subsidiary rights of illegitimate children, collateral descendants, younger children, and children by second marriages. The type of tenure and inheritance relate directly to the scarcity, productive value, and frequency of use of the land. "The least restricted forms of inheritance operate upon the land holdings for which there is the least competition" (Drucker 1977: 7). Irrigated paddy fields may be sold in exchange for sugarcane wine, death clothes, and pigs when a family must have sacrificial animals for a funeral ceremony or pay heavy fines. On such a crisis occasion, kinsmen receive preferentially lower prices (Prill-Brett 1986: 80). Neither the Igorot nor their neighbors the Ifugao were, until relatively recently, incorporated in state organizations, and their land-tenure rules were indigenous institutions within autonomous village and hamlet groupings. There were no tribal political organizations or multi-community chieftaincies.

The Ifugao combination of shifting cultivation and remarkable stone-walled, terrace pond fields supports self-sufficient populations of 77 to 154 per km² (Conklin 1980). Households exercise individual ownership rights in rice fields and in private woodlots that provide fuel, fruit, medicines, and building materials (Fig. 1.2). The inheritance of the prized agricultural terraces is by weighted bilateral primogeniture (ibid.: 32). Claims of land title are traced through from eight to ten generations of ascendant kin, and previous owners are the most commonly named ancestors on almost all ritual occasions. Land is also transferred by indirect inheritance and by purchase, and new terraces are sometimes built. Some of the more important land parcels are inherited in association with heirlooms such as stoneware jars, bronze gongs, gold neck pieces, shell belts, and carved granary idols (ibid.). Swiddens producing sweet potatoes, legumes, and vegetables on steep woodland or caneland slopes are not permanently owned but cultivated by individuals on the community common lands. They furnish the bulk of the food consumed by most families except the wealthy. The Ifugao also raise pigs, chickens, and other forms of livestock, which are individually owned. Irrigation channels belong to the constructor and his descendants, and rights to irrigate particular plots are associated with title to land. The builder of a new pond field can tap into a channel by making a payment (usually pigs) and by sharing the upkeep of the channel with an ad hoc group of water receivers in the service area (ibid.: 28). No one knows how long the highland Philippine peoples have practiced intensive pond-field rice cultivation on their magnificent terraces (Barton 1919; Keesing 1962; W. H. Scott 1966), but the institutional association of swiddens with communal rights and irrigated rice fields with individualized, elaborated, heritable property rights is clearly attested.

The Gradient of Land Use and the Spectrum of Tenure

Because altitudinal zones in mountainous regions dictate land use through temperature, exposure to sun, soil, and slope gradients, communities that cut across the environmental grain will usually exhibit correspondingly different systems of tenure. In the central Andes, major vertical life zones consist of (1) a maize zone up to 3,500 m in altitude where irrigated double-cropping and specialized horticulture are possible, (2) a tuber and indigenous cereal zone of sectorial fallowing in the range of 3,500 m to 4,100 m, and (3) a pasture zone from the upper limits of agriculture to approximately 4,700 m (Brush 1976; Mayer 1979; Guillet 1981). David Guillet's contemporary sample of 17 communities in this area supports the consistent relationship of contrasting land-tenure rules with the ecological zones and their differing agricultural regimes. High yields of irrigated maize or specialized tree, vegetable, and strawberry crops are produced in lower-zone plots under private control. Rotations may include fodder crops such as alfalfa, clover, rye grass, and barley, and the fenced, irrigated fields may be manured and chemically fertilized (Mayer 1979: 430).

In the next highest zone, where potatoes and other rainfall-dependent native tubers have long been grown, lengthy fallowing must be practiced to allow for soil regeneration and the dying out of populations of round cyst nematodes, which can reduce the yields of successive potato crops (Orlove and Godoy 1986). Village lands are divided into sectors, and most households own individual plots in all of the sectors. The community regulates which sector is to be cropped in a particular year, when planting and harvesting are to take place, and how many years of fallow must intervene before the sector is farmed again. In one sectoral fallowing system of the Peruvian highlands, there were ten sectors, each used in rotation, with a year of potatoes followed by a year of other Andean tubers, and then eight years of fallow (Orlove and Godoy 1986: 171–72). Other communities may use quinoa (*Chenopodium quinoa*) and barley in the sequence with potatoes, and fallow periods may vary. At lower altitudes, where grains are dominant, there may be as many as eight cropping years, followed by three years of fallow (Mayer 1979: 41). During the fallow period, all households have access and grazing rights to the entire sector. The system combines household property in individual tuber plots with community control over when these plots can be farmed and communitywide grazing of all plots during the fallow period. Community authorities also ratify the ownership rights of particular households and can redistribute cultivated plots that are abandoned or left vacant by owners who have died without heirs (Orlove and Godoy 1986: 171).

Grazing by llama and alpaca in the highest pasture zones, and by sheep and goats at lower elevations, is on land under communal control, but member households have indivisible use rights. In the subhumid, semifrigid climate of these high altitude grasslands, there may be scattered plots of bitter potatoes for the production of freeze-dried *chuno* (Mayer 1979: 40). Guillet (1981: 144) found highly significant associations (*p* <.001) between the intensive irrigated-maize regime and private tenure, between sectoral fallowing of tubers and communal control with individual plot-cropping rights, and between grazing and common pasture rights. In fact, the nature of corporate communal property rights held in certain resources and land-use types may be as distinctive to smallholder communities as the presence of individualized household tenure in the same communities.

The Properties of Private and Common Rights: Swiss Alpine Land and Water

The practice of intensive agriculture both correlates with and eventually *requires* private property rights. Those scarce resources that have relatively high productive potential, that yield frequently and reliably, and where outputs can be increased by the application of labor, capital, and management, will be held in socially recognized ways by individuals or households. Rights of use will be sufficiently exclusive, continuous, and transmissible so that at least a significant portion of the benefits generated by coordinated household labor, investment, and planning will accrue to those who make these efforts. The higher the long-term subsistence and / or market value of such intensively used resources, the more likely that they will be subject to detailed rules of ownership, exchange, and inheritance. Manured grainfields and kitchen gardens, irrigated rice and corn plots, terraced tobacco fields, vineyards and orchards, when used by smallholders, are protected from arbitrary appropriation or reallocation by well-defined and vigorously defended systems of private rights. The converse proposition, that extensive land use always corresponds to the absence of private property in land, can obviously not be supported, since universalistic judicial and political systems may assign rights of tenure with no regard to use. But smallholders, like those in the highland Philippines and in the Andes, frequently match different types of tenure to different resources and contrasting farming systems in the same community.

Whereas individuals and households assert specific private property rights in resources under intensive cultivation, common-pool resources such as unimproved grazing grounds, surface water, and forests may be held as common property. Designation as common or communal property does not, however, mean that there is open access to anyone who desires it.[6] We often find common property rights applied to agricultural resources of relatively low value per unit where it would be too costly to enforce exclusive private rights (Oakerson 1986; Berkes et al. 1989; Ostrom 1990), but that *can* be restricted to and defended by a local community of users. Common property is owned and regulated by a corporate group, often a residential community of smallholders, who cooperate in excluding non-members and derive joint benefits from the resource. Because such resources have the characteristic of subtractability—that is, each user has the capacity of individually appropriating resource units, which are thereby subtracted from the goods available to others (Ostrom 1990)—common property with rules controlling access and use is a means for maximizing the present value of aggregate returns to all members and, incidentally, conserving the resource owned in common. The commons is a pool of resources from which individuals take a portion for their use. But

appropriation affects production, or more precisely, the rate at which individuals appropriate affects the rate at which the resource can produce or replenish a supply. Without coordination, individuals may in the aggregate use too much too fast, causing the rate of production to fall. Sharing without collective consumption—the commons situation—requires restraint, which in turn requires coordination among users. Otherwise individuals continue to consume without regard to the diminishing marginal product of the commons as a whole. (Oakerson n.d.: 2)

The Swiss alpine community of Törbel has a long, documented history of the coexistence of individual and common property rights (Net-

6. In one of the most memorable statements of the new environmentalist movement, Garrett Hardin's "The Tragedy of the Commons" (1968) uses the scenario of a rational herdsman who increases the number of his livestock without limit, eventually destroying the resource on which he and his fellows depend. Because the herdsman gains the full benefits of each additional animal while sharing the costs of overgrazing with all the other cattle owners, his rational decisions add up to an irrational dilemma (McCay and Acheson 1987: 2–15). The finite grazing resource is open to all comers, and the possibility that a group of rational herdsmen, observing the degeneration of their environment, and communicating about cause and effect (as humans often do), might establish boundaries against outsiders

ting 1976, 1981). Well-preserved parchment rolls written in medieval Latin and dating from the thirteenth and fourteenth centuries make it plain that hay meadows, grainfields, vineyards, gardens, houses, barns, and granaries were owned by individuals as representatives of households. Bills of sale and mortgages were written by notaries, witnessed by fellow villagers, and testified to by family members. Field locations were designated by surrounding plots belonging to named owners, and there might be accompanying specified rights in irrigation water. Payments mentioned were substantial and in cash, indicating that Törbel farmland already had the high price remarked on by observers in this century (Stebler 1922). Partible inheritance, with each child receiving an equal share in the estate at the death or retirement of a parent, was the rule in Valais (Partsch 1955), and it continues to be observed, with the heirs agreeing on the composition of equivalent shares comprising parcels of land and buildings, and then drawing lots for them (Netting 1981: 172–74). Though we cannot estimate the village population before about 1700,[7] it is apparent that pressure on limited arable lands had already by the medieval period led to the intensive practices of terracing vineyards, grainfields, and gardens; manuring and irrigating meadows; and stall-feeding livestock, thereby putting a premium on possession of sufficient resources to support a household.[8]

---

and jointly control their own herd sizes, seems not to have occurred to those who accept the biological model. In fact, "commons" in the sense of the long-established community grazing areas of Europe embody clear and well-enforced property rights. As such, they are an unfortunate analogy for the real tragedies of overpopulation, atmospheric pollution, groundwater exhaustion, and overfishing in open-access resources that Hardin so presciently conceptualized.

7. Based on church registers of baptism, marriage, and burial, along with a village genealogy that allowed family reconstitution, the Törbel population fluctuated between 241 and 294 in the years 1700–1775 (Netting 1981: 96, 114). The village has occupied its current territory since at least the eleventh century, and there are indications that it was settled by Celtic speakers well before the eighth-century Allemanic invasions (Netting 1981: 8–9). A formal charter of 24 statutes pertaining to community membership and regulation of the alp was signed in 1531 by 60 named males representing a total of 69 families (Netting 1981: 62). The population of 300 to 350 was substantially reduced by the 1533 plague epidemic, but it is not unreasonable to envision a community of 250 to 350 members that supported itself for centuries on the same land base of 1,545 ha, including 967 ha of farmed land (Netting 1984).

8. It is hard to overestimate the salience of property in the lives of European peasants. David Sabean's magisterial and definitive reconstruction of the social history of a South German farming community from 1700 to 1850 suggests the richness of property relationships "between people about things": "Boundaries mediate between people. Land, houses, and tools are things which are held, managed, and argued about and are the stuff around which people in a village like Neckarhausen shaped their lives in concert with each other. In order to understand the trajectories of individual lives, the dynamics of particular families, the strategies of alliance and reciprocity, and the effects of state intervention and economic

At the same time, higher-altitude or rocky cliff and ravine areas were just as clearly demarcated as communal property. A charter dated February 1, 1483, specifically forbade any foreigner (*Fremde*) who bought or otherwise occupied land in Törbel from acquiring any right in the communal alp, common lands, or grazing places, as well as denying permission to cut wood in the village forest (Fig. 1.1). Anyone could purchase real estate in Törbel, but only a citizen, defined as a person descended in the male line from a legitimate Törbel resident, and his household, could send his cattle and sheep to the alp in summer and cut fuel for cooking and winter heat in the forest. Limiting these economically vital activities to community citizens and making new membership dependent on large cash payments *and* formal agreement by the current citizenry excluded outsiders so effectively that no new family lines became settled in Törbel after 1700 (Netting 1981: 76–82). In 1517 further alp-use rights specified that no citizen could send more cows to the alp than he could feed during the winter, thus effectively restricting households to the number of animals that their own hay meadows could support, and severely fining them for any attempt to appropriate a larger share of community grazing privileges. The total stocking of the alp was linked to the village supply of hay, though the proportions of individual household cattle might vary with their privately held meadowlands. The woodlands were similarly regulated, with trees marked annually by the elected community council and households drawing lots for their equalized shares in the timber to be cut (ibid.: 67–68). Only fallen branches and dead wood could be freely gathered from the forest.

Rights of common property that had probably existed long before they were written down effectively excluded outsiders from competing for alp and forest resources with citizens and regulated the amount of productive use that each household could make of the commons. Annual meetings of all livestock owners enforced the alp rules, appointed the paid workers who herded, milked, and made cheese over the summer season, and monitored the physical condition of the pasture. Those who sent animals to the alp and received cheese in proportion to the milk given by their cows (and carefully measured by an official several times during the summer) also had the responsibility of maintaining the common property. Work parties on the alp cleaned the springs, repaired avalanche damage to paths and walls, raked twigs from beneath the trees, and spread the dung left by the herd. Each household supplied labor in proportion to the

---

differentiation on village social practices, we have to examine the details of property rules, structures and codes. . . . Land and other goods were part of a wider set of exchanges and reciprocities through which people were disciplined. And land and its exploitation provided a focus for socialization, character formation, emotional commitment, and the long apprenticeship which instilled obligation" (Sabean 1990: 18).

TABLE 6.1
Long-term Swiss Alpine Land Use and Tenure

Nature of land use	Land-tenure type	
	Communal	Individual
Value of production per unit area	Low	High
Frequency and dependability of use or yield	Low	High
Possibility of improvement or intensification	Low	High
Area required for effective use	Large	Small
Labor- and capital-investing groups	Large[a]	Small[b]

SOURCE: Netting 1976: 144.
[a]Voluntary association or community.
[b]Individual or family.

number of cattle it had pastured. The elected commission of cattle owners who supervised alp operations and the annual assembly of all citizen users saw to it that regulations were obeyed and "free riders" apprehended and fined. In a small, face-to-face community, it was impossible for anyone to get away with fattening an extra steer on the alp and then selling it in the fall or failing to show up for the *Gemeinwerk*, the mandatory communal work days for the upkeep of the high pastures. Though the group that operated the alp was clearly corporate, with a continuing existence in relation to its collective property, its members, unlike a modern business corporation, were both managers and workers, combining ownership with use for subsistence purposes, and asserting control by egalitarian rather than hierarchical mechanisms (Picht and Agrawal 1989). The alp association resembled a cooperative in its form and function, but the institution was coterminous with the local community of citizens.

Both the alpine pastures and the high forests are extensive resources with low or slow productivity per unit area and little potential for increasing yields (Table 6.1). Forest soils were too thin and rocky to cultivate, and their altitude meant the season was too short for reliable cropping. The larch and pine trees took years to mature. Similarly, the natural grasses of the high pastures could not be irrigated for double crops of hay. Both resources were most efficiently used in indivisible form. The herd of the entire village could be moved among the pastures according to need, which varied depending on the particular year's grass growth and precipitation. When villagers visited their animals on Sunday, they looked closely for signs of overgrazing or poor condition in the livestock. Dividing the alp among the cattle owners would have required much more labor in herding and dairying, as well as possibly fencing, which would have been costly and interfered with optimal grazing movements. Privatization might also have hindered the adjustment of pasturage to the variation in household herd sizes over time. Because mountain conifers grow slowly, and yet every household needed wood fuel for cooking and heating, dividing the forest into private woodlots might have tempted some owners to overcut their tracts while other lots were not harvested optimally. A market for firewood could also have led to the same effects if some individuals sought short-term returns and the transaction costs of annually redistributing access to fuel went up. The forest had an ecologically important role for the whole village in serving as a barrier to destructive avalanches as well as conserving water and preventing erosion on steep slopes. Only planned, restricted cutting of the communal resource could provide sustained yields, equitably divided among community members, while at the same time conserving vital environmental protections.

The group that exercises joint property rights and manages a common-pool resource need not, of course, be a community or a residentially defined population unit. Törbel's three irrigation systems were formerly operated as separate associations (*Geteilschaften*), each comprising all of the owners of land whose meadows were watered from a single network of channels. The two lower and probably very ancient systems, the Springerin and the Felderin, tap the Törbelbach, the major stream that drains the village territory and provides a critical common-pool resource. Those with rights to timed and rotated periods of the flow within one system are responsible for cleaning it in the spring, paying for any major repairs in proportion to their individual water shares, and monitoring the traditional succession of use periods during each 16-day cycle (Netting 1974a). One who purchases or inherits land with its accompanying water rights is automatically a member of the association, with its responsibilities. The uppermost system, the Augstborderin, which brings water from outside the village territory, serves both Törbel and the neighboring village of Zeneggen, and they cooperate in annual maintenance.[9] A formally chartered user association in Törbel has a supervising official (the *Niventeiler*, divider of the new canal [waters]), whose position rotates each year. In the past, when the main channel was an open conduit, the association hired a guard (*Wasserhüter*) each year to patrol the ditch from its source and check for leaks and obstructions. There is even a very small group of irrigators (but with a written constitution and bylaws) that cap-

9. One account states that water rights in the Embdbach were purchased from the valley town of St. Niklaus in 1270, and documents show that the irrigation system was independently controlled by an association of Törbel and Zeneggen residents in 1343 (Stebler 1922: 71–72; Netting 1974a). The canal, cleverly engineered along the mountainside with a minimal gradient for 6 km to Törbel and another 4 km to Zeneggen, is still known as "the new one" (*die Niwa*), and it opened up for hay cultivation a high-altitude tier of meadows that had previously been forest clearings (Netting 1981: 44). The meadow areas still bear names like *Riedfluh* (rocky clearing) and *Schwendi* (ring-barking trees) that refer to their earlier, less intensive uses (Netting 1981: 50).

tures the flow of a spring in a shallow, dammed pond (*Wier*), and then releases this "common-pool resource" twice a day from a central drain onto some contiguous meadows (Stebler 1922: 69–70).

In this notably dry area of southern Switzerland, intensified management of hay meadows to produce two crops a year plus some grazing would not be possible without artificial watering techniques. The distribution of the scarce resource requires an initial investment, especially when the water source must be purchased and a long canal, plus a network of smaller channels and sluice gates, constructed. Large-scale irrigation involves capital and labor costs that are beyond the reach of most individual cultivators and can be provided more economically by cooperative effort or financed by a private owner, who then contracts with water users (Spooner 1974). To make sure that the system is properly maintained, that distribution is conducted without waste of water or time, and that members receive the shares to which they are entitled, corporate organization must actively supervise distribution and supply maintenance. Group members must also monitor each other and resolve the conflicts that inevitably arise over the use of the scarce resource.[10] Indeed, some systems of communal management and dispute settlement only come into operation during that part of the year when water is in shortest supply and when it is most in demand for growing crops.[11]

Irrigation Management: Corporate Cooperation or Oriental Despotism?

Irrigation systems become even more complex and vital to intensification when they serve wet-rice cultivation. But despite systems of dams, flood embankments, and long canals, which require large, specific investments and typically are planned and constructed by the state or other

10. Elinor Ostrom (1990: 90) has recently synthesized a set of "design principles" that characterize enduring, self-governing common-property resource institutions like those that organize mountain grazing or irrigation in a great many traditional communities. These similarities include: (1) clearly defined memberships and territorial boundaries; (2) congruence between appropriation rules and local conditions; (3) collective-choice arrangements for modifying the operational rules; (4) monitoring; (5) graduated sanctions against those who violate appropriation rules; (6) conflict-resolution mechanisms; and (7) minimal recognition of members' right to organize by external governmental authorities. The fact that rules for the commons are made, enforced, and modified by local groups whose members are the users of these resources is particularly significant. Successful institutions have a great variety of specific forms and constitutions, but local assemblies and officials retain considerable autonomy in decision-making.

11. Downing (1974) describes a Oaxacan community where official regulation of the irrigation and dispute settlement are activated only when scarce water supplies coincide with the moisture-sensitive period of the developing maize plants.

large political entities, local irrigation continues to be administered by small groups of cultivators that manage a common-pool resource with common property institutions. Although rice terraces in Bali are individually owned, and cultivation is carried on by private proprietors, the irrigation that makes this farming system possible is under the control of a *subak* irrigation society (Geertz 1972). It comprises a named, contiguous area of terraces (a "wet village"), and all people with freehold tenure in these fields are members, regardless of their residence, caste, kinship, or wealth. As the canal approaches the society's fields, it is precisely divided several times, with the proportions of water specified according to a written palm-leaf subak constitution. The society regulates the time of planting for its members and delegates the work groups that maintain the system. As a corporate group, the subak elects a chief and other officials by a council of members, each having one vote, regardless of size of holding. The society collects taxes and disburses money for improvements, fines members for infractions, and appoints the priests who conduct rituals at subak shrines that schedule the various activities of the rice-cropping calendar. Members retain all rights to sell, rent, or tenant their own land and to cultivate it as they wish. The subak society does not engage directly in production or marketing. Though Clifford Geertz (1972) refers to the society as a cooperatively owned public utility rather than a collective farm, its structure is more like that of a cooperative that provides a certain agricultural input for its members.

The irrigation society is neither a branch of government nor a private business enterprise, and its operation appears to supply benefits to farmers that neither of these institutions could provide, and that could not be distributed by individual smallholders working independently. As a cooperative with elected leaders, the subak society could create and maintain the waterworks, allot the water among users by discussion and group consensus, regulate the timing of the cultivation cycle, and control the planting of rice (Geertz 1980). Although it did not possess land as common property, the society could give or withhold permission for new terrace construction, and it legitimated transfers of terrace ownership. Water in an entire regional drainage area was apportioned among the autonomous local subak units, not by hierarchical governmental or bureaucratic mechanisms, but by religiously sanctioned integration. Ceremonies of the rice goddess cult, beginning at the river temple at the volcanic lake at the top of the system, timed the release of water to each society area in turn (Lansing 1987). Because wet rice requires the highest water input at or just after planting, followed by decreasing water levels until the harvest on a dry field, subak cycles had to be successive rather than coincident to make optimum use of the limited water resource (Geertz

1972). The higher irrigation societies received water first at the beginning of the ritual cycle in December, and each terrace neighborhood in the Balinese river valley was flooded successively until the coastal fields were reached in April. The vital temporal integration of the cellular irrigation' units and the sequence of agricultural tasks within each subak were scheduled and meshed by priests and ritual observances.

Among intensively cultivating smallholders, the same scarce, productive, improvable resources that make private property rights adaptive also provide the impetus for the corporate organizations that protect private tenure and both establish and administer common property. While the Balinese subak manages delivery of public water supplies that make individually owned rice terraces produce dependable crops, the residential hamlet has a separate corporate role in community life. Under a jurisdiction often set out in a written constitution, the hamlet is responsible for public facilities (roads, meeting houses, markets), local security (the night watch and the suppression of violence), and the settlement of civic disputes (inheritance, arguments about traditional rights). It has power to confer and withdraw citizenship, organize collective work, and control access to house land (Geertz 1980). The hamlet also allocates various public goods necessary for the protection and enjoyment of household farm property in a particular local area. In pursuit of these legitimate activities, the community can tax and fine its members, and it can own property collectively and invest in commercial ventures. Though the Balinese hamlet performs a variety of essentially social tasks, such as legitimizing marriage and divorce, enforcing sumptuary laws, and sponsoring feasts (ibid.), its major involvement with matters of property is a salient characteristic.

Corporate Institutions for Smallholder Individualists

The peasant "closed corporate community," a seminal concept proposed by Eric Wolf (1955, 1957, 1966), conveys the historical importance of corporate relationships to property but neglects some of the distinctive features of smallholder corporate groups. Using examples from Java and Mexico, Wolf (1957) stresses community closure in terms of restricted citizenship, "outright communal tenure" of village land, and unitary responsibility for tribute and corvée labor to outside powerholders as means of both accommodating to and resisting the political and economic demands of colonial conquerors. The corporation, "an enduring organization of rights and duties held by a stable membership" (Wolf 1966: 86), meets onerous external exactions by equally distributing rent in labor, produce, and money among its members. Such communities

maintain the internal order for this defensive posture by forbidding alienation of village land to outsiders and by periodic reallotments of land that
have the effect, along with ceremonial distributions such as fiestas, of leveling economic differences among members (Wolf 1966: 86).[12] That the
most economically valuable lands in such communities often remain in
local private ownership, that there may be substantial and persistent
property inequalities among households, and that the community may
be defending its common resources as much against neighboring farm
communities as against exploitation by outsiders are facts somewhat neglected in this view of the economic marginalization and subjection of
peasants.[13]

The functional links between land use and land tenure have also been
partially obscured in Geertz's (1963: 90–91) discussion of the communal
ownership systems under which the Javanese village as a corporate body
exercised various kinds of residual rights of control over fields. He suggests that the "collective apportionment procedures of traditional communal tenure" by which the village periodically rotated and redivided
plots among qualified families (Geertz 1963: 91) had been more recently
applied to the wet-rice sawah terraces that were requisitioned as blocks
by colonial sugarcane enterprises. Within the village, poverty was further
"shared" by granting communally held village rice lands to needy or
landless farm families. Although sharecropping and renting arrangements indeed distributed rights of land use to non-owners, recent scholarship makes clear the existence of firmly individualized private rights to
almost all high-value irrigated land. The high percentage of the landless
in many communities showed that many cultivators received no appor-

---

12. Wolf (1981) now feels that the cargo system did redistribute surpluses without leveling class differences. Indeed, economic support of ceremonies and public festivities might
strengthen authority within the community and reinforce inequality (Cancian 1965; Greenberg 1981). A somewhat narrower definition of the corporate community, and one that
makes no assumptions as to form of tenure or degree of internal equality, is that of Tom
Sheridan. He terms it "an organization of peasant households that controls certain basic
natural resources, and that preserves its corporate identity through time" (Sheridan 1988:
xxiii). It is primarily a community of interest rather than of place, and it may not be iso-
morphic with an actual geographic community.

13. The simple binary ideological opposition of communal and private rights was fostered by the nineteenth-century political conflict pitting highland Mesoamerican and Andean communities with sparsely populated, extensively used hinterlands against liberal governments trying "to disestablish Indian corporate jurisdiction over land in favor of private
property rights, to throw the privately owned plots on the market, and thus to open the
[lands] to colonization and seizure by non-resident outsiders" (Wolf 1981: 326). Modern
Mexican communities that eagerly espouse communal *ejido* rights in desert cattle range land
may resist government attempts to extend group tenure (and possible reallocation of use
rights) to irrigated alluvial bottom lands (Sheridan 1988).

tionment of collectively held village land. Rice lands labeled communal were in fact granted for personal use over long periods to village officials, who were the richest members of the community (W. Collier 1981; Gillian Hart 1986; Alexander and Alexander 1982). Though the corporate community may have asserted common rights in swidden and grazing lands, organized production for tax and rent revenues to the state, and administered irrigation, its control did not imply communal as opposed to household private property rights in intensively tilled lands.

Throughout our discussion of smallholder rights in resources, a continuing difficulty has been the tendency to binary distinction, focusing on essentially dichotomous variables. Tenure is not either private or communal; property does not parse neatly into open access, common, and private; groups are not either closed-corporate or open-atomistic. Rights in the same physical field may be partitioned among private owners, temporary cultivators, possessors of trees or buildings on the land, those with rights of easement to travel across the land, and a whole community permitted to graze their animals on the crop stubble. Where private property rights have great importance, as they do among smallholders, they can become legally complex and richly diversified. The several types of property use, holding, inheritance, transfer, and administration that are actively present, known, and enforced in a community of intensive cultivators (as opposed to the laws on the books and the official regulations of the state) represent a careful adjustment of social rules and practices to ecological facts.

We can move beyond the gross categories of property classification and corporate institutions by looking for regular variations in situations of controlled comparison. Among villages of upland south India in a single area with irrigated and dryland cultivation, similar crops and agricultural technology, common ethnicity, religion, and caste composition, Robert Wade (1988) has uncovered significant differences in institutionalized corporate activities. He examines the organization of open-field grazing and irrigation to illustrate the circumstances under which people solve collectively pressing problems of common-pool resources. Irrigation for paddy rice fields in a village near the end of a 20-mile canal is conducted by 12 common irrigators, who are appointed by the community council. After the farmers have transplanted the rice seedlings in their own fields, the irrigator crews distribute the water from the main canal, apply it to each paddy, help bring more water down the distributary, prevent higher villages from blocking off the water supply, and make minor repairs to field access roads. They are hired only when water becomes scarce and when farmers start to quarrel because of the tendency for top-end farmers closest to the water source to use water that should go to those with fields at the tail end of the block.[14] The work is not highly skilled, because it involves flooding the paddies rather than conducting water down multiple-field furrows, but its systematic performance by the hired labor teams saves the travel time farmers would otherwise expend in irrigating their own distant, dispersed fields. It also prevents loss of water to other, competing villages, reduces the wasting of water by managing the sequence and amount of irrigation better than individuals could (or, if left to their own devices, would), and provides a rotation schedule that consistently improves the water supply for tail-enders (Wade 1988: 77–79).

Rice is more subject to stress than other crops if soil moisture falls below the saturation point, but it is not sensitive to overwatering, and farmers who can secure more than their fair share of water may use it to retard weed growth and save on their own labor. Rice is also a highly dependable subsistence crop that needs little fertilizer. Wages for the common irrigators, which they themselves collect from individual farmers in proportion to the field sizes, are small in proportion to harvest value, and the benefits appear to justify the cost to farmers. Villages with more abundant water supplies toward the head end of the system grow more double-cropped rice and do so without common irrigators, apparently because there is less risk of crop-endangering water shortage and less competition among cultivators for an adequate water supply (Wade 1988: 161–65).

Tail-end villages with less irrigation water also grow more rain-fed crops of sorghum and cotton. When the sorghum is harvested, good livestock grazing is available on the fallow, but unfenced sesame and cotton, along with dry-season irrigated crops, are potentially subject to damage by the herds. Some 10,000 head of sheep and goats enter village lands at this time, and their herders are paid by individual farmers in return for folding the animals within a temporary fence for several nights on a field and thus manuring it. Half the fee for manuring goes to the village council, who use it to hire field guards. A civic institution thus allows for effective common grazing of otherwise unused plant growth in large, unfenced fallow areas, the manuring of individual farmers' fields, and the protection of the interspersed standing crops (Wade 1988: 60–68). The village council publicly reads the regulations governing common grazing to the assembled farmers and migrant herders every year. The field guards enforce the rules, taking straying animals to the village pound and collecting fines, some of which they divide among themselves. They also attempt to prevent crop thefts. Tail-end villages have a higher percentage

14. For what happens when theft, intimidation, and physical force replace corporate local organization and state bureaucratic administration in the allocation of irrigation, see a Sri Lankan case described by Berit Fladby (1983: 191–99).

of finely textured, deep, moisture-retentive soils than those in top-end communities that monocrop rice, and this means that there is a greater supply of fodder for a longer period after the rains. Dryland crops also require more manure than does irrigated rice. Without field guards, farmers could not secure both the benefits of large-scale manuring (what Wade refers to as "sheep-shit economics") and avoid damage to (and a great deal of conflict over) their standing crops.

With a population density of 159/km², which has almost doubled in the past 80 years (Wade 1988: 58), the village has higher risks of irrigation-water shortage and of damage by grazing livestock than top-end villages with more adequate and reliable water and a much smaller density of herd animals (Wade 1988: 184). Both the canal water and the stubble fallow grazing are common-pool resources, used jointly and with subtractive consumption. The institutions of common irrigators and field guards, administered by the corporate community, occur most frequently in villages in the bottom third of the irrigation system, where the ecological risks of crop stress from water scarcity, conflict over water, dry-crop manure needs, and livestock depredations are highest. "Any resource characterized by joint use and subtractive benefits is *potentially* subject to crowding, depletion and degradation," the so-called tragedy of the commons. But only "where joint use and subtractive benefits are coupled with scarcity, and where in consequence joint users start to interfere with each other's use do you have a commons dilemma. . . . Corporate organization is found only in villages where commons situations have become commons dilemmas" (Wade 1988: 184).

Common property and the costs of controlling and monitoring its use are an example of Boserup's more general theory of agricultural intensification.[15] For the Indian village, enclosure of the fallow or full privatization of irrigation water are not viable options. People will pay the costs of corporate organization and the administration of common-pool resources only when it becomes profitable and the risks for agricultural production of doing nothing become too high. Deliberately concerted, corporate action takes place only when net material benefits to be pro-

---

15. Margaret McKean cogently points out that population pressure and increasing agricultural intensification in Japan in the fourteenth and fifteenth centuries led to more frequent use of uncultivated hill lands for fertilizer, fodder, fuel, and building materials: "More systematic use of the commons increased the need to manage it well, define eligible users and uses, and exclude ineligible users and uses. Sound resource management required cooperation by all villagers, and became the impetus to solidary (and occasionally democratic!) self-government by village units. Thus the development of secure private property rights to arable lands simultaneously stimulated the use of commons, led to a richer and therefore more assertive peasantry organized into self-governing villages, and led to the assertion of village ownership of the commons" (McKean 1991: 4).

vided to all or most cultivators are high—when without it they would face continual collision and substantial risk of crop loss (Wade 1988: 186).

Does Land Have Its Price—And Should It?

Just as common property resources managed by indigenous corporate institutions are not a quaint holdover of socialistic traditional communities, so the market in individually owned land is not an imposition of modern capitalism that inevitably destroys smallholder society. When farmland has a money price and when rights to it can be transferred freely with permanent, legally binding alienation and acquisition, smallholders are often seen to be occupying qualitatively different roles in externally dominated economic systems. If land is a market commodity, farmers may be unable to purchase enough of this basic resource to provide reliable household subsistence, and what land they have may be lost to those with superior political power and influence. Ownership rights in fee simple and state systems of land registration are lauded by economists as providing the necessary security of tenure to promote investment, a sale value that allocates land to its most productive use, and a collateral value that increases the supply of credit (Binswanger and McIntire 1987; Feder and Noronha 1987; Shipton 1988). Others see great dangers as exchange values in land displace use values,[16] with commoditization bringing in its train land concentration, debt relationships and eviction, speculation at the expense of small farmers, and the polarization of rural society (Watts 1983, Downs and Reyna 1988, Shipton 1988). The views from both right and left suggest that land tenure is determined, for good or ill, by the action of forces *outside* of rural society. In fact, agricultural intensification in situations of population pressure increases the volume of land sales if other factors are held constant. With rising population density among the Nigerian Hausa, the proportion of land purchased and loaned has gone up, although ethnic group, farming system, religious law, capitalist market economy, and state legal codes remain the same (Table 6.2); frequent sale is an indication of land scarcity.

As the smallholder commercializes, he is increasingly threatened by competition for land and resulting disputes, though such conflict is also present among less market-oriented intensive cultivators (Netting 1972). Similarly a variety of means to transfer use rights through rent, sharecropping, and mortgaging, as well as ownership rights through sale, improve the opportunity to bring appropriate amounts of land together with labor and capital for increasing production per unit of land (Feder

16. Classical economists such as Smith and Ricardo are the last to insist on the distinction between use and exchange values, because at the margin, exchange and use values are the same (Louis De Alessi, pers. comm.).

TABLE 6.2

Hausa Population and Land Acquisition, 1964–68

	Two Zaria villages	Three Sokoto villages	Outlying Kano	Central Kano
Population/mi²	250	500	439	609
Purchased land	3.7%	22.1%	24%	31%
Loaned land	5.7%	10.4%	–	–
Other (gift, inheritance, allocation)	90.6%	67.5%	–	–

SOURCE: Goddard et al. 1975: 324, 332.

and Noronha 1987).[17] Paradoxically, government attempts to codify and enforce freehold tenure and register secure land titles may add to transaction costs in the individually tailored agreements for loan, rental, barter, swap, and sale that proliferate in Third World smallholder areas (Shipton 1988). An individualization[18] of property rights may be more appropriate and welcomed "in crowded rural areas where valuable cash crops have raised competition for land, where boundary disputes have become most dangerous, and where litigation has become most costly and time consuming" (ibid.: 122), but central government bureaucracies may have trouble implementing workable universal rules and procedures.

In a stable, intensively tilled area, land purchases, although perfectly legal, may be relatively rare because of high prices and the prevalence of long-term inheritance and temporary use rights. Sales when they do occur may be of individual fields, gardens, or meadows that are both small and dispersed. Notwithstanding that they may be eagerly sought by smallholders who are attempting to expand, such fragmented plots may not lend themselves to consolidation or enclosure, and the mere presence of a land market does not mean that household farmers will necessarily be dispossessed. Loans may also be sought without pledging land as collateral. Many peasants are notably reluctant to mortgage land, and they may only do so as a last resort in times of drought or sickness, when all other options of migrant labor, craft work, and sale of domestic animals have been exercised. Credit schemes financed by international development agencies have often been distinctly unpopular in the countryside. While land transfers in both formal and informal markets may be ex-

17. Studies in Latin America have shown positive correlations between the degree of ownership security and farm investment per unit of land (Feder and Noronha 1987: 160–61), and titled farmers have a higher incidence of permanent crops than untitled farmers.

18. *Privatization* of property rights might be a more appropriate term than Shipton's *individualization*. An individual owning property rights may choose to join others and form a cooperative or a corporation in which management decisions are not exercised by the individual (Louis De Alessi, pers. comm.).

tremely important to smallholder intensive agriculturalists, the possibility of holding title to land, selling it, and using it for collateral does not mean a consequent decline in the prevalence of small farming.

This picture of smallholders possessing significant rights of private property and participating in local institutions that democratically administer common property resources may seem to some to be at best a romantic half-truth. "Peasants" are, after all, defined by their subordinate status. Don't outside elites either own the land and extort crushing rents or claim political control that entitles them to collect confiscatory taxes or tithes?

It is apparent, of course, that many intensively cultivating smallholders are indeed sharecroppers or renters, although they may also have land of their own. Where a part of farm production is drawn off in lease payments, crop divisions, land taxes, or loan interest, we need, however, to inquire more particularly just what proportion of income is given up by the smallholder and what the regularity and enforcement of the demand is. As we see in the case of South China (Chapter 8), fixed rents may be low and a variety of rights, including subleasing, inheritance, and sale, may legally belong to the tenant.

How frequently are smallholder occupants dispossessed or progressively impoverished by their terms of tenure? As far as I know, we lack reliable comparative measures of the terms and costs of land acquisition in a range of smallholder communities. What is persuasive, however, is the intensive cultivator's need to receive some perceptible return on the skills, the management, and the heavy work that this type of farming requires. If an acceptable proportion of the surplus generated by intensive agriculture does not reward the farmer, and if everything beyond mere subsistence is appropriated in rent and taxes, the smallholder's incentives are compromised. Perhaps the most consistent goal of sharecroppers, tenants, and the holders of very small parcels of land is the acquisition of resources in land and livestock that will repay added expenditures of effort and provide the security of a productive farmstead. The degree of exploitation of farmers by landlords and the state cannot be presumed—it is a matter for case-by-case empirical determination.

This chapter contends that smallholder households have consistent, socially recognized rights in private property pertaining to permanent, scarce resources whose agricultural use has been intensified. Such rights include inheritance, temporary or permanent transfer, and administration. Smallholders also characteristically have shared, institutionalized, carefully bounded rights in common property. Communal and individual rights do not represent evolutionary stages of culture or political precipitates from larger legal or political entities. Rather they are grounded in land use and in the positive utility of intensive or extensive agricultural practices. Diverse and variable systems of tenure have evolved to meet the needs of specific groups of smallholders, and they form the crucial social institutions by which farm households relate to their environment, their neighbors, and the other members of their larger societies.

CHAPTER SEVEN

Inequality, Stratification, and Polarization

AT ANY POINT in time, smallholders practicing intensive cultivation and exercising heritable private property rights in land, livestock, and buildings display measurable differences in access to important resources within the community. Some households, as represented by their heads, will be better endowed or richer than others, and this inequality will be known, remarked on, and, within certain cultural limits, considered legitimate by the local society. Where the farmers are engaged in a market economy, and where they form part of a state that includes cities and nonagricultural occupations, the differences in wealth and power between peasant smallholders and both commercial and governmental elites on the one hand and landless wage laborers on the other may well be even more pronounced. Persisting differences in the relations of a whole group to the means of production may in this case merit designation as class differences.

The nature and function of inequality, both within agricultural communities and between farmers and others in the larger society, have frequently been obscured by the assumptions and deductive models of evolutionary theory and of various political ideologies. The widely accepted view that precapitalist social formations are characterized by communal labor organization, communal land tenure, and collective utilization of resources within an egalitarian community does not fit the facts of smallholder households of permanent cultivators in either stateless or state societies. Smallholding farmers everywhere almost always produce for exchange as well as use, take part in the market economy, use some wage labor, and possess private as well as common property rights. These practices are not solely imposed on them by political domination, external economic exploitation, or the capitalist world system. Densely settled rural societies never inhabited a self-sufficient, egalitarian Eden. The postulation of inevitable polarization between a few wealthy, large-scale landowners with hired labor and the mass of landless, impoverished rural proletarian workers is contradicted by the historic and contemporary evidence of smallholder continuity and vitality in areas of dense population and high levels of intensification. The inability of large estates, agribusinesses, and collectives to match the productivity and flexibility of household farms when land, energy, and capital become scarce further confirms the durability of a middle stratum of landholding small farmers.

The great tradition of dissent on the political left has always asserted the "natural" equality of all people and the inalienable rights with which they are endowed. Since Rousseau, this has been the basis for an attack on hierarchies of wealth and power, unearned income extracted from the labor of others, and the accumulation of property and capital in the hands of an idle, irresponsible elite. It has also prompted conjecture about an original human condition of equality in a state of nature, an evolutionarily earlier social type of noble savages or primitive communism whose fundamental values have been lost to civilization, slavery, feudalism, or capitalism.

Smallholder households do not, however, fit easily or neatly into such typologies. Intensive agriculture may, in fact, be seen as a mode of production distinctively uniting the forces of agricultural production—human labor, scarce land, and the methods and tools that create high, sustainable yields—and the *necessary* social relations of *such* production (*including* the family household as the unit of production and consumption, *and* the right to hold and transfer property). This mode also displays important functional consistencies, although its forces may be as different as hoes, ox plows, and garden tractors, and it may be interconnected or articulated in social formations as varied as those of kin-based, chiefly, state, and imperial societies (cf. Crummey and Stewart 1981).

Intensive cultivators in acephalous societies like those of the Kofyar and the Ifugao, autarchic peasant subsistence economies such as that of the alpine Swiss, or complex market-based state systems like the Chinese may all be said to embody Marx's natural economy. The production of use values predominates, although there is an exchange of surpluses at a basic level (Bernstein 1981). The natural economy is based on agriculture, complemented by domestic handicrafts and manufacturing, and "a very insignificant portion" of the product enters into the process of circulation (Marx 1967 [1894]: vol. 3, 786–87). Although intensive land use may begin with fulfilling the subsistence demands of household maintenance and reproduction, the dynamic of increasing production from limited resources is capable of producing a surplus, and smallholders are both willing and able to grow crops for the market. Low-energy agricultural and craft techniques are not so rudimentary that they cannot produce a surplus for sale, even in "Black Africa" (cf. Coquery-Vidrovitch 1972: 43). Commodity exchange is not absent (cf. Amin 1976), and the degree of market participation can be variable, with considerable changes to and from self-provisioning and specialized market production by specific households according to short-term political and economic conditions.

Because smallholders are not fully dependent on the market, and because they continue to produce largely by means of unpaid family labor rather than hired workers, they might be seen as fighting a conservative and ultimately fruitless rearguard action against capitalism, but that would ignore the economic and organizational adaptive features of the household enterprise (Chapter 2). Wage labor, supplementary occupations in trade and crafts, and the diversity of the part- and full-time earnings of household members from nonfarm sources do not put smallholders out of business; such strategies may, on the contrary, allow them to continue farming. Intensification by means of the household organization of production neither ends nor begins with capitalism:

> For Marx, the capitalist mode came into being when monetary wealth was enabled to buy labor power. . . . [but] As long as people can lay their hands on the means of production (tools, resources, land) and use these to supply their own sustenance—under whatever social arrangements—there is no compelling reason for them to sell their capacity to work to someone else. For labor power to be offered for sale, the tie between producers and the means of production has to be severed for good. Thus, holders of wealth must be able to acquire the means of production and deny access, except on their own terms, to all who want to operate them. (E. R. Wolf 1982: 77)

The wage-labor watershed, regardless of how accurately it reflects the transition from yeoman countryfolk to hired hands in the dark, satanic mills of the Industrial Revolution, does not critically encapsulate what happened to the farmer who kept one foot on the land, even when it provided only a part of the food and cash for other goods he needed. I shall suggest some of the reasons for this stubborn resistance to alienation from the means of production, as well as for the failure of the capitalist farmer to create viable rural farm-factory enterprises with wage labor. This denies neither the increasing stratification of society as a whole with economic development nor the permanent division of society into classes of surplus takers and surplus producers with the advent of state coercion in both the tributary and capitalist modes of production (E. R. Wolf 1982: 99). Rather, I am claiming that elements of autonomy in agricultural decision making, household labor mobilization, and land tenure protected the smallholder from some aspects of the assertion of unequal power and dominance over the landless and dependent wage laborer. *Within* the smallholder category or class, there *were* overt differences in rights to resources, but there was also mobility between generations and within individual life courses both up and down the local ladder of wealth and possessions.

Communal Equality and the Lineage Mode of Production

The antithesis of externally generated capitalist inequality in rural so-ciety is the model of the "lineage mode of production" put forward prin-cipally by French neo-Marxist anthropologists and sociologists using West African ethnographic materials (Terray 1972; Meillassoux 1981; Co-query-Vidrovitch 1975; Rey 1975). It postulates an undifferentiated, tech-nologically primitive mode of hoe cultivation, with crop production for use, group labor, shared consumption, and resources held in common. Meillassoux (1981: 3) links the concept to Marx and Engels's characteri-zation of the agricultural domestic community as "composed of individ-uals who (a) practice self-sustaining agriculture, (b) produce and con-sume together on common land, access to which is subordinated to membership of the community, and (c) are linked together by unequal ties of personal dependence. Within this community only use values oc-cur." The cooperating group or community is a unilineal descent group headed by an elder male and consisting of his wives and children and younger adult males (cadets), some of whom may themselves be married and have children, but who live as dependents within their father's com-pound. It is unclear whether the large, multiple-family household occu-pies a single residential compound or whether several clustered house-holds may unite as a lineage for most productive and distributive activities.

Though the local polity need have no chiefly or other superordinate authority, the lineage elder, with his greater experience and knowledge of local farming techniques, decides on the area to be farmed, schedules the activities of the farming year, mobilizes the workers from the kin group, and supervises the operations. He also oversees the storage of the harvest in his own granaries and doles out the grain to the constituent, small fam-ily cooking groups. Elders own iron tools, control the livestock, which is communal rather than private property, and dominate the trade in guns, kola, and other prestige goods (Terray 1972). Given the simplicity of agricultural methods and the usual lack of surplus, the elders had weak functional authority in agriculture (Meillassoux 1981; Geschiere 1985). They did, however, assert control over the processes of reproduction by assembling, negotiating, and receiving bridewealth, thus governing the circulation of nubile women from one lineage to another. The basic in-equality in lineage societies was between elders and cadets, women, and children, whose surplus product, along with that of slaves and clients who had been integrated in the lineage, was appropriated by the elders (Jewsiewicki 1981). Exchanges of elite goods outside the group, warfare, and the rites of ancestor worship further strengthened the elders' power

(Meillassoux 1978). Although the relations of production—what Donald Donham (1990) calls "productive inequalities"—are apparent in the structure of the lineage, they do not coincide with separate, landholding households or with labor and food distribution limits at a lower level than the lineage commune.

Something approximating a lineage mode of production may be found among groups of shifting cultivators of the savanna with dispersed, long-fallowed cereal fields on land that is sufficiently abundant for short-term usufruct to be readily available to all. Large work teams can accomplish the limited clearing and cultivation required with speed and promptness, and the distribution of food from a communal granary can mitigate the environmental fluctuations and demographic variations to which the small independent household is subject. Mahir Saul (1989) describes matrilineage rights of permanent access to tracts of land among the southern Bobo of Burkina Faso. Goats and other forms of wealth, including currency, were managed by lineage elders, and individuals were not allowed to own them. There is, however, a great deal of variation among lineage-based societies of shifting cultivators. Women often operate as quite autonomous cultivators, as they do among the Maka of Cameroun, and neither corporate descent groups nor elders organize production or store and distribute subsistence goods (Geschiere 1985). In Gbaya society, based on the swidden cultivation of cassava and maize in central Cameroun, there is almost no corporate control either of labor or of land and other means of production, and farming is more individualistic than it is among smallholder households:

No corporate structures or other mechanisms exist in Gbaya society through which certain men may gain significantly at the expense of others via control over labour supplies. Older men, richer men, or village head men or canton chiefs have no effective means of exploiting the labour of younger, less powerful men, and relations between elders and juniors in particular tend to be more egalitarian than gerontocratic. . . . In short, no man commands another's labour. . . . Individual workers, both men and women, work on their own tasks and are rewarded by ownership of the resultant product. . . . In maize farming, if children and mother help father with the farm, their various contributions to the work are explicitly recognized by an allocation of profits proportional to work contributed or by gifts of clothes at the end of the season. (Burnham 1980: 167–68)

Gbaya lineages do not control land or cattle, and food and other property are distributed personally among a wide circle of co-resident kin and neighbors. The particular forms of authority, internal inequality, and corporate control of labor, land, and stored food associated with the lineage mode of production do not occur regularly among shifting cultivators, and they are entirely absent where intensive agriculture is carried on by household labor on smallholdings.

Balanced Exchange and Unequal Units

Suprahousehold unilineal lineages or bilateral descent groups may indeed exist among intensive cultivators, and they may perform important corporate functions in regulating and organizing marriage, settling internal disputes, and handling external conflict by legal negotiation or armed combat. The kin group may be a political body and a religious congregation. But the constituent smallholder households do not produce or consume in common, and they do not redistribute their crops, livestock, or property to approximate some social ideal of equality. This does not imply either any lack of interhousehold cooperation or the presence of some atomistic, laissez-faire household independence. Indeed, smallholder communities, where labor demands are often heavy, where seasonal bottlenecks may be severe, and where certain tasks and projects may far exceed the capacity of any household work force, are rife with examples of voluntary labor groups. But these socially recognized forms of labor exchange are what Marshall Sahlins has called balanced rather than generalized reciprocity, and they produce goods and services that are by and large consumed or retained by individual households. Individuals or households who perform labor for others do not do so in an altruistic fashion, where "the material side of the transaction is repressed by the social," where the counter obligation "is not stipulated by time, quantity, or quality," or where "the expectation of reciprocity is indefinite" (Sahlins 1972: 194). Rather, the exchange transaction stipulates "returns of commensurate worth or utility within a finite and narrow period and is subject to a more or less precise reckoning" (ibid.: 195). Although exchanges *within* the family household may emphasize sharing, and occasionally sustained, one-way flows of resources, relations among households of smallholders are usually much more consciously and calculatedly balanced.

Kofyar work groups play a major role in certain agricultural and domestic tasks, but people maintain a conscious balance between labor given and received. Many individuals belong to exchange labor groups or clubs, called *wuk*, which work on the household or individual fields of each member in turn. These groups of 8 to 20 people keep careful track of work received and owed, and if they work for cash, the money is distributed equally or pooled for group use. This type of balanced exchange accounts for 4.2 percent of Kofyar farm work (Netting et al. 1989: 307). The largest ad hoc labor groups are the festive "farming for beer" (*mar muos*) parties of 30 to 80 men, women, and adolescent children from the same village or plains cash-cropping neighborhood. They assemble on a prearranged day, work as a group at a particular job, and are entertained by the host household with millet beer when the work has been completed. The heavy work of making ridges for millet and sorghum crops or yam mounding is done so as to complete an entire field at one time, rather than over a period of weeks, as would be required by two or three workers. Bundling and storing millet on a raised wooden rack under a protective thatch must be done rapidly during a lull in the rains after the millet harvest in July (G. D. Stone et al. 1990). While women gather thatching grass and make it into mats, men twist palm-leaf ropes, tie the bundles, toss them onto the rack, and cover them. The work party may go to several homesteads in one day.

Though both males and females participate in hoeing, their labor may also be complementary, as in millet storage and in beer brewing, where neighborhood women help those in the host household to grind, cook, and ferment up to 400 pounds of millet for a work party. The group labor parties spur individual effort with drummers in the field, public competition in rapid hoeing, especially among young men, and the nutritious, mildly alcoholic beer, which is provided at the rate of over a gallon per person. The beer is considered a necessary and sufficient return for the work, and if because of sickness the host cannot furnish it to the workers, the obligation must be made up before the next farming season begins. There is also an expectation that every household that has a work party will send at least one of its own members to the "fields of beer" (usually one or two per farming season) of its neighbors. Absences are noted and publicly discussed. If a household does not maintain its labor reciprocity, it is made to pay a fine in beer, and repeated failures to send a worker are punished with shunning and social ostracism.

Beer-party work on household or individual fields amounts to 8.2 percent of all agrarian labor among Kofyar cash-croppers, and another 7.6 percent goes into cooperative beer-brewing activities (Netting et al. 1989). In a sample of 799 households, two-thirds had used beer-party labor on at least one occasion in the preceding year, but larger households with bigger fields usually had greater numbers of workers and proportionately more beer. Somewhat over a third of all households also hired labor for cash, but this represented only 1 percent of all work days expended (G. D. Stone et al. 1990), and it was concentrated in richer households, which also held more beer parties. Though voluntary labor groups without cash payments provided social rewards as well as the immediate material inducement of abundant beer to participants, and though they also carried an obligation to participate in work for neighbors, they were not examples of generalized reciprocity and they did not operate to equalize access to labor by all households. Differences in Kofyar household and farm size, number of beer-party workers, and amount of paid labor contributed to pronounced differences in marketed production and cash income.

The stated preference of Kofyar householders for beer-party or exchange labor as opposed to wage workers is based on economic rather than social or redistributive criteria. The cash costs of paid laborers at approximately ₦5.41 plus food per day are much higher than that of beer, which would only be ₦1.30, even if the millet were purchased. Furthermore, hired workers are thought to be demanding, undependable, and less careful in agricultural operations than neighbors, and because they are migrants, there are not enough of them available during seasonal bottleneck periods to complete the crucial tasks (Netting et al. 1989). A decrease in relative labor costs owing to a larger supply of migrants (Saul 1983) or the presence of landless Kofyar who relied on wage work might well reduce the frequency of exchange labor.

Collective and exchange labor in peasant communities seems also to follow principles of balanced reciprocity, being mobilized for tasks that exceed the capacity of the household, receiving the immediate reward of festive food and/or drink, carrying an obligation of equivalent labor return, and resulting in products that belong entirely to the host household. A statute in the sixteenth-century Törbel charter required all adult males to help in the building of a dwelling house by any member of the community. Barn raising is an obligatory communal task among contemporary Amish smallholders. Grain threshing, with its simultaneous needs for hauling in the sheaves, threshing with hand flails, draft animals, or powered machines, winnowing, and bagging or storage has often been a group task.

Work days or cash contributions in lieu of labor to the establishment and maintenance of common property resources are usually carefully calibrated to the use made by each member of the corporately owned good. Labor on the Törbel alp is proportionate to the number of cows pastured there, and the money to pay the alp staff is assessed on the amount of cheese that each owner received. The tasks of cleaning irrigation ditches and monitoring water flow are generally assigned on the basis of amount of land or water flow to which each member of the local group is entitled. All these methods recognize potential inequality of property and corresponding duties. There may be a small amount of agricultural activity that is truly communal. Törbel has traditionally kept up a community vineyard (*Gemeindereben*), for which each household was responsible to provide a pack basket of manure and several labor days at cultivation, pruning, and harvest. The wine was dispensed for civic purposes, such as at the two annual feasts of St. Stephan's Day and Corpus Christi, when each adult male citizen was given a precisely measured equal number of cups; at the exercises of the fire brigade; at parades, when band and choir members were given a drink; and at the meetings of the village council. Wine for individual burial feasts could be purchased from the community cellar. Cooperative production and communal distribution in equal shares to all group members are practiced in many smallholder communities, but they tend to be insignificant in proportion to differentiated household activities, and they do little to mitigate or level economic differences among households.

All Smallholders Are Not Created Equal . . . Or Immobile

The paradox of smallholder inequality is that, viewed from within the local community, it is both economically inevitable and constrained. Households at any point in time will have significantly different endowments of land, livestock, buildings, tools, stored agricultural products, and other forms of wealth. The community is thus economically stratified to the extent that there is an unequal distribution of socially valued goods among its member households. But the operation of the internal economic and social system impedes the concentration of resources and accumulation of wealth in the hands of a permanent upper stratum, while promoting mobility both up and down the local ladder of relative prosperity. Among intensive cultivators, such a *system of inequality with mobility* may be modified or even destroyed by external forces that appropriate revenue, land, and labor while denying a fair return to the primary producers, and there is no doubt that military, governmental, or urban mercantile elites have far greater access to wealth and coercive power, which allows them to do just that. It is also clear that the exercise of naked force in the service of exploitation can seriously damage rural society, reduce population, and cut production per unit of land. Insofar as smallholders can retreat to subsistence and survive the attacks of avaricious landlords or violent warlords, they will attempt to resist such subjugation.

The combination of scarce but highly productive resources with permanent use and investment and with transferable, heritable property rights means that smallholder farms will be inherently unequal. The land resources themselves are subject to microenvironmental differences in soil moisture, slope and insolation, resulting in predictable (over the long but not necessarily the short run) differences in yield and use. The possibilities of improvement also differ, according to, for instance, distance over which manure must be hauled from the farmstead or situation at the head versus the tail end of an irrigation network. Household demographic processes, dividing land up among a number of siblings, concentrating it in the possession of a single heir, or increasing it with a fortunate marriage, can reduce or increase a particular homestead in each generation. The developmental cycle of a single family involves changes in labor availability and dependency that can greatly affect farm productivity. To these may be added the differential effects of market prices on inputs and outputs and exactions for taxes and rent from outside the community. Given these conditions of the household farm enterprise, it would be remarkable if either rough equality or long-term stability in the distribution of resources among households could be maintained.

One of our most durable preconceptions about stateless societies with no wage labor and with only peripheral involvement in a market economy is that they are fundamentally egalitarian (Hirschman 1981; Hyden 1980; Rostow 1960; Sahlins 1960). When we examine acephalous groups of smallholders with intensive cultivation and household property rights, we in fact find measurable differences in access to resources. The highly valued pond-field rice terraces of the Ifugao, along with private forests, house sites, domestic animals, including expensive water buffalo for sacrificial purposes, and heirloom wealth items are not distributed equally among households. Ifugao accept the presence of inequality as an ecological and social fact, while acknowledging that individuals can change their own wealth in land and goods:

Certain general principles underlie Ifugao interpretation of environment, culture, and society as they impinge on the everyday workings of this flexible montane tropical agricultural system. First it is generally assumed that all resources and units of time or space are distributed unequally. Size of individual holdings, concentration of district land forms, and access to goods and services vary tremendously. . . . Second, most such differences are ranked. Third, many ranked statuses of person, property, and natural phenomena are significantly modifiable by intentional human activity. And fourth, such modification is most effectively achieved by skilled long-range calculation and competitive action. (Conklin 1980: 36)

Although the population densities of 36–200+ /km² of arable land found among Kofyar hill farmers are somewhat lower than the Ifugao range of 165–360/km² (Conklin 1980: 6), household differences in the production of cereal grain and in domestic animal ownership are also evident in this case. Table 7.1 suggests that in four Kofyar regions, containing a total of 11 villages, population density and household size were directly related, while average staple grain production and index of domestic animals per household tended to decline as population density increased. It appeared that greater land availability in the valley and Bong areas allowed people to establish new household farms more easily and produce from them more crops and livestock per household and considerably more per capita than in more crowded neighboring areas. The Gini index is a measure of relative inequality for any variable, such that a Gini of 0 indicates a perfectly equal distribution and a Gini of 1.00 indi-

TABLE 7.1
Kofyar Population, Households, and Inequality, 1961

Region[a]	Population /km²	Households[b]	Mean household size	Crop index[c]	Crop Gini	Animal index[d]	Animal Gini
Plains	200+	103	5.42	18.25	.414	7.41	.479
Hilltop	92	60	4.78	18.43	.341	12.88	.425
Bong	36	76	4.05	21.22	.403	8.43	.435
Valley	37	109	3.47	21.76	.271	10.56	.393

SOURCE: G. D. Stone et al. 1984: 95, 101.

[a]Within each region, population density figures reflect only arable land as measured from aerial photos.

[b]Households of traditional nonmigrant farmers.

[c]Crop index based on bundles of millet and sorghum harvested from both homestead and local bush farms.

[d]Animal index computed as goats + sheep + pigs + (cows × 5) + (horses × 5).

cates a total concentration of wealth in a single case (Blau 1977; Allison 1978). For both crops and domestic animals (formerly a major mark of wealth among Kofyar), the Gini indices show the most equal distributions of agricultural goods among the small valley households with abundant land resources, whereas the greatest inequality is present in the large land-scarce households of the plains (G. D. Stone et al. 1984).

Plains farmers from the crowded communities of Kwa and Kwang at the base of the Jos Plateau escarpment were among the first to migrate to vacant lands south of Namu and begin production of cash crops. Entry into the market economy might be expected to produce growing inequality, but with land readily available on the frontier, cash incomes did not contribute to rapid differentiation. The Gini indices on migrant farmer money returns on crop sales were .422 for 1961 and .479 for 1966, little different than the inequality of nonmigrant plains subsistence farmers (G. D. Stone et al. 1984: 102). Moreover, the longer farmers participated in the cash economy, the less unequal their cash incomes became, going down to a Gini of .392 for those who had spent seven or more years producing for the market (ibid.: 104). Though comparable data are rare, we may speculate that smallholders, even in the absence of strong market influences, will display appreciable economic inequality, that this may be exacerbated as the ratio of population to resources increases, and that participation in a cash economy will not necessarily increase the degree of differentiation.

Do peasants who have produced cash crops and made money payments for centuries show greater intracommunity inequality and more permanent wealth stratification? On the evidence of rural Hausaland and Karnataka (in south India), Polly Hill (1982: 55) claims that "laissez-faire rural economies, where most interhousehold transactions and services involve cash, and where land resembles a saleable commodity, are necessarily innately inegalitarian; such rural economies invariably operate in a way which tends to favour richer households at the expense of poorer." The implication is that inequality is an inevitable outcome of the market and that it introduces injustice, exploitation, and poverty within the village community.

Longitudinal data from annual household farm tax valuations in Törbel over the period 1851–1915 show the persistence of unequal distribution of landed property, but the Gini coefficients for all village individual taxpayers are not stable over time, varying from a low of .340 to a high of .495 (McGuire and Netting 1982: 273). Because some households included more than one (possibly celibate) adult tax-paying property owner, the Gini index for households ranged from .360 to .424. Though the cantonal tax lists do not include measures of nonagricultural income or livestock ownership, we may presume that local smallholder enterprises based on access to agricultural land and buildings continued to correspond closely with relative household wealth.

Perhaps the most striking characteristic of inequality in the Swiss case is that it existed within limits—there were no landless citizens of Törbel, and neither were there any farmers with so much property that they could live without working the land. The distribution of milk cows was similar, with practically every household represented; the average number of cows per owner varied only from 2.00 to 2.50 in the years 1844–1969, and the largest single herd in 1844 totaled 7 animals (Netting 1981: 26–27). Differentiation must always have been clearly evident, with 10 percent of all taxpayers controlling 30 percent of the wealth and the lower 50 percent of taxpayers having only 18 to 20 percent of assessed property (McGuire and Netting 1982: 273). Since subsistence farming was necessary to household support, and Törbel residents were citizens with long lines of descent within the community, they characteristically possessed inherited land as well as common property rights in the alp and the forest. Without the security of a diversified, viable peasant holding and nontransferable citizenship rights in local resources, one could not survive, form a family, or participate in village social life. Over the long term, low returns were balanced by low risks. Although land could be bought and sold, its historically very high price in relation to its annual return in hay, rye, or grapes made it a poor investment in strictly financial terms. Little land entered the market, many residents competed for the odd few hundred square meters that were sold, and loans from the corporate community or from locally endowed religious foundations to purchase farmland might take half a lifetime to repay. The viable smallholding accumulated gradually from various pieces inherited by a husband and wife, purchases from siblings who migrated, and the rare auction represented a small but secure basis for founding and supporting a household, but the same property converted into cash could perhaps have bought a larger farm or a town business elsewhere in Switzerland.

Little real surplus could be produced from the alpine meadows, and it is apparent that the medieval petty nobility who had more extensive land-holdings in some mountain villages almost invariably divested themselves of these estates and took up the organization of mercenary military service or trading expeditions, which offered higher cash returns. Just as it is expensive in cash and transaction costs to accumulate land, and almost impossible to acquire a consolidated holding among the dispersed plots of peasant proprietors, so the skilled, diversified labor of intensive cultivation, with its low marginal returns, is unlikely to be supplied by a large landholder who must hire and supervise workers. Whereas big farms with extensive grazing or cereal monocropping lend themselves to this type of management, intensive cultivation on high-cost land makes the acquisition of fields that are beyond the capacity of the household to farm a poor business proposition. The extremes of wealthy estate ownership and a landless rural proletariat go against the economic grain of small-scale mountain agriculture.

By looking at land ownership over time in Törbel at a finer level of resolution, we can see not only the presence of economically rational bounded inequality but also the mobility of individuals and their households up and down the scale of local prosperity. If inheritance were especially significant in the transmission of wealth, we would expect richer farmers to transmit larger amounts of property to their children. But a comparison of parental household property with that of the children's households shows that parents' wealth accounted for less than 4 percent of the variance in the child's wealth (McGuire and Netting 1982: 276). Stories were told of only children who benefited from the undivided estate of both their parents, but even factoring in the size of the sibling set of heirs did not make for more accurate estimates of their eventual landed worth. It appears that parental endowment was so much influenced by its timing in the life course, the demographic and social characteristics of its division among siblings, and the individual enterprise and luck of the inheritor that simple predictions were impossible. The average two-hectare Törbel holding was composed of some eight meadows, three pastures, six grainfields, and three gardens (Netting 1981: 17), and it might have been inherited in parts on the death or retirement of a number of relatives over a long period of time. Yet given the rules and practice of partible inheritance, siblings would seem to be good candidates for equality with one another. Even though the evidence of equitable shares going to all children is strong, siblings rapidly diverge in wealth. Some may buy certain plots from each other almost at once, as well as purchasing additional land with their own off-farm earnings. Their spouses may be differentially endowed with land, and the size and conscientiousness of their household labor pools may be varied. As individuals, they may display different levels of work effort, management skills, and business acumen.

Even in the nineteenth century, there were a few nonagricultural jobs such as cattle trading and school teaching in Törbel, and men increasingly took seasonal employment as mule drivers for the developing tourist trade of the Vispertal, as herders and cheesemakers on the alps of other communities, and as laborers on railroad, tunnel-building, and flood-control projects (Netting 1981: 55, 104). This income might be differentially invested in farm property. In any event, siblings often came to occupy very different levels in the village wealth hierarchy. Cases of adversity, such as the death of a man with a young family while he was still in debt, might lead to his children hiring out as shepherds while his own brother acquired the land at a low price but did not provide for the widow. Neither wealth nor misfortune is necessarily shared among smallholders, even when they are close kin. Personal religiously sanctioned charity or community welfare institutions such as Törbel's provision for cooperatively feeding and housing its elderly indigent members (McGuire and Netting 1982) may mitigate individual hardship, but smallholders do not appear to level obvious inequalities in household wealth and property purposely.

Personal Factors in Smallholders' Economic Mobility

The scarcity of land and other resources in smallholder societies, reflected in private property and the existence of a market for land and labor, establishes a long-term competitive situation and the possibility of household mobility, either up or down the local economic scale. Inequality exists, and processes of differentiation seem also to separate co-villagers, neighbors, and even close kin, even when individual households are originally endowed with similar wealth, property, education, farming experience, and economic opportunities. When members of farming communities talk privately about each other, mixing shrewd appraisal, gossip, and humorous or tragic anecdotes that may span several generations of a family, they are apt to account for relative success or failure in moral terms. Above all else, Törbel farmers valued hard work and frugality among their peers, and they saw clever trading, planning for the future, self-denial, and efficient household management as traits that might enable even the poorest to increase their holdings of land and cattle, rise in esteem, and occupy village political posts. Such accounts may be tinged with envy or rivalry, and people were aware that intelligence could be employed in selfish, underhanded dealing or seizing an unfair advantage. Nevertheless, the most frequent reasons given for the declining fortunes of an individual were laziness, drunkenness, and being a spendthrift. The smallholder virtues and vices may have a universal ring, but like the Protestant ethic of dedicated work, individual responsibility, and deferred rewards, they are peculiarly appropriate to the tasks of intensive agriculture in an autonomous farm household.

Jeff Bentley's description of the rise of a poor farmer in a northern Portuguese community of smallholders illustrates these themes and suggests the frequent combination of business enterprises with agriculture:

One man in Penabranca was often cited as *the* best example of success through hard work. He was known as *O Grilo* (The Cricket). The Grilo was born in 1919 and married in 1944. His wife and he were both from small farm families. Between the two of them they inherited only half a hectare of land, a house and a little forest parcel; but by the time I knew him in 1983–84 he had nearly 4 hectares of fields and over 2 hectares of forest, in 20 different parcels, as well as 9 dairy cows and a thriving business making and selling *bagaceira*, the local brandy distilled from grape skins.

One villager explained that the Grilo started his socioeconomic climb with a nanny goat. The goat got pregnant and had two kids. The Grilo sold the three goats at a fair and bought a calf. He raised the calf and when it produced another calf he raised it and trained them both to work as a team. He did custom plowing and other work with the cows, saving the money. While most Minhoto peasants are tremendously hard workers, the Grilo was outstanding even by local standards. He sold firewood in the regional capital of Braga. They say he would load his cart with firewood in the daytime, then get the team to pull it the 15 kilometers to Braga while he slept on the load, so that he could work even as he slept.

In addition to his hard work, the Grilo's success was based on his legendary negotiating skills, which earned respect and resentment in the community. The Grilo's little house garden was next to his next-door neighbor's garden, so that the Grilo had a right-of-way through the neighbor's plot. The neighbor also had a field two kilometers away that was three or four times the size of the Grilo's garden. The Grilo convinced him to swap the field for the Grilo's garden, pointing out that he would not only be able to enlarge his garden with the addition of the Grilo's but that he would now be able to plow up the Grilo's right-of-way as well.

The Grilo invested all his money in production: land, distilleries and a sawmill (which he later sold). It wasn't until 1981 that he used some of his savings to repair the old house. (Jeffery Bentley, pers. comm.; cf. 1992: 9–10)

Farmers in small, corporate communities often appear to deny inequality, speaking of their relative poverty in comparison with city people and proclaiming, "Here we are all the same" (Bailey 1971). Widely ramifying kinship and life-long acquaintance with fellow villagers, along with community institutions for common property use and redistributing wealth are often seen as promoting an egalitarian structure. Even the market economy can be seen as less impersonal in a Bangladesh village where 24 percent of land transactions are between brothers, 29 percent are in the same patrilineal descent group, 27 percent are within the village, and only 21 percent are with outsiders (Jansen 1987: 123). Sales on the open market could bring in 30–40 percent more than sales to relatives, but the generalized mutual economic, social, and political support existing among kin is not willingly forgone (ibid.: 124). Debt dependence on local rich men also establishes many-stranded sharecropping and patron–client relationships. But household histories reveal diverging economic trajectories, including unequal class and welfare status, even among brothers.

Eirik Jansen (1987: 113–36) recounts the chronicle of two Bangladeshi brothers, Eynuddin and Joinuddin (henceforth E. and J.), who set up separate households in 1965. E., who was six years older than J., had two sons, aged 14 and 16, and a daughter aged 8, and he had a reputation in the village as being the more clever and hardworking of the brothers. The younger brother, J., had a boy of 2 and daughters of 4 and 6. Each brother inherited 0.2 ha of farmland, and they split the homestead site of 0.16 ha. But E. continued to sharecrop an additional 0.51 ha they had formerly used jointly, and he had a pair of cows for plowing, whereas J. had a cow and calf. J. produced only enough rice for three or four months, and he was reluctant to take up wage labor, which his father and grandfather had never done. He went into debt to finance his former standard of living, including gifts to beggars, and finally borrowed from E., who took a mortgage on part of J.'s land, which the latter then received back to sharecrop. J. finally entered the labor market, but he had to borrow more money to finance his daughters' dowries and marriage ceremonies. After 15 years, J. possessed 0.04 ha, which was sharecropped out to his brother, he had sold his cattle and sheet-iron roof, and his son would be a landless laborer. E. in the meantime had educated a son as a police constable, who sent money home frequently. E. acquired land until he owned 0.81 ha, held 0.4 ha as mortgagee, and sharecropped 1.01 ha with the help of his sons and some hired labor at peak seasons. He had given loans to his brother and others. By 1980 he owned four cattle and two plows and had a prosperous household of 13 members. Though E. had begun with access to somewhat more land and a more favorable household dependency ratio than his brother (Jansen 1987: 113), the difference in their economic statuses had been significantly widened by their management activities over time.

Differentiation as an ongoing process among smallholders in the same community points up the importance of individual differences in capability and management, both in regards to the agricultural activities and nonfarm enterprises that figure so strongly in diversified local economies. Hans Ruthenberg (1968: 328–29) has pointed out that differences in Tanzanian smallholders' gross cash returns are poorly correlated with differences in acreage planted and in available household labor supplies. Only about one-third of the variation in Tanzanian Sukumaland cotton production or Ukara Island intensive cassava cropping is predicted by land and labor, suggesting that the unmeasured "differing entrepreneurial qualities of the farmers" are essential factors. Specialized knowledge of one's own soil and water resources, skill in timing agricultural operations with seasonal changes in climate, choices of crop mixes, attention to livestock husbandry, mobilization and complementarity of household labor, and astute marketing decisions are abilities that are not equally distributed among householders. The cumulative result of a great many daily choices and calculations shows up in the economic performance of the small farm and the comparative income of the family.

Peter Matlon's (1981) study of income disparity in Nigerian Hausa farming households showed that factor endowments of land and family labor power had a relatively minor effect, and off-farm enterprises contributed somewhat more. The potentially critical factor in explaining differential income was resource productivity as affected by management. There were important differences in land preparation methods, the timing and manner of planting, local seed varieties selected, density of intercropping, rotation practices, and the timing and intensity of weeding (Matlon 1981: 358). In the three income classes, there was a 49 percent difference in production per ha between the high and low groups. High-income farmers applied 21 percent more labor per ha and 27 percent more fertilizer. Productivity declined rapidly with a delay in planting sorghum and millet as well as a delay in successive weedings. Poorer farmers had lower-quality soils and a higher proportion of rented land; and they also applied less manure. It is striking that the agricultural practices most regularly associated with intensification and higher, more permanent, crop yields are those that contribute to the higher incomes of more successful farmers.

Economic mobility and an inequality linked to differential achievement become visible where recent settlement of an arable area with the potential for intensive production occurs along with an expanding market. James Eder has traced the careers of migrant swidden farmers from Cuyo Island who moved to Palawan in the Philippines in the 1930s and 1940s, and who produced fruit, vegetables, and livestock for sale in the marketplace of Puerto Principesa City. The pioneer men and women farmers did not differ significantly from one another in age, education, previous occupation, or start-up capital. They could acquire land and begin to produce subsistence crops with only minor cash costs, and they were not dependent on loans or extension services to embark on their truck gardening (Eder 1982: 195). With low transport costs, a nearby market, and a demand for their perishable fruit and vegetables, they were able to sell a

TABLE 7.2
Philippine Smallholder Property Ownership
(Pesos)

	Amount			
	Upper group	Middle group	Lower group	Mean
Productive property				
Land	25,974	7,137	7,091	10,698
Agricultural and other equipment	1,585	746	251	713
Livestock	2,245	991	456	1,020
Bank savings and investments	2,226	260	0	532
Subtotal	32,030	9,134	7,798	12,963
Nonproductive property	5,422	2,007	619	2,116
Grand Total	37,452	11,141	8,417	15,078[a]
Household income				
Cash	6,499	2,526	1,278	2,824
Subsistence	1,556	1,019	1,070	1,141
Total	8,055	3,615	2,346	3,965[b]

SOURCE: Eder 1982: 20, table 2.2.
[a]Equivalent to U.S. $2,250 per household, or U.S. $418 per consumption unit.
[b]Equivalent to U.S. $592 per household, or U.S. $110 per consumption unit.

variety of crops at attractive prices. It was possible for independent operators to establish smallholdings, to enter trade, service, or artisanship occupations, or to work for wages. Large areas were not needed for their labor-intensive cultivation, and, though there were rising land prices after 1960, it was possible to earn a livelihood from a relatively small, permanently tilled plot. Under these circumstances, there was a marked divergence in landholdings, incomes, security, consumption patterns, and lifestyles in the community. Households were ranked in three groups by community members and independently on property and income by the investigator (ibid.: 20–21). Household subsistence income showed little variation among the three groups, but property ownership, especially in land, livestock, bank savings, and houses, along with cash income, decisively differentiated the upper from the middle and lower groups. (See Table 7.2.)

   This pronounced economic inequality is credited to differences in competence, personality, and motivation, highly adaptive personal traits that may be explained by "diverse and ultimately fortuitous learning experiences during childhood, adolescence, and early adulthood" (Eder 1982: 4). In the life histories of successful and unsuccessful migrants, Eder (ibid.: 180) was able to identify birth order and early home environment, occupational and geographical mobility, and marriage relationships as the principal life experiences that were associated with these personal attributes. "Some individuals appeared to find, by virtue of their personali-

ties, the individualizing requirements of successful market gardening—
the relative isolation from other workers, the repetitiveness of tasks, the
attention to detail—particularly congenial" (ibid.: 187). The work, the
application, the planning and management of the smallholder household,
both on and off the farm, may be intimately related to the success of the
enterprise and to the family's position in an emergent structure of inequality.

Richard Wilk provocatively notes different strategies of household
management by Kekchi Maya farmers, some of whom pooled money
and labor in productive investments such as beekeeping or pig raising,
while others did not. These differences could not be reduced to individual
personalities, exposure to external norms, or articulation to capitalist
forms of production. Rather, it appeared that patriarchally managed
households failed to provide sufficient incentives to members, while
those with a more participatory form of family budgeting could plan and
act together, with members willingly contributing more time, effort, and
attention to the project (Wilk 1989: 43). Successful smallholder households often coordinate subsistence production and cash income generation through informed task sharing based on mutual interest. De facto as
opposed to de jure hierarchy in the household may be bad for its corporate economic health.

Differentiation in Java

What may appear to be a limited system of relatively benign internal
stratification with opportunities for mobility in somewhat isolated and
economically autonomous communities like those of the Kofyar and the
Törbel Swiss can always be called unrepresentative, marginal, and not
subject to the severe population and market pressures felt by other smallholders. No technology of intensification can maintain reliable subsistence production when farm sizes fall below a certain level, and in the
heightened competition for land, no social system can protect some of
the population from landlessness. What then is the degree of inequality in
areas of very high rural population density? How is it represented in land
ownership, land use, and household income from both agricultural and
nonfarm occupations? Is there mobility within the smallholder community, and under what circumstances does movement up and down the
local economic scale take place? Is the process of differentiation one of
polarizing landowners with increasing access to property and wealth and
a class of landless wage workers, or is it one of stratification, with many
closely ranked gradations of income and considerable mobility?

With a remarkably dense and still growing farm population and an irrigated rice technology so intensive as to be called "involuted," Java has

provided anthropologists and economists with excellent empirical material on inequality. The community of Kali Loro in central Java has 714 people per km², and only 63 percent of village households own the high-value wet-rice land (Benjamin White 1976). The wealthiest 6 percent of the population own half of all this irrigated area, and those who own no sawah may rent or sharecrop small amounts or enter into patron-client relationships that may include wage work for landed households. Fully 90 percent of the population own some garden land, which produces an important supply of vegetable and tree crops (Stoler 1981). The average adult can work only about 40 days per year on the scarce wet-rice lands, and only about 2.5 hours per day are devoted by men to all agricultural activities, but as Table 3.7 and the discussion in Chapter 3 point out, other wage work, crafts, and trade activities with much lower returns per hour supplement farm production for those with inadequate land resources. A five-person household in Kali Loro needs at least 0.2 ha of double-cropped rice fields for bare subsistence, and most families fall below this minimum. Inequality may be exacerbated by differential access to economic opportunities outside the smallholder sphere:

Landless, near-landless, and small-farm and large-farm households obtain significant proportions of their income from non-agricultural activities, but it must be remembered that they do so for different reasons; the landless and small farm . . . "agricultural deficit" households must supplement agricultural incomes with relatively open-access occupations requiring little or no capital and offering very low returns. . . . On the other hand, the large-farm and landowning . . . "agricultural surplus" households are able to invest this surplus in relatively high-capital, high-return activities from which the capital-starved, low-income groups are excluded—rice hullers, pickup trucks, cassava and other processing industries, shopkeeping, "armchair" trading with large amounts of capital, money-lending, etc. (Benjamin White 1979: 101)

A survey of the coastal Javanese village of Sukudono by Gillian Hart (1980, 1986) shows similar patterns. In a population of 735/km², 33.7 percent control no land, either through ownership, rent, or sharecropping; 19.2 percent control (that is, derive household income from) <0.2 ha; 26.7 percent control 0.2–0.5 ha; 10.5 percent control 0.5–1.0 ha; and 9.9 percent control >1.0 ha (Gillian Hart 1986: 96). The upper tenth of the population have over 58 percent of the land. If only control of wet-rice holdings is considered, the Gini index for the village is .589 excluding the landless, or .727 including the landless, as compared to .414 to .469 for other regions in Java (ibid.). The proportion of the population with no access to cultivable land goes up with local density: 10 percent at 750/km², 22 percent at 1,290/km², and 40 percent at 1,958/km² (ibid.: 99). To obtain an adequate level of income for a family of five, a household should produce at least 240 kg of milled rice equivalent per person annually, of which 120 kg is for staple food alone. This requires control of at least 0.5 ha of rice fields, and only 20 percent of the village population fall into this category. Home garden and cattle ownership correlate directly with rice land in Class I (> 0.5 ha rice or a fishpond), Class II (0.2–0.5 ha rice), and Class III (landless or <0.2 ha rice), and the three classes have household possessions averaging Rp 133,000, Rp 21,000, and Rp 6,000 respectively (Gillian Hart 1980: 198). There are similar disparities in house size, value, and quality. Table 3.6 indicates the way in which the rich spend more time per person on their own farm production and on housework, while the poor must devote more hours to wage work. The middle Class II has the highest total of income-earning household work hours, suggesting a combination of intensive agricultural labor on limited land and a significant amount of wage work (ibid.: 202).

While landless people sought income stability through pooling returns from the diverse activities of their members of working age, they were largely dependent on purchased food grains and subject to large and unpredictable fluctuations in prices. Small landowners systematically gained access to the better-paying jobs within the village, while the majority of the landless were relegated to inferior jobs, many requiring travel outside the village (Gillian Hart 1986: 164–65). It is probable that the poor also suffer from a significantly lower nutrition and health status (ibid.: 215). The ability of the rural elite to accumulate wealth and take advantage of educational opportunities has been in part owing to their status as favored clients of the state. Supravillage authorities provide them with preferential access to agricultural inputs and credit and to a range of highly remunerative commercial activities in transportation, rural cooperatives, and large-scale trade (ibid.: 199–200). Inequality in access to land, in returns on labor, and in market participation is clearly marked in Java, and it especially affects the life chances of the landless poor and the rich whose accumulation takes place outside of agriculture. It would be important to determine whether smallholders in Class II are losing control of land, being forced by debt to become tenants or sharecroppers (Gillian Hart 1980: 197).

Smallholder Persistence and the Absence of Land Concentration in China

If farmland is an extremely high-value good that can be sold, mortgaged, rented, sharecropped, and worked by either household or wage labor, we might expect that it would tend to become concentrated in the hands of a wealthy landowner class. To determine whether this process is in fact taking place, we need historical evidence from a specific region where intensive cultivation has been practiced for centuries. Longitudinal information on the growth and decline of particular farm properties will also be useful. Chinese materials provide admirable documentation of smallholder persistence in areas of dense, permanent rural population. On the dry North China plain of Shandong province, spring crops of sorghum, millet, and, more recently, maize are rotated with winter wheat and summer-sown soybeans, and cotton has long been planted as a cash crop (P. C. C. Huang 1985: 58–62). Rapid cultivation, interplanting, and fertilization have allowed intensive use of a land base that already supported more than one person per acre in Han times, around the beginning of the Christian era. The characteristic landholding unit has been the small family farm, and such smallholdings occupied 84 percent of the land in the 1930s. A survey of almost 400,000 farms found that just 1.05 percent were larger than 100 *mu* (6.7 ha, slightly less than 20 acres), and these included less than 10 percent of the total cultivated area. Similar land distribution, with big landowners making up only 1.6 percent, was recorded in 1725–50 (P. C. C. Huang 1985: 79–103). At that time, about one-quarter of the rural population was made up of landless tenants and wage laborers, and another one-third owned 10 mu (0.67 ha, or 2 acres) or less.

In the earlier part of this century, only 14–17 percent of farm work was done by hired laborers, two-thirds of whom were hired for about 200 days, while the rest were casual laborers who worked 40–50 days a year (P. C. C. Huang 1985: 81). Middle and rich peasants with fragmented farms of less that 100 mu hired some labor, but they and their families did at least half the work. Landowners with 100–200 mu practiced "managerial" farming, using some hired labor and several teams of draft animals, but in three-quarters of all cases continuing to supply some household labor and direct supervision. Unlike the land-short family farmer, who could not fire seasonally surplus laborers, the managerial farmer could bring together an optimal combination of land and labor and did not need to tolerate underemployment or low individual productivity (ibid.: 70, 171). Using the same crops, cropping frequency, and intensive techniques as their smallholder neighbors, the bigger farmers had no consistent pattern of either lower or higher yields per unit of land (ibid.: 140). Although the gross income per mu of the small farms was almost 30 percent higher than that of the managerial farms, this intensification (see Chapter 5) was achieved at a cost of 75 percent more labor/day/mu (ibid.: 158).

With a larger and more flexibly utilized land and labor base, managerial farmers had a thin margin of profitability over the neighboring smallholders, but even this was eroded when the farmer ceased regularly monitoring his workers in the field (P. C. C. Huang 1985: 298). Beyond about 200 mu, the economic advantages of managerial farming were lost to the transaction costs of divided work teams, more supervision, and travel to more distant, dispersed fields. At this point, landowners began to lease out land, although the net income per mu was only about half that of the managerial farmers (ibid.: 173). Rents that could be used to initiate such commercial activities as moneylending, land pawning, and the buying, storing, and reselling of produce were used by these land-lords to make higher profits. Educating sons for a degree and civil service employment was another means to move into an upper tier of Chinese society with higher rewards (ibid.: 169). A member of the bureaucracy or a rich urban merchant might have an income easily ten times that of a rich peasant, but attaining such prosperity usually meant moving out of the agricultural milieu. Managerial farmers who divided their property among sons also did not maintain large working farms for generations. "Managerial agriculture . . . tended to return via landlordism to the same small farming from which it arose" (ibid.). Land concentration was bal-anced by dispersion owing to partible inheritance and sale, and one sur-vey showed three-quarters of the North Chinese villages with no resident landlord in possession of more than 100 mu (ibid.: 223). Even holdings so small that household labor returned a very low marginal product could support a high proportion of owner cultivators and a low incidence of tenancy with production of commercial crops supplemented by cottage industries such as cotton spinning and weaving, straw braiding, and sheepskin processing (ibid.: 192).

The levels of wealth stratification in a smallholder community of inten-sive cultivators resemble less a ranked series of fixed and impermeable classes (landlords, managerial farmers, rich, middle, and poor peasants, landless laborers and craftsmen) than a ladder of minutely graded rungs, with households moving up and down over time. What first appears to be a static hierarchy of access to land and wealth, a condition that may indeed be more closely approximated in a caste, slave, or plantation-based society, is actually characterized by mobility and changing status. Of 10 rich households in 9 Chinese villages of the 1930s, only half could be traced back clearly for three generations in the wealthy category. The rest had risen in the present or preceding generation with profits from commercial farming or urban employment. Downward mobility, largely stemming from the ancient Chinese custom of partible inheritance among sons, left only 3 of 19 households identified as rich in the 1890s with wealthy descendants in the third generation (P. C. C. Huang 1985: 78). The local community-stratification ladder is, of course, not self-contained and encapsulated. Those few Chinese farmers who climbed the rungs of the peasant ladder to managerial farming sometimes left the vil-lage system permanently to become absentee landlords and then urban merchants or degree-holding gentry (ibid.: 178). The landless poor, without the tie of farm property for subsistence, were much more vulnerable to years of crop failure or economic depression and therefore to employment outside of the community, emigration, and forced refugee status. Those smallholders who could keep at least one foot lodged on the land knew that their relative position in the community could improve or decline, given the imponderables of climate, markets, and illness, dependent on the results of their own labor, skill, and management abilities, and conditioned by the demographic fortunes of their farm households.

Inequality, Mobility, and the Household Developmental Cycle

For Chayanov, a certain amount of peasant economic inequality was owing to demographic differentiation based on the typical sequence of changes in peasant-family consumer/worker numbers that determine cyclical trends in family income (Chayanov 1986: 56–60; Greenhalgh 1985). Because the number, age, and gender of household members are major (though not the sole) influences on small farmers' production and consumption, stages in the developmental cycle of the family will interact with the initial endowment of land and the other factors of production, contributing either to mobility or to the maintenance of the family's economic position. The age of the head of the household, indicating the family's stage in the developmental cycle, may regularly be associated with property. Peter Matlon's (1981) survey of Hausa farmers showed that the poorest 30 percent of households included families headed by persons of 60 or older, where the extended family may have begun to disintegrate, and families with a head under 25, who may have lacked operating capital, adequate inherited land, and management experience. Founding a smallholder household at marriage or at the point of fission from a preexisting farm-family enterprise often entails using less land and equipment than were present on the mature parental farm and operating under quite different dependency ratios. Acquiring rights to more land, as first increased numbers of family consumers and then a larger available labor force demand it, may define a trajectory of growth for the smallholding, followed by gradual dispersion in dowries, transfers to children at retirement, and eventually inheritance.

Smallholders are engaged in a ceaseless struggle to reach a better fit between their limited land, labor, and capital resources. Any impression of rural stability and internal balance dims as soon as we look at farms and households over time. In any given farming system, with specific conditions of soil, water, and terrain, there will be some farm areas that are too small to be viable, others that fall into an optimum range for household support, and still others large enough to be divided. Chayanov

(1986: 249) presented longitudinal quantitative data on the growth and decline of farm size, traced in part to the demographics of the household consumer/worker ratio but also to market forces and population pressure. In one district over the period 1882–1911, farms with sown areas at the small end of the continuum (0–3 ha) showed the highest percentage of disappearance (51.9 percent) as owners died out or migrated, but of the small farms that remained undivided, 71.6 percent increased their areas (ibid.: 246, 247). Around half of all farms in the 3–6 and 6–9 ha range remained undivided, and fewer than 20 percent of these increased their sizes, suggesting that such holdings were adequate for the average farm family. The biggest farms of more than 12 ha were divided in 57.6 percent of all cases, and of the undivided ones, 58.0 percent reduced their areas. Young families with undivided small farms tended to expand their farm areas, while older, and presumably larger, complex families with more than 10 ha showed declining land areas, division, and economic breakdown (ibid.: 248). Chayanov believed that if, over time, there were a tendency for more individual farms to pass from larger to smaller size categories than in the reverse direction, there would be reduced levels of well-being, but he did not consider the potential of smallholders to intensify production from shrinking farm areas. It is, however, obvious that accretion and dispersion of land over the life course of a farm household may substantially alter its productive potential and its relative wealth.

In looking at households as the salient production and consumption units of smallholder agriculture, I pointed out in Chapter 2 the direct correlations between household size and farm size or other measures of farm-family wealth and affluence. Susan Greenhalgh observes that births, marriages, and deaths in the context of cultural preferences as to household type give rise to phases such as expansion, maturation, consolidation, and division, with accompanying changes in household membership. Although she is referring to developmental cycles among various occupational groups on Taiwan, her comments can be applied directly to mainland Chinese smallholders:

Income varies over the family cycle because of quantitative changes in consumers (*needs*) and workers (*resources*) to meet these needs. Less obvious but equally important are qualitative changes in the *organizational capabilities*, and thus economic *strategies*, of the family as it moves from simple (nuclear) to complex (stem and joint) stages of the cycle. For example, in the area of *labor*, the larger working force of complex families increases their ability to diversify the family economy and disperse workers to new economic niches. With two or more adult males, complex families can more easily obtain *credit*, for creditors have greater assurance that their loans will be repaid. Large families can achieve scale economies in *consumption* and enforce policies of low consumption, thus increasing the share of the profits for reinvestment in the family enterprise. With several adult women, a few can perform all the *household tasks* and release the others to work in income-generating activities. (Greenhalgh 1985: 575–76)

Family enlargement often brings with it both increases in net wealth and economies of scale, even if income per capita does not rise as rapidly. In a Taiwanese sample, 85 percent of all losses of land were related to the family cycle, chiefly occurring through division of the estate, but 31 percent of gains also came with or in anticipation of changes in household membership (Greenhalgh 1985: 588). Though the sample is not limited to farm families, it is suggestive that two-fifths to one-half of total inequality in the Taiwanese case is accounted for by differences between the phases of the family cycle. Mobility also followed family change, income rising during expansion and falling during contraction phases (ibid.: 577). Social class position in terms of access to the means of production accounted for much less inequality and mobility than the family cycle, and it would be interesting to know if inequality and mobility within the farmer group were similarly influenced by stage of household development. The fact that both father's wealth and initial property endowment were poor predictors of the mobility and eventual maximum wealth of Törbel sons (McGuire and Netting 1982) suggests that smallholder agriculture does not perpetuate fixed strata of property ownership and wealth.

Inevitable Polarization or Increasing Stratification? Does the Smallholder Drop into the Landless Rural Proletariat?

The theory of inevitable polarization is the most imperious and potentially damaging challenge to any postulation of a smallholder society characterized by a relatively stable distribution of farm sizes, inequality within limits, and considerable mobility over the course of the household developmental cycle and between generations. Lenin contended that the penetration of capitalism into the countryside would result in concentration of landholdings by rich peasants and absentee urban landlords, while poorer farmers were forced into debt, dispossessed, and converted into a class of landless rural proletarian wage laborers.[1] Along with the neoclassical economists, Marxists assume that traditional egalitarian village institutions and practices have been replaced by a competitive labor market with impersonal wage relationships (Gillian Hart 1986: 5). This process of commercialization intensified labor exploitation, exacerbated social tensions, and created classes with radically opposed economic interests.

---

1. Lenin's vision of the differentiation of the peasantry into two classes is "currently the most important scheme that uses linear evolutionary logic" (Cancian 1992: 3). This "primordialist" view of the breakdown of an originally self-sufficient, closed, egalitarian peasant community is shared by Lenin and modernization theorists (ibid.).

Big farmers, who could achieve an economic advantage by introducing large-scale, capital-intensive, mechanized agriculture, gradually destroyed the small farms that could not compete with the lower costs of production and the lower commodity prices of the larger operations. Like the modern advocates of industrial farming, Lenin believed that scientific and mechanical methods would concentrate production and land ownership until agriculture became "merely a branch of industry" (Shanin 1971: 249). Class polarization was as much an inexorable law of progress in capitalist agriculture as it was in manufacturing industry (Kautsky 1983: 1–5).[2] Large farmers could adopt new agricultural technology more rapidly through better access to information, credit, and the financial capacity to purchase inputs. They could afford the indivisible ("lumpy") capital inputs for large machinery, and their profits from the more productive enterprises could be used for land purchase and tenant eviction, accelerating polarization through economies of scale (Hayami and Kikuchi 1982: 133). The penetration of the market economy, with its price fluctuations, and the irresistible commercialization of technological inputs accompanying the Green Revolution are usually seen as furthering land accumulation and the impoverishment of wage laborers. The underlying conception of differentiation is similar, based on "a specific secular trajectory of economic change whereby the market in products leads to a market in land which in turn leads to the *concentration* of landholdings that finally leads to *differentiation* between land owners and wage laborers" (Kahn 1981: 556).[3]

Even without technological innovation, growing population pressure on limited land resources may lead to decreasing returns to labor relative

2. Kautsky "thought the destruction of household handicraft production by competition from more efficient industrial enterprises would force peasants to purchase on the market what they formerly produced at home. This, along with increased taxes caused by gentry and landlords who themselves would require more money for the purchase of new manufactures, would force peasants to switch from production for home use to the production of agricultural commodities for sale on the market to earn money. Penetration of the market into peasant societies, however, would make farmers prey to price fluctuations and the manipulation of middle men. Their greater need for money would force them to increase the scale of their agriculture to survive, but land would often not be available. Instead, sons and daughters, who no longer would have winter employment in handicrafts, would become surplus and would have to migrate from the farm to become wage laborers. Back on the farm this would create seasonal shortages of labor that would only be overcome by hiring temporary wage workers" (Sorensen 1988: 10). In Korea, on the contrary, land reform, massive migration to the cities, and a rise in agricultural wages have led to larger and more prosperous peasant smallholdings instead of proletarianization and immiseration (ibid.: 12).

3. Joel Kahn (1981) points out that this standard analysis conflates land and labor markets, ignoring the possibility of a market in nonagricultural products and services that employs landless peasants. Indeed, increasing the degree of inequality does not necessarily result in land concentrations or the absolute impoverishment of wage laborers.

to returns to land, and this has been cited as the "basic economic force underlying increasing inequalities in income and asset distribution in rural Asia" (Hayami and Kikuchi 1982: 59). A widely accepted academic and political model of change in Asia links population growth to inequality and immiseration in an invariant causal sequence. Strong population pressure results in "increased parcelling of the land into dwarf holdings, in the accumulation of large holdings wherever land-tenure practices allowed it, in the formation of a class of landless laborers, in widespread indebtedness, in the growth of tenancy, in the decline of rural incomes, and in poverty, malnutrition, and food shortage" (Pelzer 1948: 165, cited in Hayami and Kikuchi 1982). In fact, the logic of agricultural intensification is based on the dynamic of increasing population density regularly associated with higher production per unit of land and either stable or declining marginal production of labor. But does it also result in growing inequity? Is a smallholder population necessarily polarized into large commercial farmers and landless workers with falling per capita income and increasing misery?

An alternative model advanced by Yujiro Hayami and Masao Kikuchi (1982: 60) is that of peasant stratification where there is increasing differentiation in a continuous spectrum ranging from landless laborers to noncultivating landlords. People remain tied in multistranded personalized relations, and all community members have some claims to the output of the land (e.g., workers may receive a share in the harvest, as they do in Indonesia). Semisubsistence peasants will survive, though the majority may become poorer, with smaller farms. The fact that smallholders with household labor achieve greater output per ha without the transaction costs of wage labor on large commercial farms (Chapter 5) means that the small farmers retain some competitive advantage. There is some evidence that stratification reflects the process of change in traditional, ongoing communities of intensive cultivators more closely than does polarization. During the 1960s, most countries in Southeast Asia showed no change in farm-size distribution, and Gini indices actually declined in Thailand, the Philippines, Indonesia, and Pakistan (Hayami and Kikuchi 1982: 60). The distribution of land among growers of irrigated sugar cane in a Maharashtran village of India between 1930 and 1970 showed a statistical tendency for the rich to move down while the poor moved up: "The overall trend is toward less inequality" (Attwood 1992: 162). Even where there has been a rapid increase of agricultural commodity production for national and international markets, an economy of small household producers has survived with a remarkable degree of control over family labor and land (Swindell 1985).

The process of rural stratification can be followed over time in the village of East Laguna in coastal Luzon in the Philippines (Hayami and Ki-

TABLE 7.3
Village Population Growth, Density, and Access to Land,
East Laguna, Philippines, 1903–76

Year	Population	Land area (ha irrigated rice)	Persons/ha	Farmers (%)	Landless (%)
1903	94	52	1.8	n.a.	n.a.
1960	349	104	3.4	n.a.	n.a.
1966	393	104	3.8	70%	30%
1974	549	111	4.9	58	42
1976	644	108	6.0	50	50

SOURCE: Hayami and Kikuchi 1982: 103, 105, tables 5.3 and 5.4.

kuchi 1982: 103–23). Village population increased from 94 in 1903 to 644 in 1976 owing both to a sharp rise in the birthrate after 1966 and an inflow of adult migrants in response to the high demand for labor following an expansion of irrigated rice land and the subsequent introduction of Green Revolution technology. The land area of irrigated paddy fields doubled between 1903 and 1960, but then ceased to grow, and the ratio of people to land has more than tripled in 76 years. Landlessness has increased from 30 percent of the local population in 1966 to 50 percent in 1976. (See Table 7.3.)

Yields of rice went up from 2.2 to 6.8 metric tons/ha over the period 1956–76 in East Laguna as a result of the employment of modern semi-dwarf varieties, chemical fertilizers, intensive weeding, and straight-row planting. Though average farm size declined from 2.3 to 2.0 ha between 1966 and 1976, and the number of farms of less than 2 ha increased from 43 to 61 percent, the size distribution of farms remained relatively constant (Hayami and Kikuchi 1982: 105). The largest farm was only 8 ha, and there was little discernible tendency toward polarization or the disappearance of middle-sized operations.

Most of the farmers are now tenant sharecroppers with considerable security and rights of inheritance (Hayami and Kikuchi 1982: 109). Larger tenants may sub-rent portions of their holdings to the landless, and patron-client relationships involving harvest share have been maintained or strengthened between farmers and workers. The traditional system of giving anyone who works in the harvest and threshing one-sixth of the rice that person harvested has been modified to require weeding of anyone who wants to establish a right to participate in the harvest. Though this reduces the cost of labor, it maintains the customary system of crop sharing as opposed to impersonal wage relations. The structure of smallholder households employing other community members for certain agricultural tasks has persisted, while the strata of peasant sub-classes has multiplied (ibid.: 122).

Polarization, on the other hand, is seen on the large estates and haciendas of inner central Luzon, where settlement was more recent, population density is lower, and the indigenous rights of local communities were not clearly established. With an expansion in demand for sugar, tobacco, and indigo in the nineteenth century, and with an official land registry set up by the Spanish colonial and U.S. governments, it was both profitable and legally possible for local elites to claim large private holdings, often grabbing the land of smallholders (Hayami and Kikuchi 1982: 71–74). Hacienda owners monopolized the most productive areas, extracting high rents and paying fixed daily wages, while protecting their properties with police and private armies. Political and economic controls prevented the creation of smallholdings, resulting in a "deformed agrarian structure." "Pervasive landlordism in the Philippines was a heritage of colonialism resulting from the imposition of private property rights in land at a stage in the economy when technology and resource endowments were still consistent with communal ownership" (ibid.: 75). Where peasant communities have not been broken up or dispossessed by external elites with state support, the tendency has not been for smallholders to accumulate land and become commercial entrepreneurs using machinery and wage laborers. If stratification with mobility is the more dominant pattern, we might expect that declining farm size and an increasing landless population would not necessarily lead to polarization and hardening class boundaries.

Even where one might expect to see permanent landlessness and impoverishment making class lines more impermeable with every increase of rural population in rural Bangladesh, polarization cannot be demonstrated. In a village studied by Michael Harris (1989), 434 plots in 1922 had increased to 788 in 1987, with 73 percent having become smaller through subdivision and only 27 percent remaining the same size or growing. The average farmer had 1.34 acres in five or six plots, and about one-third of the cultivators were landless. Of the landless group, over half had inherited land that they had sold or lost through a dispute, and more people had moved out of the landless category than into it. There appeared to be considerable circulation of land, with 42 percent of all households having at some time sold and 45 percent having bought land.

Relatively intense population pressure coinciding with a high degree of agricultural intensification and small landholdings as in Bangladesh results in a system that paradoxically restrains the growth of inequality and resists tendencies toward increased rural polarization. Lower population density, more extensive techniques, and larger farms in a market economy may, on the contrary, further inequality and promote the Leninist division between wealthy landowners and poor wage workers. The contrast drawn by Carl Gotsch between Comilla District in Bangladesh, where the median irrigated multicropped holding was 1.0–2.5 acres, and

Sahiwal District in Pakistan, where median farm size was 7.5–12.5 acres, is instructive. Though access to land is unequal in both cases, concentration of ownership is higher in Pakistan. There 6 percent of farmers own 54 percent of the land and 43 percent of the farmers own only 5 percent of the land and operate a total of 9 percent (Gotsch 1972: 333). The richest 8 percent of the Bangladeshi farmers own 16 percent of the land, while the poorest 26 percent of owners own 3 percent of the land. The Bangladesh case shows the characteristic smallholder unimodal, bell-shaped curve of land distribution, with ownership and operation of farms closely matched, whereas the much larger farms in the Pakistan sample show a strongly skewed distribution in favor of landlords, with actual agricultural operations carried on by tenants (Fig. 7.1). While 50 percent of all farm units in the Pakistan case are operated by full tenants, the owner-cum-tenant class in Bangladesh comprises less than 25 percent of the total farm population (Gotsch 1972: 332–33).

The introduction of the same tube-well irrigation technology in both cases, making possible the adoption of the same improved rice seed and fertilizer package, contributed to further differentiation between big and small farmers in Pakistan, while the benefits were more equitably divided in Bangladesh. Under the arid conditions of the Sahiwal District in Pakistan, the water from one well could irrigate 60–80 acres, producing not only rice but sugarcane and cotton and raising net revenues 45 percent. Farmers with landholdings of more than 25 acres, who also knew the bureaucratic procedures and wielded political influence allowing them to tap government institutions for privileged access to extension, credit, and canal irrigation services, installed 75 percent of all tube wells in Sahiwal District. Those who owned wells were in a position to monopolize water, refusing it to smallholders and securing increased labor from landless and near-landless dependents, who were recruited into their political factions. Factional rivalry among big landowners led to cattle theft, abduction of women, the demolition of buildings, and violence by squads of hoodlum *goonda* enforcers (Gotsch 1972: 337). Apparent polarization in landholdings, access to new productive technology, income, and political power was accompanied by increased conflict and greater use of naked coercion.

In Bangladesh fewer than 1 percent of the farmers were rich enough to purchase and utilize a tube well individually or in partnership. Instead, village cooperative societies installed the wells with the help of a state agency, arranged to pay for pump rental, set up and supervised the water-distribution procedure, and determined the sale price of irrigation water for nonmembers. The higher cropping intensity that was possible with irrigated winter crops more than doubled farm income, but wide participation by small and middle-sized landholders served to maintain the preexisting more egalitarian income distribution. Higher demands for la-

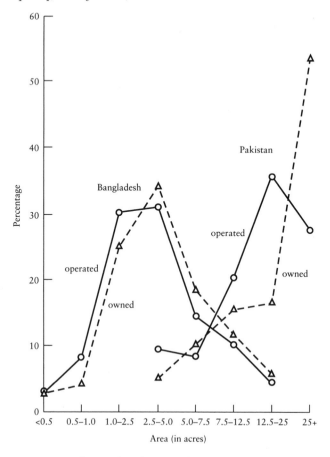

Fig. 7.1. Land owned and operated, Bangladesh and Pakistan.
(From Gotsch 1972: 332–33, tables 1, 2)

bor might also be expected to have a positive impact on the welfare of
landless and near-landless rural workers (Gotsch 1972: 336). Indeed, the
economic success of smallholders in the cooperative gave them the resources to reduce the power of the large farmer-trader-moneylender
group in the local community. Where land distribution was not originally
highly unequal and where opportunities for mobility existed, the adoption of a modern technology for agricultural intensification did not lead
to the rapid development of greater socioeconomic disparity and hierarchy, as it did in Pakistan.

Evidence of marked differentiation in the ownership of farmland is
often believed to demonstrate that a process of Leninist rural polarization
and accompanying impoverishment of the proletariat is taking place. A

survey of nine Javanese villages indeed shows a recent increase in landless-ness, with 40 percent of households lacking any cultivation rights and 9 percent of all households owning more than half of the irrigated rice sa-wah lands (White and Wiradi 1989: 291). But landless households derive only 7 percent of their incomes from agricultural wages, as opposed to 65 percent from nonfarm sources, and average total annual incomes re-main above the poverty line (based on milled-rice equivalents) of Rp 280,000 per household (ibid.: 293). Land concentration by the rich and the loss of land by the poor is not, however, proceeding, despite the ex-istence of inequality and a market economy:

On the one hand, wealthy households have many other avenues for profitable investment, and many demands for nonproductive expenditure, which compete with the alternative of land acquisition. On the other hand, the many smaller owners whose agricultural incomes do not provide reproduction at minimal lev-els . . . are able by participating in a variety of low-return nonfarm activities both inside and outside the village to achieve subsistence incomes without the distress sales of their "sublivelihood" plots. These patterns, which are certainly not unique to Java, call for interpretations of agrarian differentiation processes under conditions of commoditization and productivity growth which place the phe-nomenon of "part-time" farming and farm labor at all levels of the agrarian struc-ture in more central focus. (White and Wiradi 1989: 299)

Differentiation does not spell the demise of the smallholder, despite a lower proportion of this group in the rural population. Experienced stu-dents of political economy counsel caution in postulating polarization. "Easy generalizations about agrarian change and rural class formation processes, drawn more from standard models than from careful empiri-cal research, are both intellectually and politically irresponsible," Benja-min White observes (1989: 29).

Smallholder Persistence and the Market Economy

Even quite micro-level comparisons of areas where intensification is a recent and ongoing process demonstrate that household management of diversified agricultural production and market participation may foster both greater equality and a *higher* level of average welfare. A careful com-parison by Tom Conelly (1992, n.d.) of Masumbi, a Luo village, and Hamisi, a Luhya community, in Kenya shows that a marked difference in population density (200 vs. 700+ /km^2) and average land holdings (2.5 ha vs. 1.3 ha) correlates directly with such measures of farming intensity as the stall feeding of livestock, fodder production, manuring of field crops, chemical fertilizer use, and maize yields (Table 7.4).

The Luhya community, with the smaller, more intensively tilled hold-ings, also devotes a higher proportion of farmland to the cash crops of

TABLE 7.4
Kenyan Agricultural Intensity, 1986–87

	Kenyan villages	
	Masumbi (Luo)	Hamisi (Luhya)
Population density	200/km^2	700+/km^2
Land per household	2.5 ha	1.3 ha
Food crops	1.2 ha	0.6 ha
Cash crops	0.05 ha	0.27 ha
Fallow grazing land	1.2 ha	0.3 ha
Livestock	6.5	3.1
Stall feeding of animals[a]		
Tethered	69%	99%
Cut and carry cattle feed	1.3%	39.4%
Fodder crop production	0%	54.3%
Maize fields manured		
All of field	25%	69%
> 50% of field	13%	23%
≤ 50% of field	62%	8%
Chemical fertilizer use	0%	60%
Maize yields (kg/ha)	1,000–2,000	2,000–2,500

SOURCE: Conelly 1992: tables 1–3, 9.
[a]Percent of observations.

TABLE 7.5
Kenyan Cattle Ownership and Milk Production by Socioeconomic Group, 1986–87

	Kenyan villages							
	Masumbi (Luo)				Hamisi (Luhya)			
	Upper	Middle	Lower	All	Upper	Middle	Lower	All
Cows with milk offtake / household	1.20	0.26	0.00	0.46	0.50	0.20	0.22	0.29
Average liters milk / week / household	12.15	1.88	0.00	4.22	5.22	2.44	2.81	3.34
Kenyan shillings spent on milk / week / household	8.10	8.40	5.50	7.40	17.80	16.10	19.20	17.50
Liters milk consumed / week / household	13.40	3.30	2.20	5.80	8.70	5.10	4.80	6.00

SOURCE: Conelly n.d.: tables 3–6.

tea, coffee, French beans, and eucalyptus. Though the average number of livestock in the densely settled village is less than half that of the Luo village, milk-producing cows are more equally distributed (Table 7.5). Lactating animals vary between 0.50 and 0.20 in the three socioeconomic groups of Hamisi, while the rich of Masumbi have 1.20 cows and the poor have none (Table 7.5). Though grazing land is more abundant in Masumbi and there is less need for the fodder growing and stall feeding of Hamisi (Table 7.4), the two lower socioeconomic groups of Masumbi produce, buy, and consume less milk on the average than their counterparts in Hamisi (Table 7.5). It is apparent that with the dramatic increase in the human population of western Kenya and the growing scarcity of land in many communities, the size of livestock herds has steadily declined since the colonial era (Conelly n.d.). Inequality in livestock ownership is apparent, and it is noteworthy that the gap between haves and have-nots is more pronounced and has a larger impact on household food consumption where population density is lower and smallholder farm production is less intensive.

The remarkable persistence and self-reproduction of the distinctive intensively cultivated smallholding and its household are historically attested and remain evident even as agricultural production becomes more specialized and market-oriented and as rural society takes on a more stratified character. Neither wage labor on the farm nor urban capital invested in farm property dislodges the family household from its central role in production. Farmers who sell a larger part of what they grow and depart markedly from peasant self-sufficiency neither enlarge their operations to estates nor become non-working landlords. The economic history of the Netherlands documents local population growth combined with explosive mercantile expansion and agricultural specialization. Jan de Vries examined the Friesland district of Idaarderadeel from 1511, when it had 184 farmsteads of 11 to 30 ha, through 1640, when there were 215, to 1769, when 198 farms existed (Fig. 7.2). Indeed, the number of households using more than a few hectares had held steady at about 200 for four to five centuries (De Vries 1974: 121–25), suggesting a close relationship between the type of intensive exploitation and the range of optimum farm sizes. Neither land consolidation into a few large properties nor fragmentation into tiny peasant subsistence plots was taking place. Over this period, windmills and sluices were increasingly used to prevent flooding, large lakes were drained by companies of urban investors, and the resulting pastures fed dairy herds that grew from an average of 5 or 6 cows in the early 1500s to 26.5 in 1690. Improved transportation and a weekly market allowed increasing shipping of butter and cattle.

Reclaimed and intensively cultivated land enlarged modal farm size by about one-half between 1511 and 1769 (Fig. 7.2), but the nature of the distribution changed only by a small increase in the number of farms of 100 *pondematen* and over in the eighteenth century (De Vries 1974: 121–24). The number of farmsteads and resident households went up by only 12 percent over 150 years in the sixteenth and early seventeenth centuries, while population jumped 250 percent. By 1749, three-quarters of the lo-

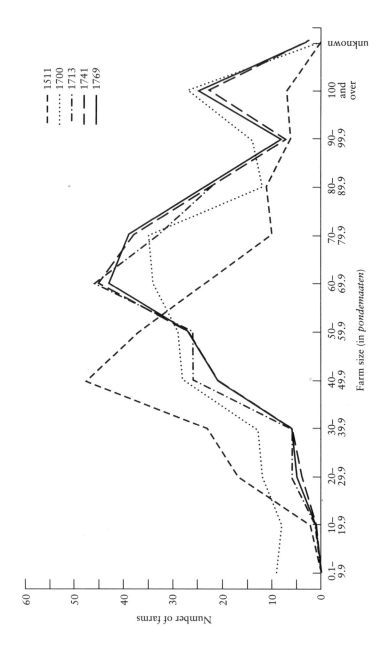

Fig. 7.2. Distribution of farm size in Idaarderadeel, Netherlands, 1511–1769 (in pondemaaten; 1 pondemaat = 0.3678 ha). (From De Vries 1974: 122)

cal households were headed by non-farmers. A Dutch economy thriving with overseas trade, manufacturing, and urbanization provided occupations for the landless in larger villages as craftsmen and traders, building ships, handling the export of farm products, and supplying the increasingly prosperous and specialized dairy farmers with clothing, leather goods, ironware, foodstuffs, furniture, clocks, glass, and candles (ibid.: 125). Workers were needed in dike and canal maintenance, and by 1749, most farmers employed both live-in farm servants and day laborers. Almost all of the cultivated land was owned by merchants or other members of the urban elite, but farmers held long-term leases on favorable terms. Economic development and the expansion of a precocious capitalism certainly characterized the Dutch republic of this period, but there is no suggestion that mercantile growth and the accumulation of wealth progressivly squeezed the smallholding intensive farmer out of existence. Indeed, the growing demand for specialized agricultural products, the supply of nonfarm goods and services, the proliferation of occupational niches, and the provision of capital for land reclamation promoted the progressive adaptation of smallholder household cultivation units to a rapidly modernizing economy.

It may, of course, be argued that real polarization is contingent upon large-scale capital investment in energy-intensive agricultural inputs and machinery at the same time that a dense population of impoverished smallholders who lack alternative employment options are losing their land. Contemporary rural Egypt would appear to provide such an example. Capitalist farmers in the Egyptian community investigated by Nicholas Hopkins rely on wage labor and invest heavily in equipment. The 15.2 percent of farmers with 5 *feddan* (2 ha) or more land hold 51.1 percent of village land. The seven largest farmers, among about 1,500 landowners, occupy 20 percent of the land, own 27 percent of the tractors, and have shares in 40 percent of the pumps (Hopkins 1987a: 161). About 45 percent of the households in the community are landless. Yet smallholders with two to five feddan of irrigated wheat and clover occupy a stable place as petty commodity producers in a complex, highly differentiated market economy. Hopkins credits this resistence to polarization in part to the resilience of the household structure and its effectiveness in organizing and supervising labor and in coordinating all the different steps and resources needed in intensive agricultural production (see Chapter 2, pp. 79–80).

The sexual division of labor in the household furthers this smallholder persistence. This is not to say that big farmers who are also machine owners and well-capitalized merchants do not play a commanding role in the local political economy. But household producers using their own labor and managerial skills may still be able to provide commodities more cheaply than either capitalist or state agricultural complexes (Swindell 1985: 58).

The second process militating against polarization is the gradual penetration of the Egyptian village by non-agrarian activities and incomes. One-third of all households are no longer dependent on the agrarian sector, and additional options for employment have been opened up by migration and education (Hopkins 1987a: 167). One might follow Hopkins (ibid.: 169) in seeing the smallholder household as "an arena of activity that is not formed by the capitalist mode of production" but is articulated with it as a path of petty commodity production in a diversified market economy. But regardless of the theoretical classification, the labor-intensive, highly productive small market farm, owned and controlled by the farmer and worked by a household that may have other occupations, survives as a significant unit even in developed capitalistic economies such as those of Germany and Japan (ibid.: 168).

Though smallholder households are by definition unable to live from rents or returns on investment the way a non-working landed gentry can, and their family members are not fully dependent for subsistence and income on wage work, as is the rural proletariat, their place in what may be a stratification system of increasing gradation is not necessarily declining, and their welfare level may in fact be improving. Without the predicted total capitalization of agriculture and absolute polarization, the smallholder household firm can be a powerful agent for economic development, and the continued existence of such farms should not be dismissed as populist romanticism. The Indian middle peasants described by Morgan Maclachlan (1987) have improved their living standards by increasing well irrigation, making possible year-round cultivation of small plots of vegetables, popcorn, grapes, and mulberry leaves for their own silk-cocoon production. Since 1983 they have become dependent on paid mechanical services, and there has been a marked increase in the proportion of land devoted to commodity production. With natural resources suitable for intensive use and a location that allows integration into a regional market economy, 60 percent of the village population have remained middle peasants, hiring little labor and rarely working for others. Their crops and cultivation techniques "required extremely careful management of a host of factors and a highly motivated, flexible, and rather skilled labor force available for work at any time, any day, year round" (Maclachlan 1987: 22). Returns are sufficient that few of these smallholders are likely to become distressed to the point of selling land, while high land prices make it difficult to increase farm size in a major way.

A "bottom up," micro-economic orientation, begining with the household as a unit of production and decision making, does not presuppose some general tendency toward equality. What Maclachlan (1987: 11) calls "intensification theory" allows for two types of inequality, an internal form having to do with the division of labor and benefits among workers and their dependents within the household, and an external form having to do with relations between agricultural households and local elites who perform little or no farm work but demand benefits on the grounds of property ownership. Private property institutions under conditions of land scarcity almost certainly ensure some degree of landlordism, with renting and sharecropping, and direct or indirect taxation of smallholders by the state is similarly a given. The question is not whether farmers' relations with elites entail some form of inequality but whether "they only require transfer payments that can be treated as an overhead cost of production and that do not impair the peasant's ability to exploit his own labor" (Maclachlan 1987: 13). If the entire increment in returns resulting from the extra labor, care, and management efforts of the intensive cultivator is appropriated by a landlord or a government instead of accruing in significant measure to the welfare and long-term reproduction of the household, the smallholders will lose the incentive to do what they do best. Exploitation that oversteps these limits may temporarily increase wealth accumulation and impoverish smallholders, but it will interfere with the intensive strategy, lead to declines in overall farm production and market activity, and risk rural conflict and disorder.

The ultimate inequality is one in which class or caste barriers introduce a permanent, heritable, hierarchical distinction between agricultural producers and elite owners and consumers, between labor and capital that appropriates the surplus products of that labor.[4] One of the first objections to the Boserup model of intensification was that the additional work effort needed to raise production per unit of land came in response to physical and political coercion rather than as an adaptation to population growth and scarcity of agrarian resources (Nell 1972). The benefits of increased farm labor could most directly and thoroughly be appropriated through military conquest and the establishment of a servile status group, slaves or serfs, with no independent rights to the means of production. My discussion of the voluntarily increased labor input, the organization of the household, the long-term investment in sustainable productive re-

---

4. Even the lines of caste separating groups of Hindu Indians do not make rural society into a stagnant, traditional hierarchy or a system in which only rich landholders can get richer (Attwood 1992: 154–55). On the contrary, "mobility is pervasive, getting more pronounced as the years go by" (ibid.: 156). In a commercializing Maharashtran village of sugar cane cultivators, small farmers of middle peasant castes did not fall into the landless proletariat, the number of landless families declined, and some of the landowners, who lacked skill and expertise in farming, became poorer. Though there was some acquisition of land by middle- and low-caste individuals, about half of all families in these groups remained landless (ibid.: 146–51).

sources, and the importance of heritable private property rights that characterize smallholder intensive agriculture has suggested that unfree cultivators as well as hired laborers would perform such tasks with much less efficiency, if at all. This is not the occasion to address the economics of chattel slavery or the medieval manor, but it is apparent that servile labor is not appropriate to the conditions of intensification:

> Where population is sparse and fertile land abundant and uncontrolled, a social hierarchy can be maintained only by direct, personal control over members of the lower class. . . . Bonded labor is a characteristic feature of communities with a hierarchic structure, but surrounded by so much uncontrolled land suitable for cultivation by long-fallow methods that it is impossible to prevent the members of the lower class from finding alternative means of subsistence unless they are made personally unfree. (Boserup 1965: 73)

The New World slave plantations, with their monocropping of cash crops, their simplified, closely supervised labor routines, and their prohibition of independent market activity could not duplicate the responsibility and productivity of the smallholder. European observers on Barbados remarked on the slack performance, undependability, and low returns on labor of the sugarcane plantation slaves (Brookfield 1984). It is little wonder that servile labor has never been competitive with the smallholder in regions of high population density and intensive agriculture.

Even where the institution of slavery exists, its large-scale imposition among intensive cultivators in their own territories does not appear to be either economically warranted or politically feasible. The Gamo are a stateless society of the south-central Ethiopian highlands reported on by Marc Abélès (1981). Living at local densities of 200 to 1,000 per km^2, they practice the intensive farming of barley, ensete (*Ensete edule*), cabbage, chickpeas, and potatoes on dispersed homesteads of regularly manured terraces. Land was individually owned and could be lent, sharecropped by tenants, inherited (unequally among siblings), and sold. Only village citizens could own and work land; non-citizen dependent castes of blacksmiths, potters, and tanners moved frequently from village to village. Each community had at least one market, and there was an active regional market network. There was stratification and hierarchy in Gamo society, though there were no kings or effective central authority. Traditional government was based on general assemblies of citizens at the neighborhood and territory levels, and each assembly designated dignitaries from its own membership. The powers of local assemblies and dignitaries were balanced with those of hereditary high priests, who conducted sacrifices against natural catastrophes, epidemics, and wars, and who guaranteed agreements with adjoining territories to end warfare (Abélès 1981: 35–47).

Warfare between neighboring territories over such issues as cattle theft and trespassing seems to have been endemic among the Gamo in the past. Members of the conquered group were neither enslaved nor forcibly removed from their homes. A victorious territory could subordinate a defeated group either as a collective vassal supplying military aid in time of war or as subjects who worked individually on the farms of citizens from the winning side. War victors did not appropriate land, but rather secured a part-time labor tribute for feeding livestock, carrying manure to the fields, and tilling the soil. The losers remained resident in their own territories and continued to cultivate their own homestead farms. Dominated territories often emancipated themselves by provoking new wars, and conquerors sometimes maintained their nominal hegemony by abolishing tribute (Abélès 1981: 42–46). There was certainly in the Gamo case institutionalized inequality among individuals, including citizen farmers and non-citizen craftsmen; commoners, dignitaries, and priests; and the victors and vanquished in war. Private property in land and livestock was necessarily unequally distributed. But even military conflict and conquest did not permanently alienate smallholders from their homestead fields, and labor services stopped far short of establishing a permanent servile status.

Smallholder Inequality Reevaluated

Smallholder intensive cultivators do not seem to go by the book or accept the conventional categories when it comes to economic inequality and differentiation. Their households possess, inherit, and transfer ownership rights in productive property, and individual households show measurable differences in land, livestock, equipment, buildings, and other forms of wealth. Smallholders neither produce in common under some idealized lineage mode of production nor abjure market exchange and the differences that emerge from competition. Their communities may administer common property resources, but they do not reallocate intensively tilled lands or effectively level apparent economic differentials. Smallholders like the Kofyar and the alpine Swiss do cooperate, exchanging labor and services, but such reciprocal contributions are carefully balanced and monitored. They do not operate to erase distinctions among households.

But smallholders live with the realistic hope that they can, by their own efforts, skills, and management abilities, change and improve the fortunes of their households. The essence of intensive cultivation is to make the most of limited resources and at the same time increase one's access to the means of production. Farmers' experience shows them that situations of relative wealth and poverty are not preordained or fixed, that inequality among smallholders at any point in time does not prevent mobility, both up and down the social ladder. The changing ratios of consumers and workers in the household developmental cycle, demographic accidents of birth, marriage, and death, the inheritance of fields, gains from off-farm employment—all these affect a life course that is not necessarily the same for parents and children, or for different siblings from the same family. Moreover smallholders are not fatalists. Economic success, although influenced by uncontrollable factors of environment, health, and chance, bears a relationship to personal abilities and moral virtues. It is a drama of striving, and a game that some play better than others.

Rural society is not isolated from the larger spheres of urban trade, state bureaucracy, and industrial production, which display occupational, class, and caste inequalities that make smallholders' intravillage distinctions look trivial. The excesses of external wealth and power that can distort and even destroy the ranks and petty properties of the countrymen are never far away. The wonder is that the same forces of capital, law, and politics that perpetuate inequality in the larger world cannot seem to convert the mass of smallholders into a landless proletariat of wage workers. Neither the absentee landlord nor the rising peasant enterpreneur regularly accumulates wealth or consolidates land to the point where well-rooted intensive farmers are dispossessed. This does not mean that there are not smallholders with minuscule farms or laborers barely subsisting in a Javanese village. But it does suggest that finely calibrated systems of stratification with considerable mobility are more characteristic of household-based intensive cultivation than are polarization and the hardening of class barriers. As historic China so clearly represents, the familiar steps of enlarging the holding until it must be worked with hired labor and expensive management lead most often to off-farm investments with a better rate of return or to division of the estate among heirs who are once again smallholders.

The evidence from Bangladesh and the Philippines shows the emergence of inequality from a dynamic that operates in part within smallholder local economies rather than being imposed on "egalitarian" peasant communities by exploitative capitalists and predatory politicians. Controlled cross-cultural comparisons of Pakistan and Bangladesh and of two areas in Kenya demonstrate that economic disparity is actually lower where population density is high and land use most intensive, and that there is a more equitable distribution of welfare in the crowded regions. It is agricultural frontiers and sparsely settled hinterlands that further wealth accumulation, dependency of the poor, and political violence.

Even where economic development and an active market flourish, as in early modern Holland, a smallholder sector may show remarkable stability and persistence without land fragmentation or rural immiseration.

Alongside large mechanized holdings in Egypt and Green Revolution landlords in India, smallholders with household labor forces provide a decent livelihood and stay in business. Their conscientious skills, diversified operations, and careful household management are not duplicated by hired laborers or by capital-intensive machines, just as smallholders were not absorbed by the slave plantations or the serf manors of the past. Smallholder inequality and long-term mobility within the limits of a functioning rural economy have shown a historic resistance to the threats of polarizing wealth and paralyzing poverty.

CHAPTER EIGHT

Chinese Smallholders

CHINA IS THE land par excellence of smallholder intensive cultivators. No other society on earth has the same unbroken history of a dense rural population practicing permanent, sustainable agriculture in the context of a great and enduring civilization. Following this remarkable tradition of land use and its associated technology through time is made possible by the existence of documents that range from agricultural treatises to imperial laws to gazetteers of local resources, climate, and demography. That this distinctive system of production has been maintained from at least the time of the Han dynasty (205 B.C.–A.D. 220) to the present is perhaps better known than the more recently adduced fact that intensive agriculture was characteristically carried on by smallholder households with well-defined rights of ownership or contractual use of land and participation in an active market economy. The historical and contemporary evidence from China for the durable association of areas of population pressure with intensively tilled, diversified farms, household labor and tenure, and a combination of subsistence and commercial production provides a paradigm of the kind of cultural-ecological system that is the focus of this volume.

But the Chinese material also affords us an opportunity, rare in the annals of social science, to observe a set of postulated functional or systemic relationships as they change through time. In one of the great political-economic experiments of any era, the postwar communist government of China collectivized agriculture, organizing production through groups such as communes, brigades, and labor teams; sharing production among resident households in the community unit; and eliminating private property in land, livestock, and equipment. The goal was to rationalize, mechanize, and thereby increase agrarian production, get rid of the capitalist inequality that came from private property and the exploitation of hired labor, and provide more efficiently the basic social services of welfare, medical care, and education. Family households continued to exist as reproductive and residential units, but agricultural labor and technology were to be managed by corporate institutions, often at the level of the village community. Direction came from the state through local official cadres and a hierarchy of district, provincial, and national bureaucratic agencies.

This institutional structure, in many ways the antithesis of smallholder farming, was largely and quite precipitously replaced after 1978 by the "Household Responsibility System," in which households received the contractual right to use formerly group-controlled resources, to make independent choices on crops and farming methods, and to dispose privately of produce beyond a stipulated amount delivered to the state. Agricultural surpluses could henceforth either be consumed or sold to the state or on the free market. The resulting rapid gains in agricultural production were accompanied by increases in nonfarm employment. Higher peasant incomes and growing inequality became apparent as Chinese agriculture reassumed many of its traditional institutional forms.

Historic China and the Boserupian Paradigm

China is in many ways a gigantic stage on which the logic of Boserupian population growth, shrinking quantities of land per capita, and compensating intensive land use has been played out over the past two millennia (Table 8.1). Although there have been pronounced fluctuations, such as the demographic rise in the eleventh century to some 108 million and subsequent fall during the period of Mongol conquest, the general tendency was for the amount of land per capita to hover in the 8–11 mu range (0.53–0.73 ha) until the beginning of the Ming dynasty in 1369. In the following 450 years, the population increased about fivefold (Eastman 1988: 4), and land availability fell to 2.8–4 mu (0.19–0.27 ha) per person.

The fundamental contrast between China and the United States in population density and in the contingent intensity of agriculture was pointed out by the American soil scientist F. H. King on the first page of his classic 1911 monograph *Farmers of Forty Centuries*. Whereas the United States at that time had "more than twenty acres to the support of every man, woman, and child," the people of China, Korea, and Japan had scarcely two acres per capita (King 1911: 1). To accommodate the population growth from 1369 to 1949, agricultural output had to increase at the same rate, with cropped area expanding fourfold and yield per unit area doubling (Perkins 1969, cited in P. C. C. Huang 1985: 10). During the same period, the institutional patterns of land tenure and productive relations changed little.

Though fallowing and more extensive agricultural techniques were still possible in many areas during the Han period, the small permanently cultivated farm was already the ideal. The sage Mencius (372–289 B.C.)

TABLE 8.1
Chinese Population, Cultivated Land, and Land per Capita

A.D.	Dynasty or regime	Population (millions)	Cultivated land	Land per capita (mu)[a]	Land per capita (ha)
2	Western Han	59	571	9.67	0.64
105	Eastern Han	53	535	10.09	0.67
145	Eastern Han	47	507	10.78	0.72
961	Northern Song	32	255	7.96	0.53
1391	Ming	60	522	8.70	0.58
1592	Ming	200	793	3.96	0.26
1776	Qing	268	886	3.30	0.22
1850	Qing	430	1,210	2.8	0.19
1934	Republic	503	1,470	2.9	0.19
1953	People's Republic	583	1,678	2.9	0.19

SOURCES: Figures through 1776 are from Chao 1986: 87, 89; those for 1850, 1934, and 1953 are from Eastman 1988: 4.
[a] 1 *mu* = 0.067 ha or 0.16 acre; 1 ha = 15 *mu*; 1 acre = 6 *mu* (Chao 1986: 231).

envisioned an eight-person stem-family household with 100 "small mu" of farm land and 5 mu for their residence, mulberry trees, and domestic animals, suggesting a total area of just under 5 ha (Hsu 1980: 10, 109). Han farms in the period A.D. 2–146 were in fact somewhat smaller than Mencius's model, with an average landholding of 68 to 79 mu (3.3–3.9 ha) and a household size averaging 4.87 to 5.76 people (ibid.: 21). A typical Han farm had to produce a ton of grain on 7.7 acres to support a family of five (Anderson 1988: 39). Chinese agriculture had already taken the path of intensification: "The method is horticulture—hand gardening—and the basic implement is the hoe. The method sacrifices the productivity of labor for the productivity of land" (Stover 1974: 68). Careful leveling or mounding of the field, weeding, and constant pulverizing of the soil became standard, suggesting that "intensive care of each individual plant had been brought almost to the level of gardening" (Hsu 1980: 127).

Archaeological evidence indicates that over the course of the Han period, residences became smaller and more tightly spaced, but iron tools doubled in frequency, more bricks were used in wells, and larger granaries were constructed, all suggesting a higher standard of living (Hsu 1980: 128). Developed land sold for 100 times the price of frontier land (ibid.: 19). Even government policy stimulated intensification. Han land tax was levied according to the size of the plot, not its production, so the "farmer would not have been inhibited from producing as much as he could" (ibid.: 110). The Qin dynasty states (A.D. 266–420) encouraged the expansion of the smallholder peasantry to achieve the high population density that provided the best source of recruits for mass armies (P. C. C. Huang 1990: 326). Fostering partible inheritance and household division

while offering land promoted the early and universal marriage that produced high fertility rates in the population (ibid.: 326–28).

   From what may have been pockets of high population density and intensive agriculture in the Han era, the smallholder pattern expanded
through the Chinese heartland of the great Huanghe, Yangzi Jiang, and
Xi river drainage basins (Stover 1974: 68). John Lossing Buck's (1937)
massive survey of land utilization in the 1930s, based on over 53,000 in-
terviews in 168 localities of 22 provinces, depicts a crowded, highly productive landscape. The farming population averaged 1,500 per square
mile of cultivated land (579 /km²), and median farm size was 3.31 acres
(1.3 ha) divided into six different plots (Buck 1956 [1937]: 9, 19). Ninety
percent of farm area was devoted to crops, compared with 42 percent on
U.S. farms of that time, and only 1.1 percent of Chinese farmland was
set aside for pasture, reflecting reliance on a vegetarian diet rather than
livestock products. Two-thirds of all land grew two or more crops a year.
Irrigated rice was grown on one-half of all land, with water pumped
largely by human-powered treadle pumps. A quarter of all land was terraced, and 5 percent was drained (ibid.: 6). Fertilization with farm manures, human excrement, oil cake, and ashes from straw and stalks was
heavy. Buck found farm households averaging 5.21 members, and 70
percent of them were nuclear families. The size of the household was
closely associated with the size of the farm, ranging from 3.96 persons on
small farms averaging 1.43 acres, through 4.52 on medium farms of 2.84
acres, 5.02 on medium large farms of 4.92 acres, to 7.31 on very large
farms of 13.02 acres (ibid.: 272, 370). It is striking that three-quarters of
the land was owned by the farmer himself and only one-quarter rented
(ibid.: 9).

   All-China figures mask differences between the drier wheat-growing
region of the north, where only 18 percent of the land was irrigated in the
1930s, and the southern rice region, with 62 percent irrigation (Eastman
1988: 63). Mean farm size of 5.63 acres in the wheat region was larger
than the 3.14 acres in the rice region, but the comparison with average
farm sizes in the Netherlands (14.28 acres), England (63.18 acres), and
the United States (156.85 acres) demonstrates the high levels of intensification that characterized Chinese agriculture in general (Buck 1956
[1937]: 268). North China often intensified land use more slowly than its
southern neighbor. Population on the plain of northwest Shandong was
devastated by the Mongol campaigns of the twelfth century, and the environment, with only 500 mm of annual rainfall and a 6-to-7½-month
frost-free growing season, was less conducive to multicropping than the
wetter, warmer south (P. C. C. Huang 1985: 58–60). Nevertheless, as
northern rural population density and market demand from the imperial

state capital at Beijing grew, a rotation of spring sorghum or millet, winter wheat, and summer soybeans was introduced. With land availability per capita dropping from 15 to 4 mu by the eighteenth century, cotton was adopted to give increased production per unit area and employ family labor more fully (ibid.: 58, 116).

Sustainability Versus Degradation

The skills, the labor, and the diversified management strategies of the smallholder balanced the short-range goal of high production per unit of land against the necessity of reducing year-to-year variability of yields in the face of climatic fluctuations and the requirement of maintaining or even increasing soil fertility. Lacking the opportunity to find unused land in the heartland except by laboriously reclaiming it from marshes or lakes, the Chinese had to make the most of their small plots. They did not have the luxury of extensive fallowing with natural biotic regeneration or of alternating arable fields with livestock pasture. Individual well-being over a farmer's life course and household security across generations made agricultural sustainability a crucial consideration. Soil degradation through erosion, leaching of nutrients, waterlogging, or salinization could destroy a smallholder's livelihood. Notwithstanding that droughts, floods, and pestilence might overwhelm even the most devoted cultivator, efforts to reduce risk and maintain stability of production could not be slackened. Chinese intensive farmers invested in the terracing, water control, drainage, green manuring, composting, intercropping, and precise tillage that must characterize any sustainable regime. Their achievement of high agricultural production without deterioration of national resources for centuries compels attention and respect.

In a land as vast as China, it is not astonishing that the sustainability of agriculture, especially in the irrigated heartland of the south, should stand in contrast to monumental land degradation on the sparsely settled peripheries. For example, rapid population growth in Hunan during the eighteenth and nineteenth centuries, coupled with state encouragement of migration to the mountainous borders and adoption of New World crops, resulted in deforestation and erosion (Perdue 1987). Extensive, slash-and-burn techniques were used for dry-rice production, and animal grazing on the stubble interfered with double-cropping. Flooding endangered intensive agriculture in river valleys downstream. In the mountainous subtropical region of Anhui, there was a similar movement beginning in the 1730s of migrant "shack people" into forested areas where maize was planted (Osborne n.d.). Soil nutrients were depleted, timber resources were reduced, lowland irrigation systems were silted up, and river transportation was impeded. The government attempted to control land use and settle land-tenure disputes, restoring some measure of ecological stability to the region (Osborne n.d.).

Frontier land use in China presents a familiar scenario. Where marginal hill and wooded land are opened for expanding agricultural land use, extensive shifting cultivation will lead to deforestation, overgrazing, and erosion in the short run. Whether increasing population densities and/or government land-use controls can install more intensive land-use methods or ensure soil conservation remains problematic.

Environmental degradation was also present in arid sand and loess areas of Shaanxi Province. Beginning in the 1950s, production brigades built bench-terraces and erosion dams, and planted trees to stablize slopes (Deqi et al. 1981). But elsewhere, during the "Great Leap Forward," a great increase in tree cutting for charcoal to smelt pig iron took place. Efforts to increase grain production in the 1957–77 period resulted in 6 million ha of forest being felled to clear fields that were subsequently abandoned, and improper land use resulted in the desertification of millions of ha more of farm and grasslands (Smil 1987: 219). It is apparent that, although some marginal lands may be brought into permanent production by heavy capital and labor inputs, there are substantial risks of environmental deterioration under extensive methods.

The Quintessential Chinese Household: Farm Work and Cottage Industry

The social institution that mobilized and organized the skilled labor of intensive agriculture in China was the smallholder household. A great deal of recent scholarship (Hsu 1980; P. C. C. Huang 1985; Bray 1986; Chao 1986) has demonstrated that the household rather than the slave estate, the feudal manor with serfs, the sharecropper plantation, or the managerial farm with hired labor has been historically and remains today the most significant unit in intensive farming systems. Moreover, the scarcity of land, long experience with particular fields and local environmental variations, ability to specialize and diversify production, and socialization to responsible job performance that have made the smallholder household such an effective farming institution (see Chapter 2) also enable it to function effectively as an enterprise in nonagricultural activities. Its members do not trudge out together in some institutional lockstep to plow or spread fertilizer or weed fields at the same time. Indeed, the division of labor by sex and age means that their tasks are usually different and complementary, with women processing crops, caring for livestock, gathering fuel, and cooking, while men carry on the multifarious activities of farm work outside the domestic sphere.

Because Chinese rural life is fundamentally tied to the larger town and urban market economy, intensification, in the sense of putting time to productive use, could go on *outside* agriculture, and the cottage industries of silkworm raising, silk and cotton spinning, and weaving could become major supports of the peasant household. Like multicropping, handicrafts spread labor more evenly over the year; women's contributions became more valuable, and the risks of total dependence on either the agricultural or the commercial sector were reduced (Eastman 1988: 73). Already in Han times, the combination of farming and textile production was regarded as distinctive of Chinese agrarian culture, in contrast to the ruder ways of living on the frontiers, and in the winter months, women wove and spun in the evening as well as during the day (Hsu 1980: 130). Though crafts could provide as much as half the household income, they might also be pursued for very low returns. With little land and few local wage-labor jobs, the opportunity cost of labor is minimal, and rural families may carry on subsidiary production at any price level for which the net revenue (price minus the cost of material) is above zero (Chao 1986: 17).

Even when, as in the 1930s, the majority of peasant farms in northern China were below the average subsistence size of 15 mu, small family farms continued to represent 80 percent of the total cultivated area (P. C. C. Huang 1985: 185). The communities with the highest percentage of owner-cultivators and the lowest incidence of tenancy supplemented their commercial farm production of cotton, rice, or winter wheat with income from cotton spinning and weaving, straw braiding, and sheepskin processing (ibid.: 192). In modern peasant households, family members, especially younger sons and daughters, may work in labor-intensive local factories like brickworks or furniture-making shops, or they may be employed at home on piecework like knitting or component assembly. Notably, however, in a Taiwanese village studied by Bernard and Rita Gallin (1982), 83 percent of households continued to farm even when the major part of household income was derived from such off-farm activities. As Francesca Bray observes:

The organisation of resources typical of a "skill-oriented" technology such as intensive rice-farming dovetails very neatly with petty commodity production, which requires very little capital to set up a family enterprise, and absorbs surplus labour without depriving the farm of workers at times of peak demand. It can be expanded, diversified or contracted to meet market demands, but the combination with the rice-farm guarantees the family's subsistence. . . . Since the owners of the enterprise also supply the labour, rural manufacturers of this type sometimes prove more competitive than larger urban industries. (Bray 1986: 135)

The flexibility of the household in mobilizing and coordinating labor for the skilled tasks of intensive agriculture, while efficiently using surplus time for profitable nonfarm occupations and then pooling joint returns for corporate consumption and investment, suggests the continued salience of the household as a functional social unit.[1] The economic behavior of the Chinese peasant "presents a composite portrait of the rational familist, who seeks to maximize not individual interests, but household corporate interests" (Nee 1986: 200).

Landlords as Ogres and Tales of Tenure

The admittedly high demand for manual labor in Chinese agriculture, the scarcity of resources and machinery, and the historical contrast between rural poverty under perpetual threat of famine on the one hand and the wealth of the urban elite on the other lend credence to the stereotyped traditional image of the Chinese peasant as victim: one whose toil is coerced and who is mercilessly exploited for the gain of others. The ideological hinge of communist revolutionary rhetoric has always been the conflict between landlord and peasant. Ownership of land, the primary means of production, and of capital for commerce and moneylending, so the story goes, allowed the landlord class to appropriate the surplus value generated by the peasant masses.

Notwithstanding the profound stratification and hierarchical nature of historical Chinese society as a whole, this model of social relationships and property may have more to do with industrial capitalism, expansion on the steppe and desert frontiers, and European feudalism than it does with the agroecosystem of China. There was servile labor in early China, but it occurred on the frontiers of the empire, where land was abundant and military lords used a conquered local population or conscript soldiers as serfs (Hsu 1980: 10, 139). Elsewhere, slaves or contract laborers were unreliable because they lacked incentive to perform the thorough, painstaking fieldwork that intensive cultivation requires (ibid.: 127).

In the early fifteenth century, during the Ming period, costly govern-

---

1. "Efficient" labor and "profitable" use of time are relative terms that may mask economically negative processes. Philip Huang (1990: 13–15) terms this process "involutionary commercialization," pointing out that rice yields in the Yangzi Delta had already reached a plateau in the southern Song and early Ming periods. The switch to more labor-intensive commercial crops like cotton and mulberry leaves for silk production and the growth of commercialized "sideline" household handicrafts both brought lower average returns per workday than food cropping: "One of the consequences of a high level of *involutionary commercialization* sustained by the *familization of rural production* [emphasis in original] was that family farming outcompeted wage labor–based managerial agriculture, for the simple reason that managerial enterprises had to rely mainly on adult male labor paid at prevailing market wages. A family farm that used spare-time and auxiliary household labor at net returns well below the prevailing market wages could sustain much higher levels of labor intensification, and hence higher gross returns from the farm, as well as higher prices (or 'rent') for the land" (P. C. C. Huang 1990: 14).

ment undertakings and high taxes drove some freeholding peasants to take refuge as bondservants with powerful landlords who could resist state exactions (Eastman 1988: 72). But as peasant cottage industry flourished and both domestic and foreign trade increased, estate owners found that commercial and manufacturing ventures had a higher rate of return on capital than managerial farming (ibid.: 73). When dynastic rule was strong, the state also encouraged smallholders at the expense of large estates, both because they were a more accessible source of tax revenues and because they were less threatening to the governing elite (P. C. C. Huang 1985: 86).

In a dense population with a relatively abundant labor supply, the cost of free labor is low, and there is little advantage in using slaves or compulsory labor, whose entire annual subsistence must be provided (Chao 1987: 108). The quality of skilled, experienced labor needed for intensive agriculture was probably never compatible with serfdom. Only in frontier zones were peasants vulnerable to exploitation because of their need for military protection and their inability, because of distance, to participate directly in the market for rice or handicrafts (Bray 1986: 177–78).[2] When, however, the interaction of population density, resource scarcity, and market demand brings agriculture to a certain level of intensity, "the family farm comes to predominate as the basic unit of production" (ibid.: 196).

Is it, however, a romantic mirage, a "mystification," to see family farmers as in any sense autonomous or self-determining when the land they work may be rented at exorbitant rates or when state taxes and tribute exact from them any gains they might secure from the soil? To what extent were smallholders in fact property owners in China? The Han imperial government claimed all uncultivated land, but gave peasants what amounted to permanent usufruct rights to manage, occupy, and transfer land that had been improved by human labor (Tai-Shuenn Yang 1989: 53). Intensive land use is contingent, not solely on landownership, but on the right to receive a significant share of what is generated by the extra labor, diligence, and management skills of the cultivator, whether the cultivator rents, sharecrops, or borrows the land.

2. Other types of social units might also have a selective advantage under frontier conditions. Maurice Freedman (1966) speculated that localized, corporate lineages of patrilineally related households may have functioned in southeastern China to organize defense of life and property and to promote cooperative clearing and irrigation of new lands. In Burton Pasternak's (1969) view, lineages became important on Taiwan at a late stage of settlement, when there was intravillage competition for available land and water resources, and when agnates could ally themselves to protect their interests against co-villagers. Though lineages in China may have held some land and other property, they seem never to have farmed as corporate groups, and there were often marked differences in household wealth and political power among lineage members (Clammer 1985).

In the 1930s, some 29 percent of Chinese farmland was rented, and ownership was more prevalent in the northern wheat region (86 percent) than in the rice region (60 percent) (Buck 1956 [1937]: 194). Although in highly productive irrigated areas near southern cities as much as 95 percent of the land was worked by tenants (Eastman 1988), there was a pronounced tendency for Chinese peasants to have some ownership rights, observable in the proportions of all Chinese farmers who were only tenants (17 percent), who both owned and rented (29 percent), and who were wholly owners (54 percent) (ibid.: 196). Under sharecropping arrangements, which were more common in the north than in the south, the landlord supplied seeds, tools, and possibly draft animals (Eastman 1988: 76). The standard distribution of shares under a sharecropping tenancy was fixed in China at approximately 50–50 for more than 2,000 years (Chao 1987: 223). Sharecropping was already declining in favor of fixed rents by the twelfth to thirteenth centuries in southern China. "As cultivation techniques became more complex and the supervision of tenants more onerous, landlords took less and less direct interest in the way their land was farmed, and tenants acquired rights to greater security of tenure or even, eventually, to permanent tenancy" (Bray 1986: 206).

Both permanent tenancy, which was in fact equivalent to joint ownership, and long-term fixed-rent systems became popular in those areas where population density was relatively high (Chao 1987: 189). Evelyn Rawski (1972), in her pioneering study of sixteenth-century Fujian (Fukien), was one of the first to recognize that favorable tenure and rent arrangements reflected the importance of the labor contribution of the tiller. Landlords supplied none of the inputs other than land, they remained responsible for taxes, and long-term contracts made it impossible to raise rents or even evict a defaulting tenant. The higher returns that might come from farming one's own land or managing hired laborers were sacrificed by landlords who opted for a stable fixed income with low risks and small transaction costs and for the prestige of rural landownership (Rawski 1972: 24). Security of tenure was also an inducement to attract new tenants to land abandoned because of civil strife or to reward the tenant's contribution in the clearing of new land (ibid.: 20; Chao 1987: 187). By the sixteenth century, the tenant could pay a fee giving him rights to sublet or sell the cultivation right, or to transfer it to his heirs.[3]

"Fixed rents provided tenants with incentive to increase yields, because these increases would be theirs and not the landlord's to enjoy" (Rawski 1972: 18). The benefits of skill and additional effort in leveling the field, properly spacing the transplanted rice, multiple-cropping, and fertiliza-

3. For a modern case of such "subsurface" rights, which might belong to a corporate lineage as well as a private landlord, see Potter and Potter 1990: 334.

tion went to the cultivator, who also profited directly from higher market prices. Under the system of permanent tenure known as "one-field, two-lords," the "surface" rights belonging to the tenant often sold for more than the landlord's residual "subsurface" rights (Eastman 1988: 77). Since it was difficult to dispossess a tenant, rent was often in arrears. An eighteenth-century rent book shows over a quarter of all tenants defaulting on payment in any single year (Chao 1987: 191). Permanent tenure was also not rare, covering 20 to 44 percent of the tenants in several eastern Chinese provinces (Eastman 1988: 77). Skilled tenants under a system of fixed rents could hope to save enough to buy land of their own, and Hsiao-tung Fei (1939) reports that even in a crowded twentieth-century Shanghai village, he met no one who had been landless all his life (Bray 1986: 207). Given the economics of agricultural intensification and the realities of fixed rent and permanent-tenancy arrangements, in China "landlordism was not synonymous with exploitation, nor tenancy with penury" (Eastman 1988: 78).

High rates of tenancy are too often explained as the outcome of forced sales of land by debt-ridden, impoverished peasant smallholders to wealthy landlords.[4] Population pressure was assumed to contribute to cycles of immiseration. In southern China, on the contrary, tenancy and land concentration rose in periods of economic prosperity when productivity per capita, yield per ha, and surplus per ha were all rising (Shepherd 1988: 403, 418). Commercialized wet-rice cultivation and market gardening stimulated both labor- and capital-intensive land reclamation, and small farmers migrated into the area to take up tenancies. On large-scale projects, the investing landlord could claim rents of 50 percent of the crop, whereas small conversions of marginal land into paddy earned the tenants permanent tenure rights and rents of only 10 to 15 percent (Shepherd 1988: 413). Intensification could proceed rapidly because of the ease of irrigation, the presence of water transport for crops and handicrafts to markets, and the availability of urban night soil for fertilization (Shepherd 1988: 414). Low-rent permanent tenancy provided incentives for farmers and an assured return for landlords without the time and energy of direct management of agricultural operations.

Though the smallholder pattern of ownership and tenancy seems to have been present in the densely populated, intensively cultivated areas of China for at least 2,000 years, there is some evidence that it has become *increasingly* dominant. Land concentration by large owners has not been the most economically profitable strategy. Population pressure and the presence of many small farm units translated into high demand for land,

---

4. The theory of debt-sales leading to high rates of tenancy and the concentration of land ownership has been especially popular among Marxists and Depression-era writers (Shepherd 1988: 407).

but land that was not cultivated by its owner was a poor investment. Even the Confucian ideal of "agricultural fundamentalism," affirming the priority of farming among all economic activities and granting it a status above the superficial occupation of the merchant (Chao 1986: 105), could not alter the economic calculations. The rate of return on capital in agriculture was a low 2.5 to 5 percent, compared with 10 to 20 percent annually in commerce. Landed wealth continued to have high prestige in the Chinese agrarian society, but the elite also invested in urban real estate, trade, moneylending, and educating their children for the imperial bureaucracy.

Commercialization of agriculture through labor-demanding cash crops like tobacco or cotton, supplemented by cottage industry, brought higher income to farm households and increased the demand for land (Chao 1986: 105). An active land market and a variety of contractual arrangements for land use meant that most transactions were for small parcels at relatively high prices, and it was difficult for the few wealthy people to purchase large amounts of land in a short time (ibid.: 108). With growing population density from the twelfth to the early twentieth century, it appears that the number of large landholders and the size of the largest holdings actually decreased sharply, and the proportion of the landless fell from 50 percent to 20 percent (ibid.: 116, 125).

The lack of polarization differentiating a few big landowners from an impoverished mass does not mean that land was evenly distributed in China's villages. In a community of 375 households surveyed in 1862, 13.4 percent were landless; 56.7 percent had less than 3 mu, or half an acre, which was below the level of self-sufficiency; 25 percent had 3 to 15 mu; and only 5.7 percent had properties in the 15–40 mu range with surplus land for leasing (Chao 1987: 117). Management of land was usually more evenly distributed than ownership, as indicated in a Yunnan village studied by Hsiao-tung Fei (Fei and Chang 1945: 76) where 31 percent were landless but only 15 percent had no land to farm. Both rich and poor cultivated rented land, including land belonging to temples, lineages, and other collective groups. The Gini coefficient of inequality derived for landownership in three eighteenth-century Hebei villages give measurements of .622, .604, and .672, with 18.2 to 25.5 percent landless (Chao 1987: 117). Forty-seven local surveys by a government land commission in the 1930s, covering almost 1.6 million households, give a Gini for the combined sample of .55, with 19 percent landless (Chao 1986: 125). Kang Chao (ibid.: 222) finds no evidence that land distribution in traditional China was more unequal than in other countries or that the distribution had become more inequitable over time. The prevailing Chinese custom of equal land division among sons meant that the large properties of richer peasants were divided among heirs. Tracing family lines across several generations shows the countervailing processes of fragmentation of large holdings and accretion by successful household farm enterprises, revealing mobility both up and down the local scale of property (P. C. C. Huang 1985: 78). If the knowledge, skills, and work effort of intensification were to be encouraged and motivated, economic rewards at least in part commensurate with these qualities had to flow to the cultivator. To the degree that the market economy and the institutions of tenure, contract, and inheritance functioned relatively freely, rural smallholders remained the norm in China.

Comes the Revolution: Collectives Replace Smallholdings

What happens when, for political, ideological, and economic reasons, the smallholder farm household is eliminated as the primary unit of production and land tenure? This truly revolutionary social experiment was mounted in the 1950s by the communist government of China. Centuries-old patterns of intensive cultivation, high and sustainable crop yields, peasant village communities, and households as units of consumption and reproduction continued in rural areas, but agricultural work was now organized and carried out in groups under general direction by the state, formerly private land and draft animals were pooled, and the production of crops and other commodities for the market was discouraged. This radical reorientation was preceded by a strongly pro-smallholder policy instituted in 1948, after the Japanese defeat, when the civil war between the Kuomintang forces and the communists was still raging.[5] In an attempt to bring rapid reconstruction to a ravaged rural economy, and to inspire the peasants to hard work and maximum tillage, the Communist Party decreed its Four Freedoms: (1) to buy, sell, or rent land, (2) to hire labor for wages, (3) to lend money at interest, and (4) to set up private enterprises for profit (Hinton 1983: 65). The slogan was "enrich yourselves," and the motivation was obviously self-interest. A sweeping land reform redistributed land, draft animals, and implements from landlords who did not directly work on the land but rented it out or hired labor. In central south China about 40 percent of all cultivated land was confiscated, and 60 percent of the population received some, averaging 0.4–0.8 ha (1–2 acres) per family (Shue 1980: 90). Though big landlordism and arbitrary exactions may have been eliminated, the popular reform left private ownership of property intact (ibid.: 99).

Although the combination of peace, the rapid growth of market activity, and increased production after land reform certainly improved peas-

5. This policy was undoubtedly a tactical move by the CPP whose eventual goal, based on Communist premises, was to move beyond smallholdings to a fully collectivized agricultural regime (Susan Greenhalgh, pers. comm.).

ant welfare, the government believed that the rejuvenation of smallholders was not the path of socialist progress and modernization. This was a
profoundly political conviction, based on a Marxist understanding of history and science and on refusal to accept temporary economic advantage
that conflicted with the radical restructuring of society. The ideological
goal of equity along with higher production, it was believed, could not
be met under a system of private property and small-scale unmechanized
farms (Hinton 1983: 114). It was felt that there were just not enough land
and draft animals to allow endless reduplication of small farms, and that
the pre-liberation desire "of every peasant, both rich and poor, to save
enough money to buy more land" (Shue 1980: 275) meant that only a few
in each community could ever become rich. Landlords and rich peasants
were by definition exploiters who hired labor, bought land, lent money
at interest, and set up small private enterprises, while the majority who
had to work for wages, sell land, and borrow money inevitably slipped
into poverty (Hinton 1983: 65). Mao contended that redistributed land in
China might result in the same polarization and growth of a rich kulak
farmer economy that had purportedly taken place in the Soviet Union
after 1917 (Hinton 1983: 66). Land reform had not eliminated wide disparities in wealth between classes within the village (Shue 1980: 282).

When voluntary sharing of labor and resources in the Mutual Aid
Teams did not remedy the negative aspects of private production and the
slow growth in output, the decision was made in 1955 to speed up socialization. In economic terms, the pooling of land would allow economies
of scale. Eliminating boundary mounds and making big fields out of
small irregular properties would increase the land under cultivation and
allow mechanization. Cropping could be more efficient if it did not serve
peasant subsistence diversity but specialized in the products best suited to
natural conditions. The scale of labor mobilization could also increase,
with large groups using off-season slack times to build reservoirs, dams,
embankments, irrigation ditches, and wells to increase productivity
(Shue 1980: 279–82). Massed labor and bigger fields would result in
larger output, with a surplus that could be invested in better seeds, fertilizer, and technology. As in the West, traditional agricultural methods and
hand labor were seen as inherently inefficient (Bray 1986: 6), and the concentration of land and labor into larger units was thought able to bring
underutilized resources into higher production (Nee 1985: 182). If rapid
modernization of agriculture could be achieved by the development of
large-scale collective institutions without a high rate of capital investment, industrial growth could also be financed.

The rhetoric of scale, mechanization, and modernization that characterizes American industrial agriculture appears just as clearly in the dis-
courses on communist collective production, though questions of prop-

erty rights and equality are very differently addressed. The distinctive features of intensive agriculture are, however, seemingly ignored, and, paradoxically, this was done by the Chinese, who created the outstanding exemplar of this farming system. Economic rationality was narrowly construed as that of the factory under state control, rather than that of the individual, the household, or the market, and both the ends and means of agricultural policy were dictated by ideology. The new suprahousehold units of production, while justified on grounds of efficiency, also provided a means to regularize extraction of agricultural surpluses by the state and thereby help to finance socialist construction (Nee 1985: 182). Although manual labor was supposed to wither away eventually, the interim goal was to organize work more correctly. "Concentration of labor may facilitate the mobilization of labor for infrastructural construction, such as roads, public buildings, and irrigation systems, but it is not clear that it actually leads to greater efficiency in wetland rice agriculture" (ibid.: 183).

The collective organization in the period 1962–79 was based on the production team, a group of about 26 families who came from the same village and often shared ties of kinship and neighborhood (Parish 1985: 5). The team owned all cultivated lands, livestock, smaller machines, and agricultural tools, and it was the unit that organized labor, assigned work points, and divided up the harvest (Khan 1984: 77). The team, led by one of its own members, who was paid out of farm receipts, was responsible for turning over certain quotas of agricultural products to the state, and the team paid into funds that supported public services. The next level of organization, the brigade, sometimes equivalent to a village, with about 1,000 people or 7.4 teams, administered smaller workshops, heavy machines, and irrigation facilities. The commune, uniting a number of villages and hamlets, was a unit of local government with workshops, enterprises, hospitals, schools, and local militia, and it comprised an average of 13.5 brigades (ibid.: 77).[6] Individual households retained their houses, household goods, and small farm and handicraft tools, and were assigned private plots, where they raised vegetables and pigs (Hinton 1983: 114). Private plots constituted 5 to 7 percent of village land,[7] and from these

---

6. Philip Huang (1985: 307) points out that the collectivized production team shared fundamental features with the family farm. As a unit of both production and consumption, it could tolerate agricultural involution and declining marginal productivity to an extent unthinkable for a capitalist enterprise. The team could not fire surplus labor. The artificial units of the team and brigade, like imperial administrative units of the past, were imposed with the recognition of the village as a natural socioeconomic unit. Efforts to shift ownership and accounting to the supra-village level during the Great Leap Forward were unsuccessful (Huang 1985: 307).

7. Policies on private plots varied considerably over time, and in some areas they were not permitted.

gardens came almost a quarter of China's fruit and vegetables and more than 70 percent of its agricultural raw materials (Perkins and Yusuf 1984: 83).[8] Spare-time work on private plots, cottage-industry sideline activities, and private commerce at local fairs were strongly attacked during the Cultural Revolution as capitalist practices, but their economic importance was recognized and encouraged after 1978 (Khan 1984: 78–79).

The achievements of collective organization have been widely recognized. In the north Chinese village of Taitou, previously studied by Martin Yang (1965), Norma Diamond (1985) found a higher standard of living, even for rich peasants, when compared with the 1940s, along with better medical care, education, some mechanization of agriculture, field consolidation, and the rechanneling of a local river to provide more arable land.[9] But in general, the projected growth in agricultural production expected to follow from economies of scale in land use and labor mobilization, modernization of the means of production, and greater social equality did not materialize.

Work Incentives That Didn't Work

It is more accurate to say that although grain production and the income from rural industries rose substantially during the Maoist period, its growth did not consistently exceed that of the population. Per capita rice production and income were stagnant or even falling between 1962 and 1978 (Potter and Potter 1990: 161–65). Peasants' standard of living was falling behind that of urban areas and bore no comparison to that of Chinese in Taiwan or Hong Kong (ibid.: 166).

The fundamental problem in collective farming is one of incentives, and in an intensive system based on skill and individual effort, this difficulty is exacerbated. How can the group make rewards to individuals commensurate with the quantity and quality of their work? Of course, if

8. Unused land along roads or canal banks or unclaimed patches in the mountains could also be appropriated for private use and their produce privately marketed: "Private land yielded well because intense effort went into it. People got up early, stayed out late, and skipped their rest at noon to till these plots. If they had any manure in the privy at home or any compost in their pigpen they took it to this private land instead of turning it over to their team for work points. They didn't do it openly, in the daytime, but late at night when most people were asleep. And they didn't take it out with carts that rumbled but walked out quietly with carrying poles. They worked so hard on private land at night that when they turned up for collective work in the daytime they were worn out. They worked slowly and found it necessary to lie down frequently for long naps" (Hinton 1983: 289).

9. The basis for comparison is, of course, a wartime period of considerable economic and political disturbance. The provision of peace and security was certainly one of the revolution's major benefits to the countryside. As I have indicated, the infrastructural measures for flood control, irrigation, and land reclamation may have contributed much more to agriculture than did mechanization and field consolidation.

individuals are motivated mainly by altruistic desires to share with a wide circle of kin or neighbors, if they produce in order to maintain long-term reciprocal obligations that guarantee their security in some type of "moral economy" (J. C. Scott 1976), or if they work principally for prestige and political power (Sahlins 1972), rational calculations of costs and benefits under different institutional arrangements might be irrelevant. But the behavior of Chinese peasants in typically working harder on their private plots, sidelines, and household chores than on the collective fields reflected a pronounced tendency to maximize individual household advantage over the interests of the collective economy (Nee 1985: 172).

Intensive agriculture, with varied tasks and a great many management decisions, performed over long periods of time for rewards deferred for seasons or even years, is poorly adapted to collective social organization. Rigid state policies and centralized planning do not promote local sustainable practices. When Chinese peasant households were combined into integrated working units and the boundaries between fields were erased, it became impossible to identify any particular ear or row of corn as having been produced by a specific household (Putterman 1985: 67). As the economist conceptualizes marginalist work incentives, how can the household's "share of team income be calculated so that the net output value of an extra ounce of effort is returned to them, inducing them to undertake the effort provided its benefits exceed their cost"? (ibid.:67) Even in relatively small, face-to-face production teams, those who worked harder felt that they might be subsidizing the lazy, a classic free-rider dilemma (Nee 1985: 172). Despite the claims of socialism, the team is not a household writ large. "The nature of agricultural work is such that, as one moves out of the organizational framework of the peasant family into that of a collective, the evaluation of performance, the institution of a system of payments, the organization of management decisions and related matters becomes exceedingly difficult" (Khan 1984: 122).

Production for the free market, which continued to exist in some form in most localities, added further calculations. For example, complex investment decisions went into when to purchase a pig for private rearing, whether to buy extra feed grain (and, if one did, how to ensure that one had enough money left over to feed the household), and how to hold off selling one's pig till the spring hunger period, when prices were highest (Nee 1985: 172). Timing could result in greater or smaller profits, given equal skills in pig raising, and the peasant household, attentive to price fluctuations in local markets, could respond flexibly to economic conditions in a way larger entities could not. Returns that went directly to the producing household provided effective incentives.

The solution to equating pay with work in collective agriculture was to award work points tabulated on the basis of number of days or hours expended, the difficulty of the task, the strength (or gender) of the worker, or some combination of the above. Earnings both in kind and cash depended on the total income of the team or other cooperative unit after deducting for production expenses, reserve fund, welfare fund, taxes, and so on (Hinton 1983: 123). Work points based on task completion were difficult to fix scientifically "because the best way and timing of performing a particular task, and the amount of labor required, will tend to vary with natural conditions, time of day, season, soil moisture, and other factors" (Putterman 1985: 70). Cataloguing thousands of distinct tasks and assigning them work points, and then supervising task completion and keeping accurate accounts required teams and their officials to devote tremendous efforts to administration alone (ibid.; Hinton 1983: 124).[10] At times there were nightly reviews of the tasks done and the performance of individuals, often leading to conflict and disagreement within the team. If points per day were fixed for an individual over longer timespans, they might reflect the member's social status or seniority (Nee 1984: 178). If women, the old, and children were assigned minimum points, they would be unenthusiastic about working for the cooperative (Shue 1980: 302). The oxen belonging to a team might be poorly tended because the job earned few points and then worked to death in plowing that accumulated high work-point totals (ibid.: 310). If points were given for tasks without careful monitoring, there might be collusion among team members to do fast, poor-quality work.

Though it has been claimed that the work-point system worked quite well, and that only lazy people complained that rates were set too high (Hinton 1983: 155), other commentators contend that the stronger, more experienced, and more capable farmers were among the most discontented team members (Nee 1985: 180). The elaborate apparatus for setting standards and assigning work points does not appear to have matched individual labor satisfactorily, either quantitatively or qualitatively, to returns, and it failed to provide meaningful economic incentives in agriculture. The Herculean efforts of the regime to supply this defi-

10. Implementing a work-point system originally developed in the Soviet Union for a very different type of agriculture was far from easy in the Chinese context: "Coop leaders found setting standards for hauling manure and compost to the field to be an especially complicated matter. There were near fields, middle distance fields, and far fields. The quality of the animal and the size of the cart also entered into the equation. They came to expect that each worker with a standard flat cart and healthy donkey would haul 25 loads of 10 baskets each to nearby fields, 18 loads of the same volume to middle distance fields, and 15 loads to far fields. A man with a carrying pole could carry only two baskets to the field at one time or one-fifth as much as a man with a donkey and a cart. Hence on this job nobody tried to earn work points with a carrying pole" (Hinton 1983: 154).

ciency with ideological education and persuasion were also inadequate. "New values about cooperation, sharing, and the power of collective effort may be taught in production meetings, but usually it is only by putting things in economic terms that interest can be generated" (Parish and Whyte 1978: 41).

The "Household Responsibility System": Smallholders Resurgent

If an objective measurement of work input is not possible, then letting the result of work directly measure the work itself would be a better approach (Putterman 1985: 71). In 1978, the Chinese recognized that the difficulties of establishing distribution of rewards according to work under unified collective management in agriculture were more or less insurmountable. The special character of intensive agriculture as "a natural process" (ibid.: 71) required the close, day-by-day adjustments in local ecosystemic relationships of environment, technology, and labor that could be provided most effectively by the household as the key productive unit. Collective cultivation was replaced by the household responsibility system under which team land, farm animals, and most equipment were divided up among member households on the basis of equal amounts per capita or per worker (Khan 1984: 88).[11] Decisions on what to plant and how to manage labor inputs were left to the household; records of work points were no longer kept, and neither were food rations dispensed. Each household had an annual quota of specified agricultural products to be sold at stipulated prices to the state, but all production beyond this amount belonged to the household for consumption or for sale, either on the market or to government. Households were also required to pay a share of team taxes and make stated contributions to team capital accumulation and welfare funds (Khan 1984: 87). The family household signed a contract with the team for exclusive rights to use land for periods that were originally 2 to 3 years but have now increased to 15 years or more (S. Huang 1989: 166). Households were also given access to credit and the right to purchase mechanized equipment, and limited labor hiring was allowed (Putterman 1987: 122). The new system was institutionally very similar to the model of Chinese smallholders in the past. It was essentially "a system of guaranteed tenancy—peasant control of land—at a fixed rent" (Khan 1984: 27), and the state and collective

11. Peasants in poor, backward areas had adopted systems of contracting output quotas to individual households in the 1950s, but the government criticized this system as a departure from the socialist path and abolished it when the communes were set up (Khan 1984: 85). Household responsibility was espoused by Deng Xiaoping in 1962 but the plan was turned down by the Central Committee (Putterman 1985: 72).

TABLE 8.2
Chinese Annual Growth Rates, 1965–84
(Percent)

	1965–78		1978–84	
	Aggregate	Per commune worker	Aggregate	Per commune worker
Gross value agricultural output	4.0%	2.0%	9.0%	6.6%
Gross value crop output	3.2	1.2	6.6	4.2
Gross value livestock, forest, fish output	3.9	1.9	9.0	6.6
Gross value sidelines, industries	10.7	8.7	18.3	15.9

SOURCE: Ghose 1987: 38. Used with permission.

functioned as benevolent landlords who distributed land equally (Parish 1985: 18).

What began in 1979 with a few production teams in Sichuan and Anhui provinces contracting with households to grow specified quantities of crops or raise a certain number of animals expanded so rapidly that it encompassed the majority of villages by 1982 and all but a small percentage of China's peasants by 1984 (Smil 1985: 118; Hartford 1985). It was an extraordinarily swift and bloodless revolution, amounting to the "de facto privatization of Chinese farming" (Smil 1985: 118). Household responsibility also represented a stunning ideological and philosophical revolution:

The restoration of peasant households to the status of production units in mainland China, largely completed between 1981 and 1983, has sent shock waves throughout the world. Mao Zedong went beyond Lenin to make his revolution one not only against capitalist property relations but also (eventually) against material self-interest as a principle of economic and social behavior. The replacement of team farming by household farming, coupled as it is on the ideological front with the replacement of such slogans as "Fight Self" and "Serve the People" by "Fight Egalitarianism" and "To Become Rich Is Glorious," is an event the significance of which is difficult to overstate. (Putterman 1987: 104)

As a spur to production, decollectivization worked. Agricultural production, which had kept slightly ahead of population growth during the Maoist period,[12] spurted ahead after 1978 (Ghose 1987). Livestock, forest, and fish production went up faster than crops, and the value of annual increases in sidelines and local industries was more than 18 percent a year (Table 8.2). Instead of requiring local self-sufficiency in grain, the regime allowed diverse, regionally specialized commodity production in response to market demands (Croll 1987).

Chinese smallholders often decreased subsistence concentration on ce-

12. While China's 1949 population went up more than 75 percent by 1978, per capita grain production rose at a somewhat faster rate (Putterman 1987: 104).

reals in favor of higher-value products. Between 1978, the last year of the old policy regime, and 1982, production of all grains went up only 16 percent, while pork, beef, and mutton climbed 58 percent, cotton rose 66 percent, and oil-bearing crops jumped 127 percent (Griffin 1984: 304). Over the same four years, diet in rural areas improved, with the consumption of the "superior" food grains, rice and wheat, going up by more than half, accompanied by increases of 57 percent in fish and meat and 212 percent in poultry consumption (ibid.: 306). Per capita incomes went up even faster than production, by 102.3 percent, and indices of living standards like average per capita floor space climbed almost a third, from 10.7 to 13.41 m², while synthetic cloth purchases rose 270 percent (ibid.: 305–6). From 1979 to 1985, a village with capitalist joint-venture factories as well as household-based agriculture and petty commerce showed prosperity in the increase in privately owned TVs, refrigerators, washing machines, motor vehicles, and new houses (Potter and Potter 1990: 327).

Not all of the credit for greater production can be given to the institutional change that replaced collective organization with the household.[13] The years immediately following 1978 had good growing weather, and the macroeconomic policy changes that increased the prices of agricultural goods turned the prevailing terms of trade to the benefit of peasants (Griffin 1984: 304). State purchase prices went up some 25 percent (Potter and Potter 1990: 332). At the same time that the government reduced taxes on grain in real terms, it also lowered the quota amounts due to the state (Croll 1987: 108) and allowed the rice quota to be met in the form of cash (Potter and Potter 1990: 332). In 1984, individual peasants and rural supply and marketing cooperatives were permitted to transport grain to and from, and buy and sell it in, markets both within and outside their counties. The state no longer monopolized the purchase of grain, and peasant households could either contract with the village government to

---

13. Modern technology also had an increasing impact during these years. In the Yangzi Delta, the state managed to supply all the fertilizer peasants could use during the late 1970s. Small hand-held tractors permitted more efficient plowing and triple-cropping. Wheat cultivation expanded with improvements in underground drainage (Huang 1990: 316). Between 1978 and 1984, grain production grew by 16.25 billion kg a year, sown area was reduced by 7.7 million ha, and output per ha increased from 2,535 kg to 3,615 kg. An official from the Ministry of Agriculture (Guo 1990: 19) cited six reasons for the higher intensity and production increases: (1) introduction of the contract-responsibility system based on the household, with remuneration linked to output; (2) raising prices for grain; (3) use of more fertilizers; (4) wider popularization of practical scientific and technical resources, such as hybrid rice and maize; (5) more water conservancy and irrigation projects and farmland capital construction conducted by the state since the late 1950s; and (6) more investments by farmers in farm production.

grow a specified amount of grain or choose to produce solely for the market (Croll 1987: 108–9).

An increased reliance on the market was further encouraged by the relaxation of policies that favored a rural subsistence economy with all areas basically self-sufficient in food grains and substituted instead the active fostering of regional comparative advantage according to environmental possibilities (Griffin 1984: 304; Putterman 1987: 105). Hilly and mountainous areas could specialize in forestry and extensive herding, while "regions such as the Pearl [Zhu Jiang] and Yangtze river deltas which are better suited to the production of cash crops such as sugarcane, flowers, vegetables, bananas, and fish farming" would develop these products for the market (Croll 1987: 112). A district in Guangdong reduced the area planted in rice and sugarcane, practically eliminated cassava and jute, and converted substantial land to the more intensive, higher-value production of fruit trees (bananas, litchis, and oranges), vegetables, and fishponds (Potter and Potter 1990: 332). Domestic handicrafts and other sideline cottage industries also expanded rapidly as peasant households were allowed to subcontract their farmlands to others and specialize on commodity production (Croll 1987: 113). Coastal and plain areas of higher population density could thus take advantage of market exchange, and a higher proportion of resident families could support themselves by providing nonagricultural products.[14] Rapid smallholder adaptation to the range of new economic niches is witnessed by a group of Shaanxi villages where in 1983 "approximately one-third of the peasant households remained in diversified farming activities, one-third specialized in industrial and sideline production or in providing services such as transport, commerce, and water conservation, and the remaining third specialized in commodity grain production" (Croll 1987: 113).

Could better agricultural prices, more options as to what to plant and how to allocate land, and greater market participation have increased production and individual welfare if the units had remained collectives rather than households? Such advantages might have increased the flexibility of work teams or brigades, but there is no evidence that the structure of individual incentives would have changed. "It is probable that the new responsibility system has affected incentives in such a way that the peasantry now works longer hours than before, works with greater intensity

---

14. It has been suggested that the household-responsibility system in the Yangzi Delta did not improve crop yields at all, but rather led to the diversification of the rural economy. In this view, "rural industrialization and sideline development, mostly under collective units" was largely responsible for the rise in peasant incomes; with expanding off-farm employment, farm labor became a relatively scarce resource, and returns to labor could rise to more nearly optimal levels (P. C. C. Huang 1990: 18, 318).

per hour than previously and also works with greater intelligence, imagination and creativity. It is impossible to quantify this effect with the existing data, but our observations in the field nonetheless suggest that it is real," Keith Griffin observes (1984: 305). Households responded to the new incentives by increasing production using the familiar methods of intensification. During 1978–83, grain output per capita increased at a rate more than ten times that of the 1957–78 period, but the area sown in grain actually declined by a total of 5 percent in these years (Putterman 1987: 105). But the marginal productivity and efficiency of labor did not fall. Substantial numbers of people in what had been thought to be a fully employed rural work force actually left farming for other occupations, and the output per farm worker, which had long been stagnant, rose by 12.7 percent from 1978 to 1981 (ibid.: 105–6). With production-team leaders no longer allocating tasks and assigning work points on the basis of time spent in the fields, peasants could fit their labor directly to crop needs, and the demand for labor in the fields declined significantly; one group of 70 that was responsible for grain and vegetable cultivation on 300 mu of land was reduced to 45 in the first year of contract and 25 by the third year (Croll 1987: 124).

The very rapid reassertion of the household as the dominant productive unit in Chinese agriculture suggests not only its traditional centrality in intensive farming systems, but also the continued dependence of individual members on the household throughout the period of collective economy. Team and brigade organizations provided a basic grain ration, modest welfare funds for the poorest villagers, and inexpensive health care, but the household produced and pooled all additional subsistence and cash income (Nee 1985: 169).[15] Incomplete households of single elderly people, widows with young children, and orphans got little informal assistance from neighbors. Victor Nee remarks on the "lack of a tradition of charity, even between kinsmen" and a minimal village welfare system that rendered "peasants all the more dependent upon their family as a source of basic security and well being" (1985: 170).

The collective division of labor required an adult woman for work on the private plot and domestic chores, and the culturally valued stem-family pattern allowed grandparents to take care of young children and look after the house, freeing the householder's wife to earn additional work points in the fields (Nee 1985: 171).[16] Under the household-respon-

15. When the road building and backyard blast furnaces of the Great Leap Forward reduced agricultural labor and crops, villagers were thrown back on household resources to get through the famine (Nee 1985: 180). This experience may have permanently disillusioned peasants with the more radical forms of collective farming.

16. The actual prevalence of the stem-family household type in pre- and post-revolutionary China as well as after rural liberalization is not clear, but one might expect

sibility system, labor could be more efficiently allocated along traditional lines, especially where there was a scarcity of arable land per capita and a surplus of labor:

The expansion of domestic sidelines and diversifying the economy by raising pigs, chickens and other animals, fish farming, the cultivation of vegetables and fruit and the establishment of handicraft industries, have all broadened the scope of women's income-generating activities so that the pattern whereby women stay at home and are engaged in various side-occupations is increasingly characteristic of households in which the males are employed in full-time agricultural field-work. (Croll 1987: 125)

With rights in land being exercised over longer spans of time at the household level and a symbiotic division of labor in which the woman's contribution is significant, corporate continuity of the household will be further emphasized. Improving household welfare may have been achieved at the expense of a decline in women's status since the early 1980s. Women's contributions (except for certain lucrative sidelines) are now *less* clearly identifiable, since they no longer earn discrete numbers of workpoints but are merely laborers in the household work force (Susan Greenhalgh, pers. comm.). Household responsibility may merely have restored patriarchal power over subordinate women and children (Nee 1986: 198; Potter and Potter 1990: 336).

The conditions of smallholder cultivation also make the replacement of family by hired labor unlikely. Though certain bottleneck tasks such as harvesting have already brought migrations of temporary labor from poorer to richer areas, the more responsible, flexible, and manageable household labor force continues to function best in situations of limited arable land. Changes in the technology of rice agriculture such as the use of walking tractors, improvement of seeds, and the utilization of chemical fertilizers and pesticides are unlikely to affect social arrangements either. "The simple fact remains that the division of labor contained in a peasant household is still adequate for handling the entire cycle of agricultural production" (Nee 1985: 185).

Possibilities of increased income and the salience of the household for reproduction and support of the elderly mean that there is an advantage

---

considerable variation related to land availability and wealth. The large joint family, in which several brothers with their wives and children inhabit the same household with their parents, may always have been rare (Gonzalez 1983: 87). According to a village study done in Guangdong Province near Hong Kong, "only those landlords and merchant families who constituted (according to brigade records) 5 percent of Zangbu's population would have had the economic resources to support such a large number of people, the status and power to effectively govern them, and the property to motivate sons to stay together under their father's authority to await their inheritance" (Potter and Potter 1990: 19). Landless laborers might be too poor to even hold a stable nuclear family together.

in having more family members. More children and additional married couples in multiple-family households provide more cultivators, diversification into domestic sidelines, and remittances from nonresident members. Owing to scarcity of housing, and with a view to reducing the liability of the collective to care for the aged, both ordinary people and Communist Party officials have advocated that widowed or single adults live with married children or a sibling's family (Gonzalez 1983: 85). The traditionally strong association between family size and wealth is reinforced as households again become the main units of production. One 1980 survey showed only 8 percent of peasant households with one or two members had per capita incomes of more than 100 yuan, while 31 percent of groups consisting of six to eight members exceeded this level (Croll 1987: 122). The desire to maximize family labor resources has led peasants to oppose the single-child policy of the government, to reinforce the customary preference for male offspring, and to withdraw children from school (ibid.: 123). Farm-family households may indeed have corporate economic interests that put them in opposition to state-level long-term policies.

Differentiation and the Dream of Equality

Household responsibility, the rapid growth of a market economy, and the privatization of the means of production are seen by convinced socialists as sounding the knell of equality in a classless society. In the "Great Reversal" after 1978, William Hinton (1990: 19) finds accelerated social polarization, meaning "class differentiation, primarily the large-scale shift from peasant smallholder (in cooperative China this meant community shareholder) to wage laborer, and at the same time, the small-scale counter shift from peasant smallholder to capitalist (mostly petty)." Prerevolutionary rural income inequalities were based largely on ownership and control of land. It is true that land reform, collectivization, and the loss of diverse off-farm sidelines eliminated most of these significant property-based inequalities. Land reform increased the average income of the poorest 20 percent of the rural population by nearly 90 percent in real terms, while that of the next 40 percent rose by perhaps 15 percent (Perkins 1978: 562). In the collectivized villages, differences in cash and work-point income were based chiefly on the Chayanovian worker/consumer ratio. The lowest-ranking households had higher proportions of young children and retired parents. The most prosperous phase of the developmental cycle was that of the family with unmarried, working children in the 15–28 age bracket who were still at home (Selden 1985: 195).

Income differentiation in rural communities in the period from 1955 to the 1970s was in the Gini coefficient range of .16–.27, and household per capita income distribution in 1952 was only .22 as compared with .33 in 1934 (Seldon 1985: 203–4). Income inequality in the China of the 1970s was somewhat lower than in India and Indonesia, and China had in addition more adequate health, education, and welfare benefits (Selden 1985: 210). The prosperity of 1978–83 led average household income to jump from 134 yuan to 310 yuan, including farm and off-farm earnings and remittances from cities. But, contrary to expectations, including those of the government, the degree of inequality in the distribution of income did not increase (Griffin 1984: 307). The Gini of household income nationwide in fact declined steadily from .28 in 1978 to .22 in 1983 (Selden 1985: 212). The change was *toward* a more egalitarian countryside rather than away from it, and diversified private-sector income may improve the distribution of income when compared to collective public income alone (Griffin 1984: 308). A study of Dahe Township in Hebei province demonstrates an opposite conclusion. There the Ginis for per capita income at a household level go from .231 in 1979 to .30 in 1985 (Putterman 1989: 277–307). Whereas differentiation at the earlier period was influenced most by the ratio of dependents to workers in the household, by 1985 occupational factors and sources of income had the dominant effect (ibid.: 310).

It is clear that areas with less fertile land, less water, and more difficult transport have not had rapid increases in per capita income, and inequality between poor and rich regions may have increased (E. B. Vermeer 1982). But there seem to be no good answers as to whether income differentials within communities over the decade of the 1980s in fact went up. Do higher average incomes reflect generally improving welfare? Teachers' salaries and medical services were becoming a problem in many communities, and families at the bottom of the wealth scale had less security. Jealousy was expressed against families with new trucks and tractors, and in some cases they were forcibly deprived of new signs of wealth (Parish 1985: 21). Village-based brick-making, truck-transport, and small manufacturing enterprises have raised the incomes of their managers and investors, who were often former team leaders, Communist Party cadres, army veterans with skills, and educated professionals (S. Huang 1989: 194). Farming may, however, lead to less differentiation than commercial and industrial activities, where economies of scale and hierarchical management are more compelling.

Legally, land still belongs to the collective and cannot be sold, though rules against renting and transferring use rights are not rigidly enforced (Griffin 1984: 311). In principle, land should be reassigned to match the changing demographic profiles of farming households. But "frequent reallocations of land would increase uncertainty of tenure and discourage peasants from investing in land" (Griffin 1984: 312). The same economic factors that prevented land concentration and dependence on hired rather than household labor for intensive smallholder cultivation in the past are demonstrably at work in the present. Limited differentiation among farm households and mobility in wealth should continue to occur, but Leninist polarization is a specter that shows no sign of materializing.

Without Cooperation, Can There Be Modernization?

Critics of rural reform point to the fact that a doctrinaire application of the household-responsibility system and dispersing of collective assets in machinery has fostered inefficiency as well as inequality and corruption. Those who choose to remain in farming may receive subsidized plowing, fertilizer, seed, and marketing services from their communes, and indirectly from profitable rural industries (Hinton 1990: 98). Some northern wheat-growing collectives have managed to stay together, conducting large-scale mechanized operations rather than splitting up their land into individualized "noodle fields." William Hinton, who spent years in northern China as a tractor technician and teacher, says he wept when he viewed this "irrational fragmentation" from the air:

"Noodle land" could only lead, in the long run, to a dead end. I could not think of any place in the world where rural smallholders were faring well, certainly not smallholders with only a fraction of a hectare to their names and that in scattered fragments. The low output of peasants farming with hoes meant that on the average each full-time laborer could produce about a ton of grain a year, one eight-hundredth of the amount I harvested farming with tractors in Pennsylvania. (Hinton 1990: 17)

But the mechanized success stories Hinton cites refer to extensive farms, settled 20 years ago on wasteland and providing 50 acres per laborer, or 10 times the average in northeastern China (Hinton 1990: 105). His conventional view of labor-saving, modern agriculture may well be appropriate to dry-wheat production with abundant land, but it denies the viability of the historic, sustainable smallholder strategy of the Chinese.

The prejudice against intensive methods also holds for students of southern double-cropped rice areas with fruit orchards, vegetables, and fishponds. Sulamith and Jack Potter characterize Guangdong's traditional agriculture in the 1930s and 1940s as very backward, with primitive technology, poorly maintained irrigation and flood-control facilities, and small farms so fragmented that they were inefficient to work (Potter and Potter 1990: 35, 176). Population pressure and involution meant, they say, that "the old mode of family production in rice agriculture was no longer viable by the beginning of the twentieth century" (ibid.: 27). As they see it, the production-responsibility system has induced a technological regression in agriculture. Tractors and motorized threshing machines are too expensive for the single household, and increased human labor is thus replacing the "economies of scale" of the erstwhile Maoist collectives (ibid.: 335–36).

There are, of course, limits to what either households or capitalistically organized business enterprises can do. The social services of health care, education, and care for the needy require some collective responsibility, and they may be sacrificed in the short run to the race for rapid agricultural growth (Parish 1985: 23). The rapid diversification of agriculture toward crops with high market value may lead to declines in basic grain production, risking shortages in bad years and dependency on international suppliers. Public work for vital irrigation and flood-control infrastructure can no longer depend on state-organized labor by millions of workers. Both investment and total irrigated land declined from 1980 to 1984 (Lucas n.d.). In fact, irrigation maintenance continues to be carried on by village-level groups (Patnaik 1988), although even water management has in some cases been decentralized from the state-level, professionally managed system (Lucas n.d.).

State extension services are now being supplemented by private contracted extension on some cash crops and small stock (Delman 1988). Whether common property institutions for distributing irrigation water and for administering fragile hill lands for sustainable grazing and firewood can be revived or built upon the framework of brigades and communes remains to be seen.[17] Certainly the housing boom and the increase in private livestock and charcoal making threaten to result in erosion and degradation of the land. Hinton asserts that the reform has "unleashed in its wake an unprecedented attack on the environment" and sent hundreds of millions out for an instant profit by attacking "anything that could be cut down, plowed up, pumped over, dug out, shot dead, or carried away" (1990: 21). This melodramatic charge appears to ignore the sustainability that has characterized Chinese intensive land use for several millennia and the restraints that keep the smallholder from destroying the property that is the heritage of the next generation. Collective control is not the only road to environmental conservation. The extent and social context of resource degradation are matters for empirical investigation.

It is also not at all clear whether specialists contracting for agricultural land-use rights will decrease food production and increase rural proletarianization. These assertions, and the claim that "privatization has created a situation that all but guarantees agricultural stagnation" (Hinton 1990: 173), contradict much of what the Chinese historical experience teaches about smallholder behavior. Perhaps, ultimately, peasant welfare is con-

17. Louis Putterman (1985: 81) sees less chance that Chinese policymakers will support the more complex communal supervision of resource use than that they will make tenure rights more permanent and transferable, emulating a capitalist system of agriculture.

ditional on the ability to apply intensive, diversified, and flexible techniques by households with secure rights to property, but always in the context of a market economy that is peaceful and well integrated by efficient networks of transport and communication. Farming households are especially dependent on the options of their members to redirect their efforts *outside* agriculture and intensify cottage industry, rural manufacturing, commerce, and part-time wage work. Smallholder agriculture has not been in the past and is not today a prescription for technological stagnation, class polarization, or rural immiseration. Collectivization of agriculture, despite its rhetoric of cooperation, sharing, and equality, has denied the incentives that lead households to plan and work for their own advancement. Farm communes organized on industrial lines impede the economic growth that follows from alternative uses of labor, land, and management skills that are open to the members of corporate households.

CHAPTER NINE

Intensive Agriculture, Population Density, Markets, and the Smallholder Adaptation

I BELIEVE THAT the most adequate intellectual framework for understanding the distinctive and cross-culturally recurrent smallholder adaptation is that of Ester Boserup, as amplified and revised by geographers, economists, and anthropologists. Boserup's synthesis, which postulates regular relationships between intensive agriculture, population density, and markets, also suggests lines of causality and processes of change different from those invoked by the classic formulations of Malthus and Marx. As developed in her major publications of 1965, 1981, and 1990, Boserup's system breaks with the more rigid theories of environmental, technological, demographic, and political-economic determinism, while attempting to account for the same set of empirical observations. As in the cautious and limited functional cultural ecology of Julian Steward (Roseberry 1989: 18), there is no effort here at a grand, inclusive schema like those of cultural evolution, cultural materialism, the capitalist mode of production, or the world system. Boserup does not propose a theory of the rise of the state, the causes of population growth, the roots of the Industrial Revolution, or the universal categories of human thought. But the correlation of farming system and rural population pressure that she offers appears to have objective validity and to make sense of patterned change through time in many unrelated societies. It also attains paradigmatic status in its implications for such social variables as the household organization of production, coexisting types of private and communal land tenure, and bounded inequality within the community of cultivators. I have tried in this book to trace these consistent associations.

The description in the preceding chapters of what smallholder intensive farmers *do*—their technology and knowledge, their means of organizing labor, and their rights to resources—has been heavy on case examples and limited populations but light on comprehensive theory. Such an approach lends itself to an empirical and largely inductive presentation, but it avoids the precise definitions, the generalizations, and the analysis of total systems that are the stuff of social science. These ideas and interpretations are also the contested grounds on which scholars wage their wars and struggle for respect and reputation. I have not dealt fully with either the positive contributions or the criticism of Boserup's theory in its historical and contemporary context. The obligatory (and often turgid) "review of the literature," followed by the issues, the arguments, and the staking of claims to turf usually takes pride of place at the head of the book. But I have chosen to lay out the evidence for a somewhat limited and possibly derivative point of view on smallholder agriculture before grounding it in the body of theory that relates these practices to their larger context in demography and economics. This chapter examines the theoretical premises of and the ongoing debates about the Boserup position.

As Chapters 1 and 5 demonstrated, smallholder farmers are not just food producers with little fields. They practice permanent *intensive* agriculture, achieving higher total output per unit of land area and time (typically per ha and year) than that produced by more temporary or *extensive* systems of land use, such as shifting cultivation or ranching (Boserup 1965; Netting 1968: 55; Turner and Doolittle 1978; Carlstein 1982: 150, 348). Intensification, more precisely defined, is a process of increasing the utilization or productivity of land currently under production, and it contrasts with expansion, that is, the extension of land under cultivation (Turner et al. 1992).[1] As Boserup (pers. comm.) points out, any part of a community territory, and not merely the land currently under produc-

---

1. The term *intensification* as used in the literature may be confusing because it is often applied not to land use but to other factors of production, such as capital or labor. Economists commonly refer to the quantity of resources that combine with a given quantity of land, so that intensive cultivation is that in which a small area of farm is combined with a large quantity of other resources (Ellis 1988: 196). If, for example, a new tool, such as the plow, or a new means of harnessing energy, such as draft animals or the tractor, are introduced, the effect may be to expand the area under cultivation rapidly and actually *decrease* output per unit area. Slash-and-burn agriculturalists may encounter more difficult terrain or make larger fields to generate a surplus for market. Their yearly total of labor hours can increase in either case, but their extensive land-use system has not necessarily changed. One can also imagine food producers moving onto initially abundant alluvial or volcanically derived soils that produce dependable annual crops with little labor and without the need to restore the nutrient status of the land artificially. Such agriculture would be intensive by this definition, but it would not entail higher capital or labor inputs. *Intensive agriculture* is a term that specifies a higher total output in the spatial and temporal context than could be produced by extensive means, and it applies particularly to land use that maintains an area in crop production for more years than it is fallowed (cf. Conklin 1961). Defining intensive agriculture in terms of yields per unit of land over time emphasizes output as the dependent variable, and it does not prejudge the effect of economic inputs of labor, capital, or technological change. Increases in these independent variables, singly or in combination, on a constant land area, *may* intensify its use (cf. Burton and White 1984; Bradley et al. 1990), but

tion, can be intensified by shorter fallows and more frequent cultivation. Even conversion of forest to pasture can intensify land use, both by directly increasing food production and by indirectly providing animal traction and manure for cultivated areas. Expansion of low-density populations on land-abundant frontiers, however, usually promotes extensive land use with low and transitory returns.

Anyone can recognize intensively tilled smallholdings in a checkerboard of walled, manured fields, in irrigated rice paddies, or in orchards and vineyards, but it was Ester Boserup (1965, 1981) who gave a new and compelling answer to the question of how this distinctive adaptation came to be and why it persisted. As a Danish economist who worked for the United Nations during the heyday of modernization theory and the flush of international development after World War II, Boserup saw in Asia a smallholder pattern that ran counter to the received wisdom of economics. She rejected the premise that agricultural change originated most significantly from technological innovation, science, and the progressive capture of new energy sources. She also denied that the natural fertility of the soil and other environmental parameters decisively limited human exploitation of any given land area, and that this "carrying capacity" constraint could only be breached by inventions or new (usually imported) farming methods.

Population Push

Ester Boserup claimed that an increase in population density or land scarcity was an independent variable that could trigger agricultural intensification. *If* population grew (and Boserup neither claimed that demographic increase was inevitable nor originally explained why it took place), and *if* people could not move to areas where resources were still plentiful, they would have to increase the amounts and the predictability of the food they produced from a limited area. Without the spur of scarcity, people do not intensify their farming for the simple economic reason that it is more work—making ridges with a hoe, grubbing up tree stumps, carving out a terrace, or digging an irrigation ditch takes more time and effort than does firing a swidden—and the return on labor may be smaller. Using a variety of cases from preindustrial agriculture, Boserup's hypothesis postulated a regular relationship between rural population density and agricultural intensification, and it further suggested why rational actors might reject the plow or cattle manure or purchased pesticides as offered by an "enlightened" extension agent. The measure of

---

this must be demonstrated by the analysis rather than assumed. The smallholder, for whom expansion of the land base is not a viable alternative, intensifies in a distinctive and predictable manner.

TABLE 9.1

Generalized Land-Use Types, Fallowing, Cropping Frequency,
and Population Density

Land-use type	Period sown	Period fallowed	Fallow vegetation	Frequency of cropping[a] (%)	Population density per/km²	per/mi²
Forest fallow	1–2 yrs	15–25 yrs	High forest	0–10%	0–14	0–10
Bush fallow	2–8 yrs	8–10 yrs	Low forest	10–40	4–16	10–41
Short fallow	≥ 2 yrs	1–2 yrs	Grass	40–80	16–64	41–166
Annual cropping	1 crop/yr	Few months	Grass or none	80–100	64–256	166–663
Multicropping	≥ 2 crops/yr	No fallow	None	200–300	256 +	663 +

SOURCES: Boserup 1965: 28–34; Boserup 1981: 19, 23; Pingali et al. 1987: 27.

[a]Frequency of cropping is cultivated coverage area as a percentage of cultivated plus fallow area. Hans Ruthenberg's (1976: 15–16, 17) *R* value, indicating the proportion of the area under cultivation in relation to the total land available for arable farming, is a similar index. $R \leq 33$ corresponds to shifting cultivation; $R = 33$–66 reflects semipermanent fallowing systems; and $R > 66$ shows permanent systems with clearly demarcated fields, with a predominance of annual or perennial crops, manuring, and possibly irrigation or tree-crop farming. The same measure is called the cropping index (*CI*) by Frederic Pryor (1985: 737).

intensification was the *frequency* of cropping (Table 9.1), arranging land-use systems along a temporal continuum (Carlstein 1982: 348) from un-cropped virgin land to continuous cultivation.[2] Smallholders have a significant portion of their land under one or more crops annually, although they may also have forested woodlots, rough grazing, and long-fallowed outfields as parts of their farming system.

The prevailing direct relationship between rural population densities and the intensity of land use is not a difficult principle to grasp:

The less available land is per capita, the more intensively it must be used. In a comparison of three Mexican communities, Palerm (1955) showed that slash-and-burn farmers use corn fields of only 1½ hectares each year, but a long fallow period requires that a total of 12 hectares be available for rotational use. In another village, fallow and crop time are approximately equal and small permanent gardens are also maintained; there, 2½ hectares are needed annually, but the total land necessary for a family is only 4½ hectares. Where irrigation permits two crops a year and the field is not rested at all, high continuous yields reduce the land requirement, both annual and total, to .86 hectare. Chinampas, raised fields in a grid of canals, support a smallholder household on only .37 ha. Thus a given amount of land can support almost 32 times as many families under a chinampa regime as the same area exploited by shifting techniques. Permanent intensive

2. Frequency of cultivation, as B. L. Turner and W. E. Doolittle (1978) point out, is a surrogate measure for the basic variable of total output per unit of area and time. Because of climatic variability, mixed cropping, different harvest schedules, and the absence of written records, reliable data on yields from fields of a particular size over a period of years are seldom available. Frequency of land use and length of fallow is more crude, but Boserup and others use the measure because of the broad correspondence between total land productivity and cultivation frequency, especially in cases where similar levels of technology are involved (Turner et al. 1992).

cultivation is a means of economizing on land as resources become scarce. (Netting 1986: 70)

Cross-cultural studies show similar population / land use correlations, despite varying local climate, rainfall, soils, crops, and tools. A statistical comparison of 29 groups of tropical subsistence cultivators (without plows and draft animals) indicated that variations in population density accounted for 58 percent of the variation in agricultural intensities (Turner et al. 1977). For West Africa, the geographers M. B. Gleave and H. P. White (1969) found that shifting cultivation, rotational bush fallowing, and permanent agriculture coincided with successively higher average local population densities—and then realized belatedly that these observations fit the Boserup model. Shifting dry-rice cultivators in Asia lived at densities averaging 80/km² (31/mi²; range 8–91), but the groups who practiced the most intensive multicropping with irrigation and transplanting averaged 2,560/km² (988/mi²; (range 260–1,300) (Hanks 1972: 57).

Density Dynamics and Modifying Variables

Far from being a static classification, the Boserup formulation postulates processes of regular change toward intensification with population pressure and in the direction of more extensive land use if population densities decline. Smallholder systems of fixed intensive cultivation under conditions of resource scarcity can emerge as the ratio of people to land increases. One of the first studies to document the Boserup hypothesis was W. C. Clarke's (1966) work on four New Guinea Highland groups having similar environmental conditions and agricultural techniques. With increasing population density came shorter fallows, the need to cut and burn grass vegetation, to turn and mound the soil, and to make drainage or rudimentary irrigation ditches. These changes taking place in areas of demographic growth have been progressive within the memories of living informants. A more recent regional comparison of the Papua New Guinea Eastern Highlands, with densities of 30/km² or less, and Western Highland groups such as the Chimbu and the Enga, whose populations exceed 100/km², indicates that more intensive cultivation of sweet potatoes, accompanied by the rearing and foddering of larger pig herds, is characteristic of the more populous west (Feil 1987). In some heavily settled Highland communities, fields are mulched or composted, gridded with drainage ditches, fenced, and farmed annually (Brown and Podolefsky 1976).

Fifty-two cases of groups from nine countries in sub-Saharan Africa compared by Prabhu Pingali, Yves Bigot, and Hans Binswanger (1987: 17–21) suggest historic change along a continuum (Fig. 9.1):

Agricultural intensification is caused by a decrease in the cultivable area per capita when the population density increases because of the growth or concentration of population in particular regions. The causes of concentration are both historic—tribal war, slave trade, and so on—and contemporary—land-use policies, lower costs of transport, and so on. Concentration of population also occurs in areas endowed with high altitude, fertile soil, and other favorable conditions.

Though the causes of increases in population density are varied, the consequences—reduction in cultivable land per capita and the intensification of the agricultural system—are fairly similar in most locations. In addition to reducing the length of fallow on land already in use, whether cultivated or fallow, growing population density also leads to the colonization of lands—such as swamplands—formerly unused for agricultural purposes. (Pingali et al. 1987: 43)

Smallholders are not so wedded to their intensive techniques that they will continue them when population / land circumstances change. When given the opportunity of cultivating the plentiful lands of southern Brazil by swidden techniques or ranching, German and Italian colonists promptly gave up crop rotation, livestock, and forage crops they had formerly used to maintain soil fertility in their homelands (Boserup 1965:

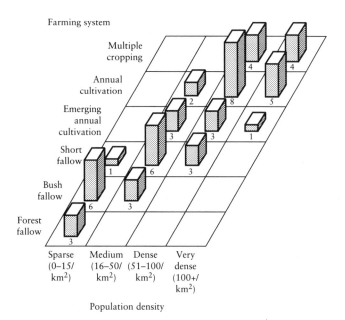

Fig. 9.1. Population density and agricultural intensification in nine African countries. Figures inside the grid represent the number of population groups in each category (total = 52). (From Pingali et al. 1987: 51)

63; Boserup 1981: 135). Experiments in introducing populations of intensive wet-rice cultivators into less densely populated areas of Vietnam and Sri Lanka failed because the colonists adopted labor-saving shifting methods (Boserup 1965: 63; Gourou 1966). Kofyar who migrated from their intensively tilled homesteads on the Jos Plateau escarpment to the sparsely settled Benue Valley frontier lands first practiced simple slash-and-burn agriculture despite the fact that they were familiar with and well practiced in the skills of manuring, intercropping, and elaborate ridging (Netting 1965, 1968). These intensive techniques, plus the addition of chemical fertilization and higher labor expenditures, only reappeared as land filled up and population density increased (Netting et al. 1989; G. D. Stone et al. 1990). The trajectory from a stable, intensive, traditional system in the Kofyar hill communities through an expansive phase of shifting cultivation to a reintensification of farming on new smallholdings took place within 30 years and one adult generation, suggesting rapid adaptability to the changing ratio of population to available land.[3]

To contend that human beings always zero in quickly on an agricultural system that maximizes benefits while minimizing labor and transport costs neglects the complexity of long-term environmental problems and the stubborn persistence of habits and culturally valued behavior. Intensification by means of microenvironmental specialization in land use, stable fence lines, optimal field-farm location patterns, and regional trade connections represents the result of experiments and investments over a period of years. Such "accrued economies," together with permanent improvements to land such as terraces, irrigation systems, and orchards, may be maintained even under conditions of falling population density, tending to fix a smallholder system very tightly to a particular resource complex and render it brittle in the face of revolutionary change (Barrau 1965; Brookfield and Hart 1971). Agriculture may, in fact, be an adjunct to other factors more immediately important in human life-support systems. A military threat to outlying intensively tilled homesteads may force settlement nucleation, requiring further intensification of fields in the defensible village and allowing perimeter areas to revert to occasional grazing, although total population density is unchanged. Kofyar villages exhibited such land-use changes during periods of warfare (Netting 1973, 1974b). A resident community may also be sited in an inhospitable environment, as were the Kachin villages in the Burmese mountains that pro-

---

3. Kofyar population densities ranged from approximately 112/km^2 in the homeland of small homestead farms around 1955 to 25 part-time residents/km^2 in the migrant bush-farming area in 1963 and back up to 99/km^2 as permanent settlement of the frontier was completed in 1984 (Netting 1992).

TABLE 9.2
Frequency of Cropping in 56 Countries, Ca. 1960

Population/km²	Technology group[a]				
	I (low)	II	III	IV	V (high)
0–4	35%	–	–	–	73%
4–16	34%	–	59%	50%	92%
16–64	39%	62%	61%	66%	91%
64–256	–	86%	82%	95%	95%
256+	99%	–	–	–	99%
Countries (N)	8	5	16	11	16

SOURCE: Boserup 1981: 19. Used with permission.
[a]For definition of technology groups, see Boserup 1981: 13.

vided military protection for trading traffic through the passes into China
(E. R. Leach 1965: 28). Intensive terraced agriculture was practiced near
the densely aggregated settlement, though overall population in the area
was low. Though there are clearly local exceptions to the association of
intensive smallholder cultivation and rural population density, the correlation that Boserup postulated appears to have general validity in any
larger-scale regional comparison.

When one examines frequency of cropping (average cultivated area expressed as a percentage of cultivated plus fallow area) in modern nations,
this index of intensification still rises with population density (Table 9.2).
Boserup (1981: 13) finds this relationship even when the 56 countries are
classified by technological level, using measures of per capita energy consumption, telephones per 1,000 population, life expectancy, and percentage of literates. Cropping frequency above 80 percent implies raising one
or more crops per year on the same land. This practice predominates in
countries with population densities above 64/km² (166/mi²), regardless
of technological level (Boserup 1981: 20).

Given local environments differing in soil nutrients and structure, precipitation, water supply, and topography and with varying crop requirements such as those of grains versus tubers, it is difficult to come up with
any "magic number" that would represent a population-density "great
divide" between extensive and intensive preindustrial systems. K. J. Pelzer (1948) saw 50/km² (130/mi²) as representing a maximum for swidden or long-fallow cultivation in Southeast Asia. Ifugao wet-rice terrace
farmers support 291/km² (Dove 1983). Tiv shifting cultivation in the favorable circumstances of the Nigerian Middle Belt can support up to
154/km² (D. E. Vermeer 1970). Hausa long-fallow savanna woodland
cultivation disappears when densities reach 58–85/km² (150–220/mi²)
and even grass fallow gives way to permanent cultivation, as in the Kano
Close-Settled Zone, when population exceeds 232/km² (600/mi²) (Grove

1961; Mortimore 1972). Kofyar intensive cultivators reached densities of 37, 67, and 92/km² of arable land in the hills and over 200/km² on the alluvial lands of the plain (G. D. Stone et al. 1984: 92). In eastern Nigeria, Ibo tropical forest horticulturalists use intensive plots of tuber, grain, and tree crops in areas with population densities of 155–386/km² (400–1,000/mi²), but can only maintain bush fallow of 10 or more years where populations are below 58/km² (150 mi²) (Morgan 1953; Udo 1965). A recent attempt to assemble African case studies of population growth and agricultural change used a density of 200/km² as distinguishing areas likely to practice intensive agriculture (Turner et al. 1992). Boserup (1981: 21) notes that peoples living at densities up to 64/km² (166/mi²) practice forest fallowing and bush fallowing predominantly, while those above this demographic line rely on more intensive cultivation. We may tentatively infer that there will be pressures to intensify land use when local densities reach 60/km² (155/mi²), and that any rural area over 200/km² (518/mi²) will have farms under annual or continuous use with minimal fallow.

Cross-cultural synchronic comparisons within and across societies support the regular association of certain density levels with degrees of agricultural intensity, but are there processes in the system that act to increase or constrain population growth? It might be claimed that rural cultivators would be motivated to have larger families both to meet the increasing labor demands of intensification and to tap alternative sources of employment until their households reached a bare subsistence level. Boserup (1990: 23) now emphasizes the way in which tenure systems in densely populated countries usually provide less encouragement to high fertility than do those in countries of sparse population. There are economic inducements for people with long-fallow farming systems in regions with communal tenure to have large families:

> The size of the area they can dispose of for cultivation is directly related to the size of their family, and most of the work, at least with food production, is done by women and children. So a man can become rich by having several wives and large numbers of children working for him. Moreover, unless he has acquired other property a man's security in old age depends on his adult children and younger wives, since he cannot mortgage or sell land in which he has only usufruct rights. (Boserup 1990: 24)

This seems to have been the case with Kofyar households on the frontier. Family restrictions are also weak in sparsely populated regions with large landholdings. Where landless or near-landless workers comprise a large share of the rural population, adult children provide security and may contribute to family income even at an early age when they work for wages on ranches or plantations.

Under intensive agricultural regimes in high-density regions, a rural

population of small and middle-sized land owners may be more motivated to preserve a smaller family size than are those with no land or
insecure tenure:

> They are less dependent upon help from adult children in emergencies and old
> age, because they can mortgage, lease, or sell land, or cultivate with hired labor.
> They may also have an interest in avoiding division of family property among
> too many heirs. If they live in areas where child labor is of little use in agriculture,
> they may have considerable economic interest in not having large families, and be
> responsive to advice and help from family planning services. (Boserup 1990: 23)

That there can be feedback influence from land scarcity to smallholder
fertility, with high land prices and delayed inheritance modifying age of
marriage and celibacy rate, was apparent in the Swiss case. Whether intensive cultivators regularly differ from their non-smallholder neighbors
in fertility is an interesting suggestion that needs to be more widely
tested.

Technology as Prime Mover

Boserup's model of intensification was revolutionary in part because it
cast doubt on evolutionary scenarios (Richards 1985: 138) that made technology the primary, indeed the sole, engine of agricultural change. In
these familiar and still compelling formulations, tools and tillage methods from the simplest digging stick through Neolithic stone hoes and
then scratch plows with animal traction succeed one another in orderly
stages, giving each cultivator the power to use more land and increase the
production of food (Hahn 1919; Childe 1951). Progress was the ability to
command larger sources of energy, rendering human labor less necessary
and more efficient as animal and mechanical power and fossil fuel provided substitute sources of energy (L. A. White 1959). Thus the smallholder was an anachronism, limited by "primitive" technology to laboriously farming a few hectares with yields presumed to be low and
unpredictable. From the pointed dibble stick on, the invention or diffusion of new technology was the key to increased food production and
bigger farms.

It is true that farming peoples can achieve more frequent and sustained
use only with a greater number and variety of agricultural techniques,
and that another surrogate measure of intensification is the presence or
absence of these methods (Brookfield and Hart 1971). But these represent
knowledge and skills as much as tools per se, and they may increase
rather than diminish labor demands. B. L. Turner and W. E. Doolittle
(1978) provide a weighted index of farming technology, including
ground and crop protection, erosion and hydraulic controls, soil-fertility

maintenance, and plant preparation. Boserup (1981: 45) lists operations according to their occurrence in a wide variety of systems (planting, scaring wild animals, harvesting) or their appearance in successively more intensive systems (weeding, soil preparation, fertilizing, watering crops, feeding domestic animals, producing fodder). Indirect labor and capital investments in future production, such as more complete clearing, leveling, and terracing of land; constructing irrigation, drainage, and flood-control facilities; training and maintaining draft animals; and providing better tools, equipment, storage, and shelter buildings become increasingly important with intensification (Boserup 1981: 45; Pryor and Maurer 1982: 327). Yet the basic means of increasing production per unit of land were all present in prehistory, and modern energy-intensive technology has often contributed to *dis*intensification and more extensive land use.

Smallholders are often chided for "peasant conservatism," resisting the adoption of high-yielding Green Revolution seed varieties, powerful tractors, or irrigation systems built with modern engineering and structural materials. Boserup never questions the ability of farmers to make rational decisions on deriving subsistence from limited resources, but she realizes that there are *costs* to intensification in terms of the declining marginal returns to labor as it encounters resource scarcity. New technology may in fact cost more in labor and capital than it returns, and it may not be readily applicable to the confined areas or mixed cultivars that are the hallmark of land-intensive agriculture. Some inventions, such as a new calendar that more closely agrees with seasonal planting periods, are adopted as soon as they are proven successful because they increase production with no additional labor. Julian Simon (1981: 200) contrasts such "invention pull" innovations with a "population push" type of change such as irrigated multicropping, which requires much more labor and hence will not be adopted until demand generated by an additional population warrants the shift.[4]

---

4. H. C. Brookfield (1984), in criticizing Boserup, contrasts "innovation," adoption of a new crop, tool, or technique that increases the productivity of labor, with "intensification," an agricultural change that reduces labor productivity. Whereas innovations bring obvious gains to the farmer, intensification is always burdensome and is adopted from necessity. *Intensification*, as I use the term, refers to the productivity of land that can be raised by technology and by labor, of various kinds and in different proportions. The efficiency of labor—that is, the return on each unit of labor input—must be empirically determined, and it does not necessarily decline under all processes of intensification. However, as I attempt to demonstrate in Chapter 3, smallholder intensification, up to the point of industrial mechanization, continues to demand a rising total labor input. In the process of agricultural intensification, "product per man need not fall since more hours are worked per person. Longer working hours is indeed one of the crucial features of the Boserup model" (Salehi-Isfahani 1987: 879).

Like the steel axe, the animal-traction plow would appear to be an obvious benefit to any cultivator. It is not, however, a technology that always saves labor and increases productivity. If that were true, all farmers who knew about plows and had suitable draft animals would universally adopt them. In fact, however, of 41 European and Asian societies meeting these prerequisites, 36 percent do not use plows (Pryor 1985: 729). Boserup says that population-mandated short fallows result in grass turf that can be broken effectively over wide areas only with animal-drawn plows. In Frederic Pryor's (1985: 736) sample of 68 Eurasian and African societies where plows are possible, a population density of 65 /km² (25 /mi²) statistically discriminates plow users from low-density non-users. Labor productivity is apparently higher in long-fallow systems, where investment in the implement and in the care, feeding, and training of draft animals is unnecessary. The ox that draws the plow for 10 weeks must be herded, fenced, or tethered for 52, and if it is likely to snow, barns must be built and hay stored for the winter. If the farmers grow "plow-positive crops," such as wheat, barley, rye, buckwheat, and teff, which require considerable surface area as an inducement to plow adoption, the predictive value of population is further increased (Pryor 1985: 737). The switch in Gambia from hoe cultivation of groundnuts to ox-plow methods with wider row spacing meant a 31 percent increase in field size but a 50 percent drop in yields, as well as higher overhead costs (Weil 1970: 251–52).

For Boserup, the dynamic element that moves extensive cultivators toward the intensive practices of the smallholder is demography rather than technology:

A growing population exhausts certain types of natural resources, such as timber, virgin land, game, and fresh water supplies, and is forced to reduce its numbers by migration or change its traditional use of resources and way of life. Increasing populations must substitute resources such as labor for the natural resources which have become scarce. They must invest labor in creation of amenities or equipment for which there was no need so long as the population was smaller. Thus, the increase of population within an area provides an incentive to replace natural resources by labor and capital. (Boserup 1981: 5)

A 48-case sample of agricultural societies from Africa shows labor use rising with farming intensity, irrespective of types of tools used (Fig. 9.2). "A switch from hand hoe to animal draft power and to the tractor significantly reduces total labor use per ha" (Pingali et al. 1987: 107), but the area under cultivation expands. Although less labor goes into land preparation, there are greater requirements for weeding and harvesting if these operations continue to be done by hand. Animal-drawn plows in West Africa raise total labor per farm while reducing it per ha (McIntire 1984, cited in Pingali et al. 1987: 106).

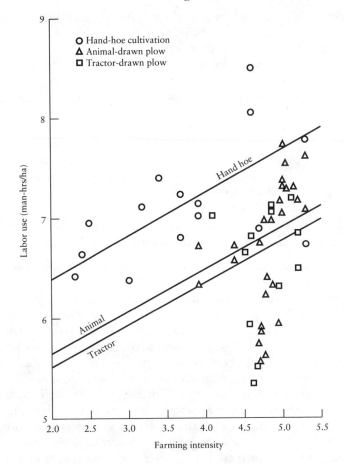

Fig. 9.2. Labor use and farming intensity: logarithmic scale. (Farming intensity is the frequency with which a plot of land is cultivated). (From Pingali et al. 1987: 108)

There seems to be no doubt that intensification of an agricultural system leads to longer and more regular working hours because of an increase in the number of operations required to maintain yields and soil fertility. Because labor input increases at a faster rate than yield per ha, labor productivity declines, as Boserup predicted (Pingali and Binswanger 1983: 11). A change in technology from hand cultivation to animal traction can reduce direct labor use by 64 percent at a given intensity of farming, but total labor, including the additional subsidiary tasks of animal care, feeding, training, and provision of replacement beasts, may reflect little absolute decline. When land is still abundant, the overhead labor costs of destumping and leveling ground under forest and bush fallow far outweigh the future benefits of plow use (ibid.: 6).

A given intensity of cultivation (based on the percentage of time in the rotation cycle devoted to cropping) can be achieved with hoe, animal traction, or tractor (Fig. 9.2), but raising the level of intensity requires longer working hours as long as the type of tool is held constant (Pingali and Binswanger 1983).[5] Technological change per se does *not* increase production per unit of land, and Pingali et al. (1987: 107) found that though yield per ha in Africa rose with farming intensity, "there was no significant association of yield with the use of animal traction or the tractor." Intensification of agriculture appears under population pressure that reduces per capita land, and it increases labor demands, even if the smallholder has access to "labor-saving" technology.[6]

Environmental Constraints on Intensification

An evolutionary way of thinking akin to conceiving technology as the prime mover of intensification is the approach emphasizing the natural environment as both opportunity and constraint. In this view, agriculture is perceived as having developed first in favorable settings, such as the rich, annually renewed alluvial soils of river valleys or the well-watered lands of an oasis (Childe 1951). The major constraint on dense permanent settlements of smallholders is environmental—that is, the inherent limitations imposed by climate, precipitation, topography, and soils (Huntington 1963). The Amazon tropical forest, for example, is thought to have had sparse populations because of low soil fertility resulting from abundant rainfall, high temperatures, leaching, and erosion (Meggers 1971; but cf. Ferdon 1959; Lathrap 1970; Roosevelt 1989; Moran 1990).

The contention that there is a definable agrarian ceiling or "carrying capacity" for each environmental zone denies the potential for intensification of production, at least in the absence of major technological change. African land use has been classified by the physical characteristics of local soils and their tendency to erode or lose fertility under prevailing

5. A 10 percent increase in farming intensity causes a 3.4 percent increase in labor use per ha (Pingali and Binswanger 1983: 9).

6. Economic historians have tended to follow agronomists in seeing agricultural progress as dependent on the adoption of new crops, new means of restoring soil nutrients, and new machines, all of which were resisted by conservative peasants (Sabean 1990: 51). The agricultural revolution in Neckarhausen (1740–1830) involved planting legumes, kale, mangolds, and potatoes, rotating crops, increasing the stall-feeding of livestock, reclaiming fields subject to flooding, and specializing in flax and hemp (ibid.: 52–60). This process of intensification had less to do with the spread of scientific knowledge and techniques than "with the patient mobilization of labor and the communal reallocation of rights and privileges" (ibid.: 51).

systems of agriculture (Allan 1965). Though the semi-arid tropics of West Africa and India have similar annual rainfall totals, Africa has a shorter growing season, greater risk of drought, and soils that are shallower, poorer in texture, more inert, and with less water-holding capacity than those in India (Matlon and Spencer 1984). Means of intensification through high-yielding Indian sorghums and millets and their associated chemical fertilizers have therefore proved ineffective in Africa.

Boserup (1981: 15) admits, of course, that certain areas with extended cold seasons or desert aridity are unsuitable for cropping. However, within a broad range of environmental potentials, she points out a process of "mutual adaptation" between food-supply systems and population density. Yet her model still appears to emphasize the plasticity of the physical environment in accommodating different degrees of land-use intensity. A short growing season and the lack (or prohibitive cost) of irrigation water may, however, prevent adoption of multiple-cropping (Grigg 1979: 77). On the other hand, a naturally flooded treeless stream valley in Kalimantan may promote wet-rice cultivation in a situation where swiddening would require more labor, and the lack of a dry season makes clearing and burning forest difficult (Padoch 1985). Yet under similar population densities, a highly constraining environment, as represented by length of dry season, can restrict intensification, while an advantageous set of conditions, like the presence of alluvial and hydromorphic soils, increases the probability of intensification (Turner et al. 1977; Turner and Brush 1987: 34).[7]

Comparisons of villages in one region with similar crops and technologies can show decisive and primary influences of the environment. In Bangladesh, extremely severe environmental constraints, such as prolonged and deep flooding that decisively limits the cropping frequency of rice, prevent raising land productivity, and this obstacle could only be alleviated by dikes and drainage systems that are too costly for either the village or the state to provide (Ali and Turner 1989). The statistical relationship between population density and intensity is strongest in those cases where environmental constraints on land quality are moderate—that is, where labor, skill, and technology can effectively raise land productivity or carrying capacity (Turner et al. 1992).

Smallholder intensification is not environmentally determined. Intensive agriculture is unlikely where population density is low or rapidly declining, regardless of the environmental potential. Nevertheless, climate, soils, and topography are critical parameters that may constrain or

7. For 29 groups of tropical subsistence cultivators without the plow, differences in population density accounted for 58 percent of the variation in agricultural intensity, while the addition of environmental variables increased the explanatory power to 79 percent (Turner et al. 1977).

contribute to intensification, as seen in the controlled comparisons of farming communities. Environmental variables are not mere *ceteris paribus* conditions. Boserup's formulation becomes richer and more informative as it becomes more ecological.

Demography and Technology as Modeled by Malthus, Marx, and Boserup

A changing ratio of population to land, caused either by population growth or by a declining resource base (deteriorating land, declining precipitation, circumscription by outsiders that denies access to formerly used resources) can elicit a variety of responses (Grigg 1979; Hammel and Howell 1987). Perhaps the most direct reaction to the stress of population pressure, even in its initial stages, is geographic expansion, either by fission and budding off when there is vacant land available or by conflict if a population is surrounded by neighbors and there is no unoccupied land (Bohannan 1954; Vayda 1961; Sahlins 1961). If such migration is successful, it can allow swidden cultivators to continue their extensive, labor-efficient land use (Freeman 1955), and population densities will remain below the levels at which the smallholder pattern typically appears.

When expansion is not possible, population itself may be altered. This can take the form of limitation, as in restricting fertility by late marriage, celibacy (Malthus's preventive checks), contraception, long lactation, or abortion, or by increasing mortality through infant neglect or infanticide. Such means of regulating local population increase, often combined with out-migration, are frequently found in smallholder communities such as those of the Swiss Alps (Netting 1981: 90–158). Population growth can also be directly halted by rising mortality owing to malnutrition, disease, and war (Malthus's positive checks). Competition for scarce resources can set in motion appropriation of resources by certain groups or classes in society and the impoverishment of others, a socioeconomic conflict emphasized by Marx. Though Boserup's predicted agricultural intensification is not mutually exclusive with the other modes of response to population pressure, it does suggest the possibility of a more positive evolutionary adaptation to stress.

The systemic interaction triad of population, environment / land, and agricultural methods /technology has been conceived and configured in different ways by Malthus, Marx, and Boserup. Whether the distinctive smallholder adaptation is a cul-de-sac, a station on the way to misery, or a step toward more adequate and reliable agrarian production that sustains a larger rural population depends on where the limits to growth lie and the direction of the causal arrows. For Malthus, land, at least in his type case of Europe, was an ultimate and inelastic constraint. Any increase in food production could enable demographic enlargement to begin. Technology was not likely to be able to improve or intensify rapidly enough to keep up with population growth, and the "positive checks" of mortality would restrain the implacable demographic overshoot of resources. Boserup tries to stand Malthus on his head, stressing the manner in which population growth impels technological changes that can significantly raise agrarian production per unit of land.

The recent economic-theory literature effectively confronts the priority of technology or population and the positive or negative aspects of population growth. A consensus is emerging that Malthus and Boserup are not contradictory but complementary, and that indeed a formal synthesis of their theories is possible (Pryor and Maurer 1982; Robinson and Schutjer 1984; Lee 1986a, 1986b).[8] It is plain that both authors recognize processes that lead to scarcities of resources, but the demographic and technological reactions to disequilibrium, the rates of change in system variables, and the nature of a new equilibrium are at issue. Neither would deny the law of diminishing returns. "The heart of all economic theory of population, from Malthus to *The Limits of Growth*, can be stated in a single sentence: The more people using a stock of resources, the lower the income per person, if all else remains equal" (Simon 1981).

For Malthus, the supposed inherent potential of human beings to increase geometrically could outrace even growing production so rapidly that income would fall to starvation levels. In 1798, during the very real expansion of eighteenth-century European populations before the benefits of the Industrial Revolution had taken hold, his theory was conceptually appropriate to a major ecological transition (Wilkinson 1973: 22). But the *inevitability* of rapid population growth and the *rates* of potential changes in food supply were conjectural and wrong. Those, like the anthropological neo-evolutionists, who were bullish on technology also portrayed human history as a Malthusian progression, but they emphasized the technological revolutions preceding the expansion of population to new and higher equilibrium levels (Lee 1986a: 121):

In a simplified Malthus-type model, we assume a constant technology and a constant amount of land. As population expands, the marginal and average productivities in agriculture fall; after some point, the birth rate begins to fall, the death rate rises, and the population growth rate falls, and the society eventually

---

8. This is not to say that there are not major differences between the positions of Boserup and Malthus, and my essay "Population, Permanent Agriculture and Politics" (Netting 1990: 32–33) may have overemphasized the areas of agreement outlined by several commentators. "Of course you can make a formal, i.e., simplified, model which combines carefully selected elements of both the theory of Malthus and my theory, but what both Malthus and I attempt to explain with our theories is what was and is typical of the development in the real world, and our explanations of that are contradictory" (Ester Boserup, pers. comm.).

achieves a relatively constant population level. Thus this type of model assumes an exogenous level of technology and an endogenous population response. Boserup proposes a much different mechanism: the population rises (exogenously) at a given rate and the technology employed in agriculture adapts to the ratio of land to population. (Pryor 1985: 731)

For Malthus, an exogenous improvement in productive technology would lead directly to more food, a higher general level of welfare, and more people. Such technological change is random, and thus no more likely to emerge if there is population pressure (Hammel and Howell 1987). Inventions raising production were rare in Malthus's day, and there was generally little surplus beyond subsistence needs (Robinson and Schutjer 1984). A difficult environment like that of the Swiss Alps closed off any possibilities of intensification:

There are no grounds less susceptible of improvement than mountainous pastures. They must necessarily be left chiefly to nature; and when they have been adequately stocked with cattle, little more can be done. The great difficulty in these parts of Switzerland, as in Norway, is to procure a sufficient quantity of fodder for the winter support of the cattle which have been fed on the mountains in the summer. . . . In Switzerland as in Norway, for the same reasons, the art of mowing seems to be carried to its highest pitch of perfection. As, however, the improvement of the lands in the valley must depend principally upon the manure arising from the stock, it is evident that the quantity of hay and the number of cattle will be mutually limited by each other, and as the population will of course be limited by the produce of the stock, it does not seem possible to increase it beyond a certain point, and that at no great distance. (Malthus 1986 [1826]: 212)

The irrigation that allowed Swiss mountain pastures to produce two crops of hay, the log barns that stored the hay for winter fodder, and the spreading of carefully husbanded manure on the meadows to maintain their fertility (Netting 1981) are all labor-intensive techniques that coincided with higher local populations of dairy cows and people. For Boserup, such technological changes were historical responses to the demands of exogenous population growth, to the obvious fact that more mouths needed to be fed. Land is not a resource of fixed quantity and quality that conditions both production and population size.[9]

If technology is held constant, Malthus and his latter-day neo-

9. A new means for securing more calories from the existing environment, such as the potato, which became an important crop in alpine Switzerland during the latter part of the eighteenth century (Netting 1981: 159–68), both could and did raise local population numbers. For Malthus, potatoes could be seen as an exogenous technological innovation allowing a farming population to break out of its previous stability (Viazzo 1989: 182–83). Boserup, on the other hand, might suggest that a period of wet summers and failure of rye crops, temporarily lowering environmental potential and exogenously raising population pressure, might have caused the Swiss to begin cultivating potatoes, though the crop may have been known to them and available for centuries (Netting 1981: 167).

Malthusian descendants would postulate a relatively inflexible environmental ceiling on population. Population density or growth would then
be governed in large measure by the relative richness of the environment
(Pryor 1986: 883). Those favored areas with the greatest natural potential
for agriculture would have the densest populations and the most intensive
cultivation systems. Frederic Pryor has tested this proposition by coding
worldwide geographic surveys for the relative favorability of soil,
weather, and topography for agriculture. Eliminating the cases where
farming is impossible, he finds that agricultural potential accounts for
only 10 percent of the variation in the importance of agriculture and only
4–5 percent of the variance in population density (Pryor 1986: 883–84).
In sharp contrast, population density explains 40 percent of the variation
in importance of agriculture and 30 percent of the index of crop frequency. Though such a correlation does not predict the direction of causality, it does cast doubt on the Malthusian proposition that environmental
potential and higher carrying capacity are major determinants of population density.

There is also disagreement between Malthus and Boserup in their predictions of how rural people will react to scarcity in their agrarian resources. Diminishing returns spelled for Malthus the doom of hunger,
disease, and war because he could not imagine an already crowded landscape producing more when previously unused labor resources were
called into service and when known techniques and tools that had previously been too costly were applied. Intensification gives a selective advantage to those who are goaded into using their slack time and inventive
capacity.

In the Boserup model of exogenous population growth, it is assumed
that the cultivators will respond to declining labor productivity by working more rather than suffering a proportional decline in consumption
(Lee 1986a: 100). With food in short supply and more costly, the payoff
from a new technique is raised. In economic terms, some capital investments now have a positive net return (Pryor and Maurer 1982: 341).
Higher population density leads, *ceteris paribus*, to lower per capita income, which raises the utility gain from an innovation (Lee 1986a: 100).

In examining the relations between increasing population density and
intensifying agricultural technology, it is not necessary that one factor or
the other be consistently, logically prior. In a population-push transformation, a Boserup trajectory (Fig. 9.3) would follow a concave path in
which demographic growth precedes technological changes, as when the
gradual increase of a growing population's food needs precipitates first
shortening of fallow (A–B) and then the upward kink of annual cultivation with fertilization (B–C). Declining returns with further population
growth and little technological change (C–D) would suggest a period of

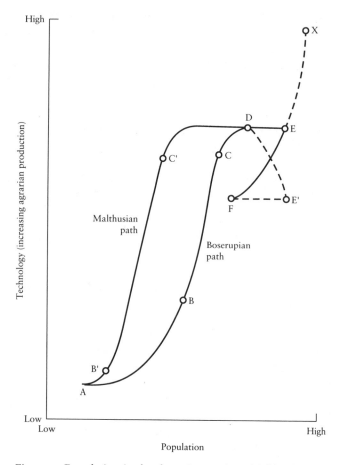

Fig. 9.3. Population/technology interaction: Malthusian and Boserupian trajectories.

involution. On the other hand, a Malthusian curve can be described (B'–C') in which the diffusion of a new crop like the sweet potato into an area with appropriate climate and tools promotes a technological leap in production. A surplus of tubers may be grown not for direct subsistence but to feed pigs for purposes of feasting and social prestige (Brookfield 1972; Allen and Crittenden 1987; J. B. Watson 1977). The new food resource, however, does lead over the long run to population increase (C'–D).

Boserup (1965: 118) does not deny that population growth rates can be so high that the necessary labor and capital investments in intensification cannot be made, and both population and technology would stabilize (D). What Ronald Lee (1986a: 120) terms a Malthusian equilibrium node

(D) can mark the point of environmental limit at which the diminishing return to fixed resources and the growing cost of technology and labor outweigh the stimulus to intensification. At this stable point of higher technology, Malthusian checks of limited resources and high population demands prevail over the Boserupian growth dynamic (ibid.: 120). "Boserupian technological change [intensification] eventually reaches a dead end" (Robinson and Schutjer 1984: 364). Further growth in population (D–E) or decline in technology (D–E′) may send the system into decline, as the Malthusian positive checks of mortality reduce population (E–F, E′–F). A number of possible growth and intensification trajectories can be traced between A and D in what Lee calls a "Boserup space," and an exogenous shock may allow the system to leap (E–X) from one technological regime such as "agriculture" to another like "industry" that is farther along the population-technology axis (Lee 1986a: 122). The Malthusian and Boserupian models are not contradictory, and in the paths of growth, the dynamics of population and technology are mutually reinforcing and either factor can take the leading role:

The critical difference between the classic Malthusian and Boserupian views is whether the appearance of cultural variation is independent of its adaptive value—whether technical or cultural change is exogenous. Malthus does not insist that technical change be exogenous, but only in the long run futile. Boserup does not insist that technical amelioration be permanent. The cycle of challenge and response extends into the long run. (Hammel and Howell 1987: 145)

Marx like Malthus saw exogenous technological development as powering historic economic growth, but the intensively cultivating smallholder was prevented from enjoying the fruits of such progress. Marx also paid little attention to the physical environment either as a limiting factor or as an element in the system that could be enhanced or degraded by human activity. Population did not figure in the Marxian model as an independent or exogenous variable in the way that it did for Boserup.

The specter of starvation for the masses was powerfully evoked by both Malthus and Marx, but they differed in fundamental ways as to its causes, and neither could measure statistically the secular variations of rural population, land use, labor expenditure, average diet, and mortality. Marx's view of the primacy of technological and scientific material progress in raising human welfare places him with the more optimistic evolutionists. Scarcity for him was a creation of an external and dysfunctional political economy.

Engels, in his "Outlines of a Critique of Political Economy," critically summarized Malthus as contending "that the earth is perennially overpopulated, when poverty, misery, distress, and immorality must prevail, that it is the lot, the eternal destiny of mankind, to exist in too great numbers, and therefore in diverse classes." Engels said that Malthus, seeing the tendency to overpopulation in both civilized and natural man, explained human misery as caused by an eternal law of nature, thus diverting attention from the misery created by class exploitation and more particularly capitalism (Meek 1953: 25).

Marx agreed that there might be a law of diminishing returns in agriculture where available land was inevitably limited. He cited Justus von Liebig to the effect that pulverizing and aerating the soil through more frequent plowing would increase the yield of the land, but that the increase could not be proportional to the additional labor expended. Marx and Engels pointed out that labor power increases with population, and that even if returns diminished, the progress of science and technology would more than compensate for this decline (Meek 1953: 30). The productivity of land could be infinitely increased by the application of capital, labor, and science, thus *reducing* human labor, but on the contrary, capital accumulation and concentration has increased the value of the means of production, raising competition to the highest pitch and creating the contradiction of starvation in the midst of wealth and abundance.

Engels insisted that every adult can produce more than he can consume, and that children will abundantly return the expenditure laid out on them. The degradation of humanity inherent in the Malthusian myth can be eliminated "by doing away with private property, competition and conflicting interests" (Engels 1844). Severely declining returns to labor need never be encountered.

Forced Work and the Agrarian Work Force: Is Intensification Coerced?

Contemporary Marxist scholars and those who adopt the perspective of political economy seem to agree that scarcity does not emerge from the normal exigencies of changing land/population ratios, and that the work effort of cultivators who provide only for the subsistence needs of their own households is moderate. A marked increase in hours of labor does not in this view seek to enhance the well-being of the farm family but results rather from economic exploitation and political force that extorts an additional surplus product from the peasant. Population pressure is in no way exogenous but it is created by society and the state limiting the access of rural workers to land and other means of production. Externally imposed tribute, rent, corvée labor, and taxation *demand* intensification by means of labor, denying the land, capital, and technology that are necessary to raise yields while simultaneously lowering the work load. An attempt to explain smallholder land-use practices as responses to the need for increasing production from a decreasing per capita land base is seen by such scholars as a naive neoclassical economic model that mystifies and obscures unequal social relationships and capitalist penetration.

Edward Nell insists that Boserup neglects the role of physical force and slavery; he says that class relationships and coercion are at least as relevant as population pressure to change in the pattern and intensity of work (Nell 1972, 1979). Involuntary servitude seems to accord poorly, however, with the self-disciplined work, responsibility, and decision making that characterize intensive agriculture. It is difficult to imagine the concern of the smallholder for land and livestock being replicated by slaves, whose meager rewards are little connected to the competence of their work. Oriental despotisms may have had armies of slaves, but they never replaced Chinese smallholder households in the practice of irrigated agriculture (Chao 1986). Modern slavery has flourished most vigorously on the frontiers of empire, where labor, not land, was in short supply, and where market monocrops could be produced by massed, unskilled, and directly supervised labor (Wallerstein 1976).

The "open resource theory" of slavery addresses the question of why slavery occurred in the sparsely populated lands of Africa but not in densely settled China and India. H. H. Nieboer observes that when agricultural land is abundant and little capital or complex technological knowledge is necessary to exploit it, "every able-bodied man can, by taking a piece of land into cultivation, provide for himself. Hence it follows that nobody voluntarily serves another; he who wants a laborer must subject him and this subjection will often assume the character of slavery" (Nieboer 1971: 298). When every piece of land is claimed by someone as his property and no open resources exist, slavery is not likely to occur. If independent farming in a situation of plentiful land earns more than the going wage, a landlord has difficulty hiring labor and may profit by slavery (Domar 1970: 60). Only unfree labor could be coerced into forgoing the opportunity to practice shifting cultivation for subsistence. But open resources alone are not significantly correlated with slavery, and the major determinants of slavery would appear to be social and political, not economic (Pryor 1977: 247).[10] Tribute, labor services, and taxes may indeed be extorted from sessile intensive cultivators with scarce, valuable resources, but direct coercion and full-time management of their farming

---

10. In circumscribed environments such as the sugar-producing islands of the Caribbean, slave population densities were high, and intensive practices such as ridging, manuring, and stall-feeding of livestock were practiced at the cost of thousands of man-hours per ha, but the slave labor was said to be highly inefficient, requiring a great deal of costly supervision and producing only a fraction of what English free laborers could do in the same time (Brookfield 1984).

may easily become counterproductive. Smallholders, despite the variety of sharecropping and tenancy relationships they may have, are neither serfs nor slaves.

The presence of smallholders and "free," as opposed to servile, labor does not, of course, guarantee the attainment of some Panglossian ecological "steady state," either of economy or polity. Conquest and naked coercion can pit the state against peasant proprietors, who may flee from their holdings or retreat into a precarious, extensive mode of subsistence. Even without direct political interference, the same demographic factors that stimulate a more productive agriculture can cause dislocation and maldistribution of resources over the *longue durée*.

The very speed at which population can grow, in contrast to the slower rate at which intensification (with all its experimentation, discipline, invented and diffused farm methods, new crops, and selection for farmers with the right stuff) develops, means that discontinuities and contradictions will necessarily appear. Where land use is still relatively extensive and where fields and gardens are interspersed with tracts of forest, marsh, and grasslands, the smallholder option of raising production through intensive methods may be blocked by the interests of the politically powerful elite in concentrating land. The best-known and therefore "classic" cases of agrarian change in Europe come from England and France, where the enclosure of common lands and the persistence of estates with wage labor contrasted with the smallholding pattern characteristic of the Netherlands and the Alps. British history shows cycles of rising population in which the standard symptoms of population pressure made rural life worse for many on the land and often heralded the approach of the real Malthusian destroyers (Grigg 1980).

Sudden swings in population changed land use in the southern French region of Languedoc. The favorable conditions of the depopulated world after the Black Death—plentiful land and low rents, labor shortages with high wages, rural diets with wheat bread and wine (LeRoy Ladurie 1974: 38–49)—were reversed under demographic pressure. With population growth, medium-sized landholdings underwent a "veritable pulverization," while larger estate owners engrossed a great deal of property, and the proliferating smallholders had to seek seasonal farm labor as a semi-proletariat (ibid.: 21, 88). Why polarization emerged in this situation and not during periods of population growth in the Low Countries or Scandinavia may require a closer look at economic and political processes as well as population growth. The rural changes chronicled for Languedoc cannot be credited solely to the expression of a single feudal mode of production, and we can appreciate Emmanuel LeRoy Ladurie's remark that he went to the history of this area in search of Marx—and he found Malthus.

There is, however, a predilection among students of political economy to discern social subjugation and hierarchical appropriation behind any increase in work load. Keith Hart contends that the labor demands of irrigated wet rice cannot be met in West Africa because the area lacks a history of politically exploited peasants:

Has the [West African] work force been sufficiently prepared by famine, impoverishment, and proletarianization to constitute a reliable, hard-working, paddy-field peasantry? Our assessment is that it has not. Agricultural intensification means getting people to work harder, and that undertaking usually requires coercion. The compulsion would have to be especially severe in the forest areas, where labor-efficient techniques of production yield up the population's food need with comparatively little effort (hence the popularity of manioc). (Hart 1982: 48)

Until we have examples of successful state conversion of extensive food producers into intensive cultivators *in the absence* of either population pressure or market incentives, the effectiveness of coercion will remain in doubt.

Intensification and Involution

The themes of population pressure on declining per capita resources, intensification of land use, and higher labor inputs with possibly declining marginal productivity are brought together with the factors of external political domination and economic exploitation in Clifford Geertz's model of agricultural involution. In a remarkable occurrence of independent but convergent discovery, Geertz in 1963 and Boserup in 1965 published parallel findings on agricultural change from shifting to permanent land use under the impetus of demand from growing population. Geertz (1963: 12, 28–35) demonstrated how Java was able to support a population averaging 480/km^2 by means of multicropped wet rice on irrigated, often terraced, sawah fields whose yields were maintained virtually undiminished over many years. Demographic increase made possible and necessary "fine-comb" techniques of water control that were increasingly labor-intensive; more thorough plowing, raking, and leveling of fields; transplanting rice seedlings from nurseries; more frequent weeding; and harvesting of each individual rice panicle with a hand-held knife (ibid.: 35). It is a classic case of progressive, sustainable agricultural intensification increasing production from each unit of land by the application of skilled labor, and as such, it reflects the process that Boserup characterizes so optimistically. But "involution" does *not* reflect economic development, and it responds to the externally generated demands of the capitalist world system as well as to exogenous population growth.

Geertz borrowed the term *involution* from its earlier use to describe patterns of decorative art that become more internally complicated and elaborate with the passage of time, and exhibit increasing technical virtuosity (Geertz 1963: 81). He implicitly contrasted involution with real qualitative technological change that would both raise food production and reduce labor inputs. Involution, he believed, was equally present in Javanese land tenure and cooperative labor. The entire culture had in some sense turned in upon itself, preserving the status quo by laborious but ultimately unrewarding effort, without true development or revolutionary change. Boserup's intensification was a recipe in this case for running faster to stay in the same place, with population growth threatening, but never quite reaching, a Malthusian crisis.

Whereas Boserup provides a synthetic, generalizing model, Geertz considers the processes of population growth and agricultural intensification in a particular geographical setting during a specific historical period. Java during much of this time was a Dutch colony, and sugarcane grown on rice sawahs was a major commodity in international trade. Geertz presciently analyzes elements of what came to be known as the world capitalist system as integral factors in the local Javanese ecosystem of intensification/involution. He saw Western intrusion as contributing to rapid population growth (Geertz 1963: 80), and the imposition of taxes payable in sugarcane for the Dutch mills as further intensifying the use of irrigated land and rural labor. These government policies did not dispossess Javanese smallholders from their lands or disrupt local irrigation communities, but they did indicate the role of state coercion in forcing higher production per unit of land than was directly necessary for subsistence.

In other colonial cases, agricultural land may be directly appropriated by invaders, taken from resident farmers for more extensive uses such as ranches, or concentrated into foreign-owned plantations for export cash crops (Carlstein 1982: 192). Though the loss of land may sometimes raise population densities and stimulate intensification in remaining areas, it may also deprive farmers of resources needed for a smallholder livelihood and make them dependent on agricultural wage labor or seasonal migration (Turner and Brush 1987: 38). It has also been argued that an increase in production *precedes* higher rural population densities and results from the extractive economy, low farm prices, and the actual domination of industrial capitalism (Warman 1983). In sum, intensification and involution describe the same fundamental process of agricultural change, but focus on different sections of the curve illustrated in Figure 9.3, give different weights to population and political/economic demands, and counterpose a positive Boserupian view of potential adaptation with a somewhat pessimistic neo-Malthusian emphasis on limits.

We might ask whether intensification represents progress or involution in Boserup's view (Paul Richards 1985: 53).[11] The remarkable ability of smallholders to match the consumption demands of a growing population against higher production per unit area has continued to be evident, even in Java. But the amount and quality of labor devoted to wet rice has been rendered more productive by the higher-yielding seed varieties, chemical fertilizers, and enlarged irrigation works of Green Revolution technology. What Geertz saw in the late 1950s as a static system was still mired in economic recession after the end of World War II and Dutch colonialism. Indonesia had not yet experienced its oil boom or the coming of scientific advances in rice cultivation (W. Collier 1981). The question of what exactly the marginal productivity of labor in agriculture was, and whether it was barely maintaining itself "by always managing to work in one more man without a serious fall in per capita income," as Geertz (1963: 80) predicted, remains difficult to answer (though see Table 3.7).

The general economic welfare of Javanese villagers has apparently improved in terms of diet, real wages, housing, schooling, and transportation, and this is related to rapid increases in rural trade, service, and construction activities as well as to higher rice and cash-crop sugarcane production (Keyfitz 1986). Yet it is clear that the agrarian dynamics of the historic Javanese smallholder system cannot be isolated from coercive colonial demands for increased production, from externally developed technologies for dramatically increasing yields, and from the proliferation of nonagricultural rural employment and commerce.

It may be that even without colonial extractive economies or parasitic national states, areas of high population density that take the path of agricultural intensification will necessarily become involuted and thus permanently impoverished. It has been claimed that this was historically true of China, where increases in total crop output were not matched by the higher output per unit of labor and higher per capita income that constitute real development:

Involution must be distinguished from modern economic development, for it does not lead to transformative change for the countryside. Small-peasant production at subsistence levels persists, becoming ever more elaborated with commercialization, intensified cropping, and household industry. As this pattern of change advances, peasant production, far from giving way to large-scale production, actually comes to obstruct the development of wage labor-based production by virtue of its ability to sustain labor input at returns that are below market

---

11. Local population increase may lead to intensification of food production or emigration, depending in part on the demand for additional food or more labor in other rural *or urban areas,* either within or outside the country. Progress or involution does not take place in isolation, and these processes are related to the macroeconomic situation, itself heavily dependent on government policies (Boserup, pers. comm.).

wages. And far from giving way to labor-saving capitalized production, it actually obstructs development in that direction by pushing change in the direction of lower-cost labor intensification and involution. (P. C. C. Huang 1990: 12–13)

It might, however, be asked whether areas with existing demographic pressure (as opposed to relatively resource-rich regions) may not achieve higher levels of welfare with smallholder cultivation combined with large-scale, capitalized urban industrial production. Development, as the case of Japan indicates, is not predicated on industrial, energy-intensive agriculture.

As a general rule, Westerners are far too ready to see Third World regions of "overpopulation" and labor-intensive land use as technologically backward, hopelessly involuted, and immiserated. Although they may accurately portray the lack of real growth in and continued dependency of many Third World nations, such views obscure the potential for gains in food supply from agricultural intensification. The enormous increases in land productivity possible among smallholders, even in areas with historically high population densities, are illustrated by data from a six-village controlled comparison from Bangladesh in the period 1950–85 (Ali and Turner 1989). Average cropping frequency went up from 112 percent to 188 percent and total land productivity from 1,618 kg/ha/yr to 3,874 kg/ha/yr in rice equivalents. Dry-season irrigation and the adoption of high-yielding rice varieties were accompanied by an increase in labor input from 202 to 425 worker days/ha/year (Ali and Turner 1989, ch. VIII, 4). Average output per worker-day over the entire period climbed from 8.05 kg to 9.11 kg, but there was a drop in productivity from 1970 to 1985, suggesting the possibility of involution or stagnation. Consumption and commodity-demand variables strongly affect land productivity. In villages without severe environmental constraints, population density explains 26.9 percent and market activity 19 percent of the variance in agricultural intensity (Ali and Turner 1989, ch. VIII, 14).

The basic and well-demonstrated Boserupian relationship between population density and agricultural intensification is seen by some as generating an inexorable movement toward Malthusian involution and stasis, but there is convincing evidence that this denouement is often avoided. The creativity of smallholder economic coping strategies, both on and off the farm, should not be negelected.

Market Pull: Economic Incentives for Intensification

The Boserup model has been most consistently criticized for its earlier neglect of the market as a major source of the incentive or requirement to intensify land use (Bronson 1972; Cowgill 1975; Grigg 1979; but cf. Boserup 1965, ch. 8). Boserup (1981: 63–90) in fact sees the growth in specialized production, infrastructure, urbanization, and exchange as reactions, like intensification, to higher population densities. There was both demand from non-farmers for food and for raw materials like wool to be processed and from dense rural populations for manufactured goods:

For commercial producers, the motivation for intensification of agriculture emerges when population growth or increasing urban incomes increase the demand for food, and push food prices up until more frequent cropping becomes profitable, in spite of increasing costs of production or need for more capital investment. By this change in sectoral terms of trade, a part of the burden of rural population increase is passed on to the urban population. . . . In many cases, a large share of the agricultural population combines subsistence production on small plots of owned or rented land with wage labor for commercial producers in the agricultural peak seasons, and this contributes to considerable flexibility in the labor market. (Boserup 1990: 14, 15)

Boserup points out that cash-crop production may lead to some decline in real wages per hour, but that increased employment in the off-season (see Chapter 3) and more work opportunities for women and children can at least partially compensate for this decline (Boserup 1990: 15). Hiring labor and leasing land in or out accompany the market integration of intensive cultivators.

Economic models based on marginalist choice by individual farmers selling produce in a monetized market have been used to explain the concentric zones of successively less-intensive land use around cities (Carlstein 1982: 212). As the Prussian geographer and agricultural economist Johann Heinrich von Thünen (1783–1850) found in mapping such zones (Fig. 9.4), using the costs of production, transport, and market prices on his own estate, it is most profitable to use the most distant lands for extensive grazing or shifting cultivation.[12] Travel to such outlying areas and transportation of products back from them is time-consuming and expensive, and only sporadic work there is necessary. In von Thünen's system, grainfields with a short fallow rotation are placed closer to the center, and the area of manured, annually tilled fields closer still, given the greater expenditure of work on them and the higher value of their products. Closest to the city are areas where perishable dairy products and garden vegetables are produced by very intensive means. There the value

12. As summarized by Carol Smith, von Thünen's thesis was "that the pattern of land use would be a function of price variation among agricultural products, with returns to production intensity diminishing as distance between a producing area and a given market or consumption center increased. By assuming an isotropic (featureless) plain with a single market center, and by taking as given the local production resources, yields, costs (primarily of labor and transport), and prices, he could calculate the returns to labor on particular plots of land, thereby deriving their economic rent. With these figures he was able to derive a formula for allocating local production resources into particular production zones for optimal returns" (C. A. Smith 1975: 6).

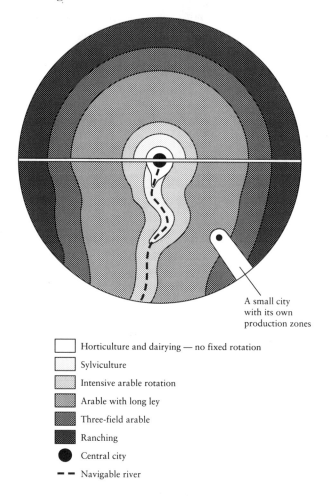

A small city
with its own
production zones

☐ Horticulture and dairying — no fixed rotation

☐ Sylviculture

▨ Intensive arable rotation

▨ Arable with long ley

▨ Three-field arable

▨ Ranching

● Central city

▬ ▬ Navigable river

Fig. 9.4. Johann Heinrich von Thünen's system of land use. The top half is the pattern on an idealized featureless plain, while the bottom shows how a river modifies the patterns of concentric rings. (From Chisholm 1970: 27)

of production per unit of land is highest and the costs of transport to market lowest. Smaller towns connected to the city by roads reproduce this separation on a smaller scale. Similar calculations determined the zonation of land use around Old Scandinavian farms, with the manured infield near the barn where livestock was kept, a farther plot of hay production or wheat and barley rotation, and a distant outfield used for rough grazing and only infrequent cultivation (Carlstein 1982: 213).

The economics of location concern both the cost of work time and transport and the relative market value of the goods produced. If farmers were on a featureless, equally fertile plain and at equal population density around a market city, they would presumably concentrate on more or less intensive agricultural production in terms of the prices of goods minus the costs of production and transport, resulting in a von Thünen patterning of activities. In fact, rural populations tend to be denser closer to cities, where market gardening is feasible and competition for land is high (Cowgill 1975). Along waterways or roads that lower transport costs to market, there will also be concentrations of farmers. Thus Boserup's population density and von Thünen's market zonation tend to coincide. Under the ideal conditions of a marketing economy in a uniform environment, both theories would apply equally: "Market centers would develop in areas of highest population density, and production intensify around them, i.e., production would intensify in those areas that were simultaneously near market centers and supporting the densest population concentrations. . . . [All else being equal], population density would vary directly with distance from major market centers so that both theories would adequately predict production intensity" (C. A. Smith 1975: 33).

In the case of western Guatemala, using a measure of production intensity that contrasts subsistence farming with producing specialized crops for the market and with nonagricultural specialization in long-distance trade or handicrafts, C. A. Smith (1975: 11) finds that population density predicts 30 percent of the variance in production intensity, while marketplace size and importance predicts 11 percent. Using distance of all townships from a central regional marketing center, where purchasing power is concentrated by a ladino upper class, 40 percent of the variance in production intensity can be explained, and, with the addition of population density and local market centrality, over 50 percent (ibid.: 21–22). This model (which is not limited to agricultural intensification alone but seems generally quite analogous to it) introduces issues of market monopolization and politically based imperfect competition that distort the influence of both population- and market-distribution variables (ibid.: 33).

But can market opportunities in and of themselves stimulate intensive land use? Do extensive cultivators adopt intensive cash-cropping while arable land still exists in abundance? Some suggestive evidence is provided by the recent history of the Kofyar, who, until the 1950s, were intensive subsistence cultivators on ancestral Jos Plateau lands and only very peripherally involved in the Nigerian market economy. As dry-season roads and truck transport made the urban food markets of Jos and Zaria accessible, the Kofyar moved into vacant bush lands in the Benue

Valley to grow yams, millet, and sorghum for sale (G. D. Stone et al. 1984; Netting et al. 1989). They lacked sufficient traditional homestead land to produce a surplus, but the frontier allowed them to farm extensively by slash-and-burn methods. Because the best new lands were rapidly claimed and occupied by an increasingly dense resident population, however, and because further expansion encountered poorly drained clay soils and greater distances to roads (G. D. Stone 1988, 1991), the Kofyar *reintensified* their agriculture. They began to subdivide fields and rapidly resumed their earlier pattern of sedentary smallholdings. Production for the market, unless it is in a specialized crop that requires irrigation, like dry-season onions, can be undertaken most efficiently by extensive techniques as long as a fresh supply of land is easily accessible. When land becomes scarce because of increasing immigration, natural population increase, or attraction of people to a market center, the desire to raise yields per unit area and the higher density of population will further the intensification process.

It is probable that the most immediate stimulus for increased agricultural production in both tribal and peasant economies is "the incentive transmitted by market forces" (Fisk 1964: 157). Once demand is no longer determined by the subsistence and security requirements of the local population but by the market for local goods and the accessibility of desirable commodities from outside the community, pressure to increase production for exchange can increase dramatically. For the Kofyar, the cost in additional labor and travel time of planting fields of yams, storing the crop, and transporting the tubers to market or selling them to truckers was willingly accepted in return for the cash to purchase cloth, kerosene, and bicycles, followed in later years by motorcycles, galvanized-metal roofing, iron bedsteads, medical services, and education (Netting et al. 1989). Market access is itself a function of interrelated population density and the level of transport technology, and improved market conditions reflect "increasing population size both in food exporting and food consuming areas" (Boserup 1987: 695). Though the Kofyar began to head-load yams to the Jos Plateau for sale to tin miners in the 1930s, major market production did not begin until lorries reached the area in the 1950s and a shorter road was constructed to link the area with Jos (Netting et al. 1989).

The major expansion of the Nigerian economy fueled by the oil boom of the 1970s (Watts 1984) led to the completion of an all-weather road from Shendam through the new bush-farm areas of Kwande and Namu to Lafia, connecting with the national network of highways (Netting et al. 1989). Cheap, state-subsidized gasoline and a proliferation of motor vehicles lowered transport costs. Essentially unrestricted trade and unregulated market prices have given opportunities to a host of independent entrepreneurs. Demand for foodstuffs also increased rapidly as the national population grew at over 3 percent per year and urbanization sky-rocketed. Because Nigeria lacks trustworthy regional production statistics on food crops produced and consumed in the country, and there are few time series of prices or truck loadings, it is almost impossible to document the local growth of agricultural production. This means that the rate and extent of increase in amounts of agricultural commodities, especially those produced by indigenous means (Paul Richards 1985), is not known. It is apparent, however, that smallholders have both expanded and intensified their cultivation in response to market demands.

In the 1960s, while continuing to provide practically all of their own subsistence needs, Kofyar farmers increased average household incomes from under £10 to over £34 by beginning migrant cash-crop farming on the frontier (Netting 1968: 216–23). By 1983, a sample of 729 cash-cropping households were selling an average of 561 yams, 3.57 bags of millet, 0.63 bags of sorghum, 0.4 bags of rice, and smaller quantities of groundnuts, locust beans, cowpeas, and vegetables for a mean annual income of ₦1,160 (Netting et al. 1989: 312). Though minimal subsistence food requirements could be met by homestead cultivation in the home villages, the expansion and intensification of farm production in the newly settled areas was motivated largely by the desire to participate in the market economy.

Among permanently situated smallholders, the degree of agricultural intensification can also increase as market demand grows. The double-cropping of irrigated rice in sixteenth-century Fujian (Fukien) occurred in the southern part of the province but not in the north. Both areas had access to the improved seed variety that made double-cropping possible; climate, topography, and supplies of irrigation water were similar in both; and population density and labor supply were *not* higher in the south (Rawski 1972: 32). Yet it was in the south where oil cake, crushed oyster shell, and urban night soil were purchased and applied as fertilizer. Waterwheels and reservoirs were installed. New crops like sugarcane and tobacco competed with rice for land. Weaving became a cottage industry, reducing female labor in agriculture. It was the market that brought about the greater intensification of agriculture and the higher labor inputs of the south. Nearness to the trading ports of coastal Fujian and cheap river transportation made it profitable for peasants in the south to intensify their land use and diversify their production (ibid.: 47). Overseas trade and market expansion induced rapid changes in cropping pattern, higher land values, and increased production per unit area in the south, whereas static markets and the limitations imposed on commerce by the high cost of overland transport in the north presented no equivalent stimulus to agriculture (ibid.: 96).

With the few historic exceptions of relatively isolated groups with high population densities but little dependence on markets, such as the pre-contact Kofyar, Philippine Ifugao, and New Guinea Enga, most contemporary smallholder intensive cultivators are adapting to the dual demands of resource scarcity based on high population density and of market participation (Turner and Brush 1987: 31–35). These might be characterized as contrasts between subsistence and market incentives, or what B. L. Turner and S. B. Brush call consumption and commodity demand themes (ibid.: 31). They unite the insights of Boserup on population pressure, of von Thünen on production and transport costs and market prices, and of the neoclassical economists who regarded farmers as rational economic agents allocating resources in response to the market so as to maximize profit (ibid.: 34). Demographic and economic factors are part of the same local ecosystems, and in general reinforce each other.

James Eder (1982) describes how the settlers of San Jose on Palawan Island in the Philippines chose a location near Puerto Princesa City so that they could have low-cost access to a market for farm products. Over the years, they switched from cash crops of cattle and bananas to the market gardening of vegetables, arboriculture, and pig and poultry raising. With more farmers and declining parcel size, the gardening became "not only a source of cash income but also a yield-increasing labor intensive innovation that economized on land" (Eder 1977: 16). Such intensification had the greatest appeal for the small farmer who lacked sufficient land for extensive use.

As Eder points out, the relative causal importance of population pressure and market incentives in any particular intensification sequence may be difficult to specify (Eder 1977: 2). Smallholders may in fact respond differently to these same factors, leading to different levels of intensification, agronomic productivity, and wealth (Eder 1982), but there is a sense in which local person/resource ratios and market conditions allow a variety of choices. Decisions on labor and land use remain to some extent voluntary, changing over the life course and different among individual cultivators. This is not to say that the state cannot influence population by its tax policy or attempt to set prices of food and terms of trade by nonmarket mechanisms. But the smallholder adapts most consistently over time by the means of agricultural intensification to the conditions of *demographic density* and the *market*. The more direct and imperative demands of state coercion or exploitation by an elite, onerous and unjust as they may be, neither create the distinctive smallholder pattern of intensive production nor sustain it in the face of conflicting social and economic forces.

CHAPTER TEN

Peasant Farming and the Chayanov Model

To ATTEMPT TO discuss smallholder agriculture and household orga-nization without reference to the theories and empirical studies of the great Russian economist A. V. Chayanov would be to ignore what is without doubt the most influential body of ideas on these topics in the social sciences. Since the translation in 1966 of his *Peasant Farm Organi-zation* (originally published in Russian in 1925) and *On the Theory of Non-Capitalist Economic Systems*, with commentaries by Daniel Thorner and Basile Kerblay, Chayanov's work has evoked a flurry of scholarly activity, stimulated in part by the brilliant and seminal, yet flawed, reading of the American anthropologist Marshall Sahlins (1971, 1972). In the period of political and intellectual ferment after the Russian revolution, Chayanov was able to seize on a splendid body of empirical materials on peasant agriculture and demography, collected by the czarist state since the 1880s (Shanin 1972), and with his colleagues at the Institute of Agricultural Economy to develop a general synthetic model of peasant family-farm organization. By beginning with the household as a unit of production and consumption and applying the tools of neoclassical economic analy-sis, he could explain peasant decisions that seemed to violate the normal expectations of a profit-making farming business. His interpretation "de-fined a particular peasant economy by the characteristics of family labor and the relative autonomy of its usage at the roots of peasant survival strategies which are systematically different from those of capitalist enter-prises" (Shanin 1986: 3).

Russian agricultural economies of scale in farm equipment and on-farm transport might provide the lowest cost for produce and the highest income at an optimum farm size of 2,025 ha for a long-fallow system and 162 ha for crop rotation, but peasant farms relying on family labor and limited capital were much smaller and showed no tendency to grow in order to maximize profit (Chayanov 1986: 90–91). With a larger labor force of adult workers or with a higher daily return, household members might work less as an alternative to earning more. Chayanov reports the ratios of payment (in rubles) per working day in agriculture to number of days worked annually by Russian peasants per worker per consumer as being 0–1.0 : 114.3; 1.0–1.25 : 100.2; 1.25–1.50 : 93.1; >1.50 : 90.1 (ibid.: 81).

The two major differences between a peasant family farm and a capitalist enterprise are that the former "relies on its own labor rather than hired labor, and it produces mainly to satisfy the family's own consumption needs, rather than to maximize profits on the market" (P. C. C. Huang 1990: 5). In the Russian economic environment, with a national market and both peasant and capitalist types of farms, peasant choices in production, land renting, and off-farm labor were different from those of capitalists, yet peasants didn't go broke. In comparison to the capitalist farm owner, the peasant family was willing to pay a higher price for land, more interest on borrowed capital, and higher rent for leasing land, while selling its produce for lower prices (Thorner 1966: xviii). "Peasant farms often work at a consistent nominally negative profit yet survive—an impossibility for capitalist farming. Maximization of total income rather than of profit or of marginal product guides in many cases the production and employment strategies of peasant family farms" (Shanin 1986: 4). Chayanov's conviction that the viability of the peasant family farm allowed it to compete successfully with large-scale capitalist or collective farms went against the mainstream of Marxist thought and contradicted the inevitable polarization into capitalist farmers and proletarian workers foreseen by Lenin (Thorner 1966).[1]

Chayanov realized that valuing unpaid family labor at the going wage in the encompassing market economy would seriously distort the cost-benefit analysis of the peasant household and produce the anomaly of viable farm families that appeared to be bankrupt. As Peggy Barlett (1980) points out, economic calculations based on accounting principles developed for capitalist firms founder on the notion of opportunity costs (the return from resources at their next best use):

By attributing a fixed wage to family labor, the calculation suggests that some farmers actually lose money. In fact, these households have many workers and are investing more labor in the household enterprise, since the labor is available.

---

1. Lenin denied the economic rationality and productivity of peasant household agriculture. The destruction of the landlord-peasant relationship and the abolition of private property in land and livestock were primary goals of the Russian Revolution. "Such a waste of human power and labor as is involved in small peasant economy cannot go on any longer. The productivity of labor and the economy of effort would be doubled and trebled in agriculture, if, from the present disjoined individual system, we could pass to one of collective tillage" (Lenin quoted in Maynard 1942: 359).

Thus the opportunity cost of their labor is too high, when set at the going wage. Since alternative opportunities for labor vary from family to family and from month to month, as well as according to the age and strength of the child and the contacts with neighbors who might hire him or her, in reality the exact figure most appropriate to understand each household's decisions is very hard to determine. Therefore, profit is best calculated by the Chayanovian "returns to labor" approach, which subtracts from the value of the harvest only paid costs. (Barlett 1982: 95)

Farmers are aware of their labor investments, but are making decisions based on "the annual product minus outlays" (Chayanov 1986: 5). The Costa Rican farmers who had increased their household labor to cultivate terraced tobacco / corn-beans crop rotations did not attribute a wage to unpaid family labor, although they knew with considerable accuracy the number of *jornals*, or six-hour work days, it would take a hired peon to do each task. "Oh, no, don't figure in the time we spend working. We don't come out making anything then," one said. "I can't count my own work on tobacco or the childrens' either; you don't make money on tobacco if you count the family's work," another farmer told Barlett in the course of her household economic survey (Barlett 1982: 94).[2] The distinguishing feature of peasant agriculture, as Chayanov astutely perceived, lies in the pivotal role of household labor, which cannot be calculated in terms of market wage rates and the profits and losses of a capitalist firm.

Chayanov's Economic (?) Model

Chayanov's model of a primarily self-sufficient subsistence farm without wage labor and with a household dedicated to its own reproduction, "in business for its health," so to speak, was particularly attractive to anthropologists. Their own focus on localized ethnographic study of social groups where kinship seemed more important than the market, and where the abstract generalizations of formal economics did not appear to hold, made the Chayanovian perspective congenial. The ideas that pre-capitalist, kin-ordered societies produced for use rather than exchange, that the domestic mode of production has created no surplus output, and that balanced, precisely reckoned reciprocity of market buying and selling is fundamentally opposed to the "generalized reciprocity" of long-term altruistic relations among kin (Sahlins 1972) seemed to fit neatly within the peasant-household framework. There was a peasant economy that provided the material goods to satisfy biological and social wants

2. In farm-management studies in Kerala, India, it was found that if an imputed value were given to land rent and family labor, returns on holdings below five acres were generally negative (Herring 1983: 244).

(Dalton 1961), but it was not a special "economizing" subsystem of culture that applied scarce means against alternative ends (Sahlins 1960) and operated by rules designed to maximize the achievement of some end or to minimize the expenditure of some means (Dalton 1961).

Some economists and anthropologists followed Karl Polanyi (1944) in opposing the application of formal economic analysis to social transactions such as gift giving, reciprocity, and redistribution that bear only a superficial resemblance to market transactions (Ben-Porath 1980). Formal economics presupposes choice and ready marketability of land, labor, and capital, all quantifiable in terms of money (P. C. C. Huang 1985: 5). The opposing "substantivist" point of view emphasized the social relationships in which nominally economic behavior was said to be "embedded." Both substantivists and Marxist scholars criticized the universalist assertions of neoclassical economics "for inappropriately generalizing capitalist forms of rationality, for not respecting cultural differences, and most importantly for not recognizing the institutionally specific patterns of non-capitalist economics" (Donham 1981: 518). In fact, Chayanov put forth an unapologetically economic model of how the peasant family allocates its limited time and resources. "An organizational analysis of peasant family economic activity is our task—a family that does not hire outside labor, has a certain area of land available to it, has its own means of production, and is sometimes obliged to expend some of its labor force on nonagricultural crafts and trades" (Chayanov 1986: 51).[3]

The misinterpretation of Chayanov by substantivist anthropologists comes both from neglecting the economic roots of behavior within households responding to market imperfections as well as to their own production and consumption needs and from incorporating into the anthropologists' idealized peasant type case the particular agronomic and historical factors that Chayanov drew from the distinctively Russian farming experience. Because Russian peasants of the early twentieth century practiced extensive cereal cultivation with long fallows on relatively abundant land, but in a context of poor transportation and an undeveloped market economy, they best exemplify a *swidden cultivator* adaptation without the demands of either scarce land or marketable surpluses that

3. Marshall Sahlins, who is self-identified as a substantivist for whom the economy does not form a separate institution with its own rules, appropriates the Chayanovian model to suggest that it is really kinship and political institutions that shape behavior and organize production. He ignores Chayanov's attempt to propose a general theory of noncapitalist household behavior. "In practice, substantivists have to develop a particular analysis for each society studied because the institutions and institutional relations vary from society to society. Within this framework it is not possible to develop a general theory of economic process" (Tannenbaum 1984a: 29).

characterize intensive agriculturalists. It is true that in the wake of the breakup of the serf estates after 1860, there was by 1900–1914 very low use of wage labor and little commodity production in rural Russia (Shanin 1986: 13). But although he took into account a wealth of material on varied farming systems, market involvement, and land scarcity within Russia, and regular differences between Russian peasants and land-constrained farmers like the Swiss, Chayanov was making the simplify-ing assumptions required for a conceptual analytic model. It is unfortu-nate that his exemplary focus on the microeconomics of the smallholder household has been used to obscure the existence of rational economic decision making at that level and the range of factors beyond basic food consumption that influence these calculations.

Ever the economist, Chayanov asserted that "any economic unit, in-cluding the peasant farm, is acquisitive—an undertaking aiming at max-imum income. In an economic unit based on hired labor, this tendency to boundless expansion is limited by capital availability and, if this in-creases, is practically boundless. But in the family farm, apart from capi-tal available expressed in means of production, this tendency is limited by the family labor force and the increasing drudgery of work if its intensity is forced up" (Chayanov 1986: 119). The process is one of reaching an equilibrium between the labor of those workers in the household who are available to produce necessary goods and the consumption needs of household members. Chayanov was not arguing that peasant families maximize profits (a concept that has meaning only in wage-labor econo-mies), but that they regularly make decisions designed to maximize gains and minimize costs (Donham 1981). Maximization here is not an empir-ical proposition, then, but rather a method of analysis (Donham 1981: 519).[4]

Peasants may lack the causally related categories of price, capital, inter-est, and rent of standard economics, but Chayanov assumes that they op-erate with their own system of economic rationality, based on satisfaction of family needs (that is, marginal utility) and the drudgery of labor (Tan-nenbaum 1984a: 27). The material, quantitative character of food sup-plies, labor hours, land availability, and mouths to feed is apparent to the householder, as is the network of long-term implicit contracts among

---

4. Donald Donham notes that Sahlins's basic dichotomy is between maximizing capital-ist societies with unlimited production for exchange and the domestic mode of production for use, where wants are finite and few, and technical means are unchanging but adequate. But maximizing is not so much wanting the most as wanting consistently—that is, seeking to realize consistent preferences or hierarchies of value. "The opposite of maximization is not, as Sahlins would have it, domestic production; it is unintelligible disorder" (Donham 1981: 527).

household members, but the family firm does not depend on a book-keeper:

> The very advantage or disadvantage of any particular economic initiative on the peasant farm is decided, not by an arithmetic calculation of income and expenditure, but most frequently by intuitively perceiving whether this initiative is economically acceptable or not. In the same way, the peasant farm's organizational plan is constructed at the present time, not by a system of connected logical structures and reckonings, but by the force of succession and imitation of the experience and *selection*, over many years and often subconsciously, of successful methods of economic work. (Chayanov 1986: 119)

The farmer does not impute some wage cost to unpaid family labor or estimate the value of a day's work as an economist would. Indeed, Chayanov insisted on regarding these factors as undifferentiable, and on taking the entire family household as a single economic unit, treating annual product minus outlays as a single return on family activity (Thorner 1966: xiv).[5] "Capitalist profit accounting cannot be applied to a peasant family farm on which there is little or no wage labor, where the family's own labor input cannot be readily disaggregated into unit labor costs, and where the farm's annual yield is a single 'labor product' that cannot be readily disaggregated into units of income" (P. C. C. Huang 1985: 4–5). Emphasizing the corporate character of the producing and consuming peasant household certainly represents a simplification, but it allows model building and analysis that would otherwise be impossible. In focusing on the smallholder household as a key socioeconomic unit whose internal behavior reflects consistent choices that make rational economic sense in the family-farm context, we are accepting and building on Chayanov's fundamental insight.

Chayanov was the first observer to point out those unique, economic characteristics that differentiate the small peasant household farm from larger or capitalist farms:

> Empirically, the small peasant continues to add labor to the production process even if the marginal return to a unit of labor is very low. In contrast, a farmer hiring labor will presumably conform to neoclassical rationality: at the point at which marginal returns from a unit of labor equal the marginal cost of that unit, the farmer will apply no more labor, as each additional unit would cost more than its return. (Herring 1983: 243)

Since the costs of consumption by the resident family and its draft animals are relatively fixed, and since nonfarm economic opportunities are

---

5. Ethnocentric, patriarchal concepts of the family have been criticized in Chayanov's overly restrictive "assumption that 'the household' is an indivisible unit under the control of a single head. There is a lot of evidence from different societies that such a formulation is plain wrong, and that it underestimates the independent roles of women and of children" (Harriss 1982: 211).

typically limited, family labor has little opportunity cost. This economic calculus makes the intensification of labor and managerial time that we have seen reflected in smallholder practice economically rational.

Land, Labor, and Consumption in the Ideal-Typical Peasant Household

The special conditions required for the Chayanov model, which are both its theoretical strength and its empirical weakness, are (1) that arable land not be in short supply and be obtainable by the household either when it is formed or gradually over the household developmental cycle in response to its needs, (2) that no hired labor or opportunity to work outside the household for wages exist, and (3) that "each peasant community has a social norm for the minimum acceptable income per person, and thus, by implication, the household as a unit has a minimum acceptable consumption level" (Ellis 1988: 107). Under these circumstances, the labor force is the limiting factor of production on the family farm, and it "is fixed by being present in the composition of the family. It cannot be increased or decreased at will, and since it is subject to the necessity of combining the factors expediently we naturally ought to put other factors of production [land, capital] in an optimal relationship to this fixed element" (Chayanov 1986: 92). At any point in time, the number of workers, based on household size and the age, gender, and individual capability of members, is "something" given, as is the number of consumers, with their intakes dependent on their biological needs. The labor time of the farm family, unlike that in other kinds of firms, is not a variable cost but a fixed stock, because "you do not hire or fire family members" (Machlachlan 1987: 4).

In a "natural" self-provisioning economy of independent family farms, the major change affecting the peasant household will arise from the "biological laws" of its own development (Chayanov 1986: 110). The crucial consumer/worker ratio can vary from 2:2 (that is, 1.0) in a newly-wed couple who both work for their own subsistence, to 5:2 (2.5) when the family includes three dependent children, to 5:4 (1.25) when two of the children are sufficiently adult to work alongside their parents, to 6:2 (3.0) when a younger couple support two elderly parents as well as two of their own young children. "*Every family*, depending on its age, is in its different phases of development a completely distinct labor machine as regards labor force, intensity of demand, consumer-worker ratio, and the possibility of applying the principles of complex cooperation" (ibid.: 60). What is now known as the "developmental cycle" (Fortes 1949; Goody 1958) was described by Chayanov, not mainly in terms of its demographic components or the arrangements of kin resulting from social

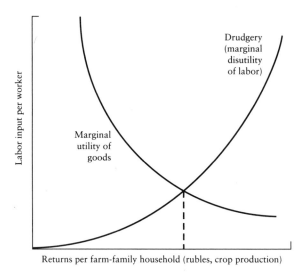

Fig. 10.1. Labor input and farm output: the Chayanov model. (From Chayanov 1986: 82)

rules of marriage and postmarital residence, but as a mechanism organizing production and consumption.

Chayanov's basic premise is that since the peasant farm has no recourse to hired labor, the composition of its labor pool and the degree of labor activity are entirely determined by the number of able-bodied family members and the number of consuming members for whom they must provide. The consumer/worker ratio affects production per worker as demonstrated by the intersection of curves describing utility and drudgery (Fig. 10.1). This is in effect a standard neoclassical analysis of the marginal utility of goods produced by each additional input of labor balanced by the marginal disutility (drudgery) of each increment of labor (Durrenberger 1979), but since utility and drudgery are completely subjective variables with no independent empirical referents, the model may have a mainly heuristic value (Herring 1984). The curve of marginal utility for a household may be raised by an increase in the number of consumers each worker must support, but it also goes up with the amount that must be reinvested in the farm to maintain its production and with any other factors that require part of the farm's product, such as taxes or loan repayments (Tannenbaum 1984b: 28). Marginal disutility, or drudgery, is a measure of the noxiousness of labor and is inversely related to productivity. Many hands make light work, and with more workers serving the same number of consumers, drudgery declines. More productive techniques, higher soil fertility, shorter distance to market, and higher prices for crops all increase the output per unit of labor and thus lower the drudgery (Tannenbaum 1984a: 28):

> On standard marginalist theory, the family farm will raise labor input until the (falling) extra product just ceases to be worth the (rising) extra unpleasantness of effort. The value to the family of extra subsistence output rises as the number of consumer units, working or not, rises. Conversely, the unpleasantness to the family of extra effort falls as the land area and quality (and hence the amount of extra effort required per unit of extra food) increase; and, above all, as the number of people able to work rises. (Connell and Lipton 1977: 50)

Chayanov's theoretical demonstration of the relation between measures of demand satisfaction and the burden of labor is quite straightforward:

> The economic activity of labor differs from any other activity in that the quantity of values that become available to the person running the farm agrees with the quantity of physical labor he has expended. But the expenditure of physical energy is by no means without limit for the human organism. After a comparatively small expenditure essential to the organism and accompanied by a feeling of satisfaction, further expenditure of energy requires an effort of will. The greater the quantity of work carried out by a man in a definite time period, the greater and greater drudgery for the man are the last (marginal) units of labor expended. On the other hand, the subjective evaluation of the values obtained by this marginal labor will depend on the extent of its marginal utility for the farm family. But since marginal utility falls with growth of the total sum of values that become available to the subject running the farm, there comes a moment at a certain level of rising labor income when the drudgery of the marginal labor expenditure will equal the subjective evaluation of the marginal utility of the sum obtained by this labor. The output of the worker on the labor farm will remain at this point of natural equilibrium, since any further increase in labor expenditure will be subjectively disadvantageous. Thus any labor farm has a natural limit to its output, determined by the proportions between intensity of annual family labor and degree of satisfaction of its demands. (Chayanov 1986: 81–82)

Minimum output is set by the requirement that the farm household meet its minimum acceptable standard of living, whereas the upper limit of production is defined by the maximum number of full working days that it is physiologically feasible for worker members of the household to perform (Ellis 1988: 108–9). Family size and composition define both the minimum and the maximum levels of output, and thus the relative weights attached to the utility of production as opposed to the drudgery of work (ibid.: 110).

Though farm returns, whether measured in crops produced, rubles earned, or (indirectly) sown areas may go up along the horizontal axis of

TABLE 10.1
Consumer / Worker Ratio, Income, and Working Days, Volokolamsk, Russia, 1910

	Consumers per worker			
	1.01–1.20	1.21–1.40	1.41–1.60	>1.60
Worker's "output" (rubles)	131.9	151.5	218.8	283.4
Working days per worker	98.8	102.3	157.2	161.3

SOURCE: Chayanov 1986: 78.

Figure 10.1, Chayanov seems to be stressing the costs or limits to this expansion both in terms of steeply rising labor drudgery and of the falling marginal benefits to the family whose demands have been satisfied. He is concerned less with the multiple reasons for increasing what he refers to as the "self-exploitation" of labor[6] than with the point at which labor ceases, perhaps because (and this is indeed speculative) peasant labor in Russian shifting cultivation was generally being exercised at a level far below its physiological potential of full working days.[7] The implicit question seems to be "Why don't Russian peasants work more?"

Whereas I have emphasized land scarcity and the compensatory application of intensive techniques as well as rising demand for market goods as factors increasing labor input, Chayanov sees demand generated from *within* the household by the growing ratio of consumers to workers. From materials collected in Volokolamsk district in 1910, where labor was recorded for each family separately, he was enabled to measure the influence of an increase in the consumer / worker ratio on the intensity of family labor directly (Table 10.1). The argument is clear and simple, as is its graphic representation (Fig. 10.2), where successively higher c/w ra-

6. Chayanov's usage of the phrase "self-exploitation of peasant labor" has given rise to much misunderstanding: "The word 'exploitation' has led Marxists and non-Marxists alike to associate the concept somehow with the notion of the 'extraction' of the 'surplus value' of labor, the classical Marxist meaning of 'exploitation.'" Chayanov (1986: 72–89) himself intended no such meaning. "It simply makes no sense to speak of a family extracting the surplus value of its own labor" (P. C. C. Huang 1990: 6).

7. The identification of additional labor input with drudgery, noxiousness, and self-exploitation conjures up an image of physically demanding, indeed exhausting toil, to which anyone would be strongly averse. The peasant farmer's work seems categorized as harder, more monotonous, *and less rewarding* than alternative kinds of exertion, and the implication seems to be present that only what Adam Smith called the "lash of hunger" (the barely nourished consumers of the household) compel this grinding labor. I have been at pains to indicate that intensive cultivators do a good bit of work, but their increased labor may start from a fairly low base in shifting cultivation, yearly totals of hours may not seem unreasonable, additional labor is voluntarily applied, and the farmer may prefer work on the land and its rewards to other available employments. None of this, of course, contradicts Chayanov's model, but the phrasing of labor costs in terms of drudgery does give one of the economic factors a strongly value-laden connotation.

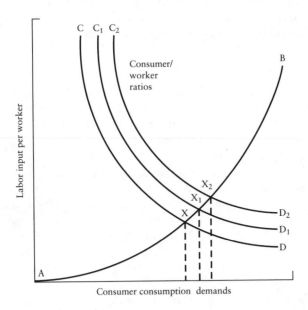

Fig. 10.2. Labor input and consumer consumption demands at different consumer/worker ratios. (From Chayanov 1986: 84)

tios push the marginal utility of goods curve *CD* up to new equilibrium points x_1 and x_2 on the drudgery curve, indicating higher labor intensity per worker. But the clarity of the model depends on all else being equal:

Looking at the table [10.1], we see that, other things being equal, the peasant worker, stimulated to work by the demands of his family, develops *greater energy* as the pressure of these demands becomes stronger. The measure of self-exploitation depends to the highest degree on how heavily the worker is burdened by the consumer demands of his family. The force of the influence of consumer demands in this case is so great that for a whole series of areas the worker, under pressure from a growing consumer demand, develops his output in strict accordance with the growing number of consumers. The volume of the family's activity depends entirely on the number of consumers and not at all on the number of workers. (Chayanov 1986: 78)

The peasant farm family enterprise is a real and significant unit, its organization hinges on economic decisions that arise from a rationality based on different factors from those of the capitalistic farm, and there are regularities that link household consumption and production. Chayanov's model stands as an original, seminal characterization of these issues in peasant economy, but its selection of relevant variables and its limitation to a type of agricultural system that is geographically and historically rare should caution us against its uncritical application.

Chayanov, Boserup, and Demography

The causal arrow that determines amount of labor per worker in the Chayanovian peasant household comes from the consumer/worker ratio—that is, from the demand generated by family size and composition. This is a *demographic* parameter and has some resemblance to the population pressure that provides the motive power for agricultural intensification in Ester Boserup's (1965) model of change (Carlstein 1982; Barlett 1982; Maclachlan 1987). Both, in fact, examine population growth, but Chayanov concerns himself with the microlevel of the household in which the critical change is in the ratio of consumers and producers over the span of the family's developmental cycle. Labor increase or intensification by individual workers is owing to the demands made on their productive time by a larger number of dependent consumers. The growth in household population is an independent variable determined by "biological" laws or "natural" dynamics, a demographic causative agent totally rejected in Sahlins's treatment of "Chayanov's rule" (Carlstein 1982: 243). Boserup's scenario of increasing population density is modeled at the level of the geographic region rather than the household (ibid.: 237), but she uses similar marginalist reasoning and shares with Chayanov the general point "that the utilization of labor in household-based farm economies was highly responsive to demographic conditions" (Maclachlan 1987: 2).

It is implicit in Boserup's argument that population growth, if it results from increased fertility or longevity or both, would lead to an increase in the ratio of children and elderly people to adult workers, creating just that situation of heightened social dependency that, according to Chayanov, increases traditional farmers' utilization of labor (Maclachlan 1987: 2–6). Where Boserup and Chayanov part company is in their respective emphases on intensification as a directional, evolutionary, systemic change in a regional geographic context as opposed to a cyclical, household-specific process that distinguishes the work effort of individuals within a population. Boserup drew material from a variety of sources, especially in South Asia and Africa, contrasting shifting and intensive cultivation and charting the broad correlation of these agricultural types with person/land ratios. Chayanov's data, though systematically collected and suitable for statistical analysis, came from a country with some of the lowest population densities and most extensive agriculture in Europe.

True slash-and-burn cultivation of the sort that persisted into this century in Finland[8] and parts of Poland characterized much of early Russian agriculture. As settlers moved into the Volga forest and the swamps of the central plain after A.D. 1100, and after frequent depopulation by warfare, they plowed small strips, which were fallowed for ten years or more, and spent much of their time hunting, fishing, and honey gathering (G. T. Robinson 1932: 9–10). Patriarchal, three-generation, multiple-family households formed communal settlements that shared the labors of land clearance (Blum 1961: 25–26).[9] Even when the medieval three-field rotation system was adopted in parts of Russia in the sixteenth century, meadows might be reallocated among village households every year after mowing (G. T. Robinson 1932: 11), and plow land might be redistributed at intervals of six years or more (Blum 1961: 525). The famous repartitional commune form may have been breaking down into individual household holdings by the seventeenth century (Blum 1961: 522), but the practice of periodic redistribution may have increased in the eighteenth century under official sponsorship in order to adjust membership of peasant families to available land, allocate equivalent state tax assessments to households, and allow landlords to ensure that domestic units could pay rent and labor services (G. T. Robinson 1932: 34–35; Blum 1961: 512–14). The very fact that it was possible to redistribute land resources[10] and the continued existence of large amounts of fallow (37 percent of plow land in the central black soil region of Russia in 1881 [G. T. Robinson 1932: 98]) suggest relative land abundance. In 1877, the average peasant allotment was 14.4 ha, compared to the 1884 French figures of 3.6 ha for *all* holdings, with three-quarters of all farms consisting of less than 2 ha. Even with wheat yields increasing by 30 percent between 1870 and 1900, Russian peasants had yields lower than those of western Europe and only half those of U.S. farmers. It was estimated that four-fifths of the rural workers were in excess of the numbers needed for

8. Terry Jordan and Matti Kaups (1989) have recently suggested that Finnish swidden technology and log cabins provided the distinctive adaptation of the North American temperate forest–frontier pioneers.

9. The joint or multiple-family household, with early, near-universal marriage and a number of married children continuing to reside in the parental household, was a distinctive feature of Russia and other Slavic areas of Europe (Hammel 1972; Halpern and Halpern 1972; Czap 1982, 1983; Laslett 1984). It was also encouraged by the owners of serf estates who were interested in the increase and management of the labor force. Though Chayanov never refers to this type of household labor organization, its large size and retention of adult workers may have meant that the multiple family could maintain a more constant consumer/worker ratio over time, without the cyclically increased labor demands that result from the formation of independent simple family households with many dependents.

10. In the absence of communitywide periodic redistribution of land, a similar effect could be achieved by young, growing families acquiring land by lease, purchase or gift from older, shrinking households (R. M. Smith 1984: 12).

cultivation of the allotment lands under the prevailing system of tillage (G. T. Robinson 1932: 97–98).

With land readily available and lacking the stimulus of large urban markets, it is understandable that peasants should reduce their labor to that necessary for household consumption, as Chayanov demonstrated that they did. The absence of thorough fertilization, deep plowing, complex crop diversification, and rotation fitted a pattern of extensive land use reflected in Chayanov's model. The lack of economic options for labor beyond the household subsistence sphere and the absence of market incentives may also have reflected the earlier feudal economy of serf estates in Russia and Poland. Extensive farming on large estates allowed the Polish nobility to produce rye for the domestic market and wheat for foreign export using the unpaid labor services and taxes in kind from resident serfs. The serfs supplied their own consumption and reproduction needs from household plots, but only a small proportion of them had access to the market, and they were not permitted to sell their labor at market prices. Monocrop grain farming had relatively little use for the labor of women and children, there were many unused work days in winter, and surplus production benefited the lord rather than the serf household (Kula 1976: 42–51). Because little money was paid to the lord and purchases were dispensable, peasants could survive bad years without buying or selling anything, and economic betterment came only with larger families, including grown children who could work for themselves as well as for the lord (ibid.: 67, 72). The focus on family self-sufficiency, on household labor, and on relative isolation from the market that characterizes the Chayanov model portrays in many respects the situation of the feudal serf, whose labor beyond the subsistence minimum was coerced and whose ability to enter the market was severely constrained by the political system.[11]

Even in Russia there were more densely settled zones near cities where

11. Chayanov's microeconomic focus on the household has also been criticized for ignoring the encompassing international capitalist system that appropriated peasant surplus production: "Chayanov's work expresses the limitations imposed upon the development of the productive forces, and seeks the constraints at the level of the specificity of agricultural techniques, rather than in the social limits to accumulation and technical change. These limits arose historically from appropriation of agricultural surplus product by the gentry and state, and an international division of labor which made Russia export grains at the expense of domestic living standards and domestic rural savings. In doing so Chayanov interpreted the pockets of Russian agriculture where accumulation and technical change were proceeding with great rapidity as an index of the potential of peasant cooperation and agricultural extension, rather than as an index of the development of capitalism (in southern wheat-farming, in the intensive agriculture of the industrial centre and the Baltic states)" (Harrison 1977: 334).

TABLE 10.2

Labor and Crop Intensification in Russia, 1898

Crop	Working days / desyatina[a]	Gross income (rubles) less outlay on materials / desyatina	Payment (rubles / working day)
Oats (Moscow)	24.7	51.41	2.08
Potatoes (Moscow)	48.9	86.20	1.76
Flax (Volokolamsk)	83.0	74.71	0.90

SOURCE: Chayanov 1966: 114.

[a]One *desyatina* equals 1.093 ha or 2.7 acres.

more intensive food and cash-crop production was evident. There were areas of beet and flax farms in the more heavily populated west of the country and intensive dairy and vegetable farms in the northern industrial provinces. Where the peasant farm had a minimum land area, there was a stimulus either to increase earnings from crafts and trade or to expand cash-crop production, depending on the local market situation. This forced labor intensity beyond its optimum limit, getting increased agricultural income at the cost of greater drudgery and lower returns on labor, "either by an intensification of work methods or by using more labor intensive crops and jobs" (Chayanov 1986: 113). As population pressure grows, labor must be substituted for land to maintain peasant living standards, and there is a tendency to go from extensive, labor-efficient crops like oats to intensive crops like potatoes and flax that require more labor and bring small returns per working day (Table 10.2).

In densely populated areas, land shortage does not permit the peasant family to use its labor power for the highest possible payment, and extra labor input is subject to diminishing returns. Chayanov cites data on Swiss farms with little land that trebled their labor intensity: "They suffered a big loss in income per labor unit, but they gained the opportunity to use their labor power fully, even on the small plot, and to sustain their families. In the same way, small farms in the north and west of Russia increased the growing of potatoes and hemp, which are often of lower profitability than oats but are more labor intensive and thereby increase the farm family's gross product" (Chayanov 1986: 7). As the amount of land per worker declined, the Russian peasant switched from grains to the labor-intensive crops of flax and potatoes (Table 10.2), reaching what Chayanov (1986: 115) calls "the almost kitchen garden growing of potatoes (101 working days per desyatina [roughly a hectare])." The same Swiss peasants with inadequate land had to pay prices for land that significantly exceeded its capitalized rent. Chayanov ranks the amount of land available for use as being of equal importance to the size of the labor fam-

TABLE 10.3

Swiss Farm Size, Rural Labor, and Rural Income, 1910

Farm size (ha)	Ha / consumer	Per ha			Gross Income (Fr)	Payment / labor-day (Fr)
		Hired labor (%)	Working days	Gross outlays, materials (Fr)		
0–5	1.2	7.4%	147	304.9	902	2.90
5–10	2.1	19.4	115	212.8	778	3.36
10–15	3.2	30.1	89	214.9	728	3.62
15–30	4.8	47.5	76	183.5	610	3.87
> 30	7.9	57.3	56	170.6	501	3.70

SOURCE: Chayanov 1986: 116.

ily in determining the intensity of cultivation, and he also mentions the less important, yet essential, social factor of the traditional standard of living (Chayanov 1986: 12).

While not ignoring the effects of environmental factors such as high soil quality, Chayanov contends that in regional terms, the main factors of the natural economy are demographic—population density and migration. "These determine land utilization, level of prosperity, and, thus, the ever-varying amount of capital accumulation and taxability of the population; the last forms the basis for organizing the nation's state and culture" (Chayanov 1986: 12). His recognition of the influence of demography on farming system and labor intensity does not appear to differ fundamentally from that of Boserup, but Chayanov's Russian materials allowed him to ignore the factor of relative land scarcity and emphasize the importance of the consumer/worker ratio in determining household labor input. He could also cite the fact that improvement of the market or a more advantageous location of the farm led to an increase in earnings but also a fall in labor units and a reduced working year (Chayanov 1986: 41). But we must remember that Chayanov was always discussing an ideal-typical family farm, which by definition was not capitalistic—that is, it hired no labor (Chayanov 1986: 117). This perhaps obscured his appreciation of the fact that on Swiss peasant farms (Table 10.3), as average farm size increased and labor/ha declined, total labor, including a growing proportion of hired labor, continued to rise. Both demographic pressure and market demand were evident in the contrasting processes of intensification and disintensification that characterized subsistence-oriented smallholders and capitalist wage-labor farms respectively.

Chayanovian Shifting Cultivators

Many of the difficulties in the application of the Chayanov model to a wide range of family-farm systems stem from its defining characteristics of land abundance, self-sufficiency, little commodity production or market participation, and no hired labor. Although rarely found in rural communities with dense populations, intensive agriculture, and permanent smallholdings,[12] the Chayanov type case, with its predicted household behavior, is represented in areas of the tropics with abundant land. Hans Binswanger and John McIntire (1987: 75–78) cite the following qualifications, derived largely from African examples, for such low-pressure regions:

1. Population density is low, cultivable land is abundant, and land has no sale price.

2. Land-use rights are available to the indigenous population at no cost or in exchange for token payments.

3. Technology is simple, consisting of hand tools and possibly draft animals. Management skills are unimportant and technical economies of scale are limited.

4. Transport and communication costs are high. The region is geographically isolated.

From these conditions, the following propositions are derived:

1. There is practically no hiring or exchanging of labor among resident farmers, even during the peak labor seasons of sowing and weeding.

2. The cultivated area per working household member is largely invariant to household size or wealth. Households may have access to different land qualities in roughly the same proportions, or land quality is uniform.

3. There is no regular output market in every year. Variability in crop output is highly covariant and does not generate *regular* opportunities for exchange between surplus and deficit households in the same year. There is a high level of self-sufficiency in agricultural and nonagricultural goods.

4. "Once an output level that sufficiently provides for current household consumption and household storage demand is reached, marginal utility of effort is very low" (Binswanger and McIntire 1987: 78). Households will tend to work fewer hours than in peasant societies where there are export opportunities and labor markets.

12. Frank Cancian (1989) places Chayanov with Eric Wolf and George Foster as "homogeneity theorists" who emphasize peasant isolation, self-sufficiency, independence from the market and lack of socioeconomic differentiation. Chayanov (1966: 111–13) noted that his model was most applicable in regions of low population density and abundant land, but he also claimed that "pure family farms" formed a very considerable social sector in India, China, and Japan (Cancian 1989: 144–45). Chayanov incorporated Asian wet-rice economies into a model derived from land-extensive systems in Europe, and Eric Wolf (1957) similarly stressed similarities between closed corporate peasant communities of Mexican shifting cultivators and Javanese irrigators. Critical differences in the social correlates of population density and land use were neglected.

In such societies of shifting cultivators with substantial isolation from the market, we might expect to find the consumer/worker ratio more closely linked to labor time per worker or production per worker.

The Kantu, a group of Iban-like swidden dry-rice cultivators from interior Kalimantan, Indonesia, increase their clearings of forest by about 0.545 ha per household producer for each increase of 1.0 in the household consumer/producer ratio. Since annual rice consumption for the average household member is 415 kg, the yield from 0.533 ha of secondary swidden, there is a close congruence between producer's work and household demand, as Chayanov predicted. The Kantu cultural ideal is the large, multiple-family household, including married children, and M. R. Dove (1984: 122) shows that large households typically have large numbers of producers, a low consumer/producer ratio, low intensity of labor, and a high return on labor.[13] Having many workers makes it possible to cultivate multiple swiddens in swamps and both secondary and primary forest, thereby spreading out household labor use over the year. A newly married couple may stay in the natal household while they have small children, but they are likely to move into an independent household when their consumer/producer ratio is lower than the joint ratio of the parental household. Households with high consumer/producer ratios do hire labor, paying fixed wages in rice, and their labor intensity falls below the level predicted by Chayanov. Those households at the other end of the continuum with low consumer/producer ratios do not stop working, perhaps because the high returns on their labor enable them to pursue "market oriented consumption plans" (ibid.: 120).

The Theory and the Fact of Smallholder Households: When Chayanov Doesn't Fit

The Kantu case may appear to follow "Chayanov's rule" that "the intensity of labor per worker will increase in direct relation to the domestic ratio of consumers to workers" (Sahlins 1972: 102), but even it involves some wage work and some market production, violating the unrealistic condition that the system be one of household production for use. Marshall Sahlins, in plotting a Chayanov slope of household labor intensity, accounts for the fact that some households overproduce while others underproduce by citing kin and political relations. Households that do not meet their consumption needs can call on relatives with a surplus to help them over periods of adversity. In the case of the Kapauku of Irian Jaya in western New Guinea, "Big Men" mobilize domestic groups to produce excess sweet potatoes and pigs as part of the competition for status and

---

13. The same advantages might have accrued to the large, multiple-family Russian peasant households mentioned in footnote 9 above.

political power (ibid.: 115–23). Deviations or "inflections" from the Chayanov slope are *social* rather than *economic*. For Sahlins, household labor intensities bear no relation to land and resource availability, to the intensification of land use, to market participation, and to nonfarm employment (as among the Tonga). While accepting the "practical reason" of the Chayanovian consumer/producer ratio, he emphasizes its culturally specific modification by social and political considerations.[14]

The somewhat slippery Chayanovian slope that is still discernible among shifting cultivators disappears entirely as shortage of resources, partial intensification, and market involvement appear. Shan cultivators in northwestern Thailand combine several cropping systems of irrigated rice, highland rice swiddens, soybeans and garlic in dry-season irrigated fields, and sesame in highland swiddens (Durrenberger 1979). Garlic and soybeans are cash crops, and surplus rice and sesame may also be sold (Durrenberger 1978). Notwithstanding that the swidden and irrigated systems have different labor requirements, labor correlates closely with farmed area in both cases, but when Paul Durrenberger carefully measured the household labor of producers,[15] the amounts produced, consumer rice requirements, and the consumer/worker ratio, he found "no significant correlation between rice harvested and the production targets derived under these assumptions" (Durrenberger 1979: 450). "For Thongmakhsan as a whole, the consumer/worker ratio does not correlate with any of the agricultural work per worker categories" (Tannenbaum 1984b: 929). One of the major factors that influence the Chayanov curve of marginal utility is the ratio of consumers to workers (Durrenberger 1984: 10), but a great many other factors affect the subjective judg-

14. In drawing the causal line from kinship to economic behavior, Sahlins (1972: 123) suggests further research on the possibility that societies using Hawaiian kinship terminology will be "more intensive economic systems" than those with Eskimo kinship, which isolates the immediate family. Hawaiian kinship should generate a greater surplus tendency by developing more social pressure to share on households with the highest ratios of workers to consumers. A counterproposal might test the proposition that intensive cultivators, regardless of their social organization, may have less interhousehold sharing of food than do local groups of shifting cultivators.

15. It is worth noting that neither Chayanov himself nor Sahlins and others who have tried to apply the model often used direct measures of labor input. Rather labor intensity was estimated by reference to proxies such as "sown area" or cash income per worker (Donham 1981: 520; Gross 1984). This negates any process of intensification in which area farmed declines while labor hours increase, both in direct and indirect productive activities. Sahlins (1972: 103) bases labor intensity on the surface cultivated per worker or the output in crops produced. Wanda Minge-Kalman's (1977) figures for a French-speaking Swiss village, based on labor information recorded daily by every member of a household, show that over seven times as many hours/ha were devoted to strawberries and raspberries as to hay production, though the hay meadow occupied 2.5 ha and the berry plot only 0.14 ha. For a ten-family sample, Minge-Kalman (1977) shows a .78 correlation between increases in labor intensity per worker and increases in the consumer/worker ratio, supporting Chayanov's hypothesis.

ment of drudgery, such as the conditions of production, the market situation, and the location of the farm relative to markets. Machines and capital that increase productivity lower the drudgery curve, and the influence of urban culture may raise the level of demand. In the Shan case, the Chayanovian model of the relation of household labor to production fits best in swidden cultivation (Durrenberger 1979: 453), but the varied production strategies for swidden, irrigated, and garden agriculture, for a number of different crops, for the mix of subsistence and cash products, and for different environmental conditions, mean that household labor and consumption *alone* do not constitute an adequate predictor of labor per worker. The Shan community in Thailand investigated by Durrenberger and Nicola Tannenbaum is not homogeneous, and households engage in three contrasting production strategies: intensified, normal, and minimal. Sahlins (1972: 102–23) refers to three similar possibilities of under-, over-, and sufficient production, but credits surplus production to the effort to establish political power. Households with insufficient production become both economically and politically dependent (Tannenbaum 1984b: 930):

In Thongmakhsan, strategy choice reflects the household's ability and willingness to support itself at the community's standard of living. Households choose production strategies based on their goals. For most households these goals are to live at the community's standard of living, able to produce enough rice to feed itself and enough money to buy necessary consumption items, some luxury goods, and to contribute to the local Buddhist temple. As long as the household members are healthy, have not had unexpected expenses, and have had sufficient rice yields, they can maintain themselves with the normal production strategy. If the household has had poor yields, illness, or unexpected expenses, it will have to intensify production to repay debts or it will not be able to support itself at the community's standard of living. Alternatively, a household may choose to intensify production to increase its well-being by building a new irrigated field or a new house. Both of these goals require more income than is required to support the household. Consequently these households use the intensified production strategy. . . . Most households at the minimum level of production have no choice in the matter; someone is sick or the household lacks the labor necessary to get it out of debt and back to the community's standard of living. A few households with adequate labor choose to live at this minimal level. Access to hill fields and the potential for developing new irrigated fields makes long-term production at the minimal level a matter of choice. In other locations, where access to productive resources is more limited, households may never have any choice about remaining at the minimal production level. (Tannenbaum 1984b: 932–33)

Production strategies that reflect differing exogenous conditions like illness or crop failure and different motivations, such as the desire to exceed community living standards (Barlett 1980), mean that there will be departures from the ideal relationship between the consumer/worker ratio and the intensity of production (Tannenbaum 1984b: 933)

Chayanov's analysis does incorporate other factors besides the consumer/worker ratio in determining the utility curve (Tannenbaum 1984b: 939), but the model's radical simplifications, by glossing over the complexities of differing household productive strategies and ignoring cash-producing activities, encourage an emphasis on the easily computed consumer/worker ratios as correlated with product per worker. The attractive shortcut to household labor input misses the complexities introduced by shortages of land and market participation that consistently enter into smallholder household decision making. Using 12 data sets covering societies as distant as Tsembaga swiddeners and Iowa farmers in 1880, Michael Chibnik found a positive correlation between consumer/worker and production/worker ratios in 11 of 12 cases, but only in 4 cases was the correlation greater than 0.3. This suggests that nondemographic variables, such as access to land and capital, value of farm implements, and location of farm in relation to transportation and market centers, are more important factors in affecting household production (Chibnik 1987: 96).

Intensification such as that found under Chinese conditions of scarce land, high population density, and greater labor intensity makes it difficult to adjust farm size as the peasant family changes demographically. Instead, it was the domestic unit of production that had to adjust its hand/mouth ratio, and, though "the biological mechanism of sex and reproduction is the ultimate source of family labor, . . . it is notoriously slow and often unreliable" (McGough 1984: 190). The demographic makeup of the household unit of production might change over the years through marriage, adoption of child brides, and sending out children as servants, and a more rapid adjustment of the consumer/worker ratio might be effected by hiring labor in and out. But labor recruitment was closely tied to landholding and to the socioeconomic status of the household, which were in part determined by legal and governmental relationships in the encompassing political economy.[16]

Land may also be unequally allocated according to differential political and economic power within the community. In England before the Black Death of the fourteenth century, the evidence does not reflect customary

16. Shu-min Huang (1984) believes that larger Chinese farm families have larger farms and also high dependency ratios and income per worker. But where dependency—that is, consumer/worker ratios—is highest, we find "slumping" net family incomes and income per worker, along with the lowest expenditure per consumer (S. Huang 1984: 177). In congested rural areas, there is no way in which poor families with many dependents can expand their holdings and thus intensify their labor.

holdings that grew or shrank in accordance with biological families' phases of expansion and contraction. In fact, some groups were expanding family land at the expense of the economically weak, and there was no cyclical mobility of families through the socioeconomic strata of customary tenantry (R. M. Smith 1984: 21).

In Sri Lanka, where farmers confront an economy of general poverty, high unemployment (both rural and urban), and an extremely tight market for land, the Chayanovian solution, expanding the size of the farm, "is an empirically verified, and fervent, peasant desire . . . , but a bootless one" (Herring 1984: 143). A quarter of the families had adults working part-time off-farm, all farms hired some labor, and family-labor farms (to the extent they had ever existed) had been replaced by small-scale commodity production with distinct capitalist features. Labor intensification could not be accounted for by number of consumers without consideration of the size of holding, opportunities for off-farm employment, the prices of inputs, outputs, and consumer goods, and farm taxes (ibid.: 145). Even in situations like that of Malay irrigated-rice growers, where size of holding and household labor and consumption formerly governed production, variation in household outputs are now determined much more by variations in the use of tractors and herbicides (Kahn 1981).

Among Kenyan Embu cultivators, though there was a low positive correlation ($R = .3$) between harvest size and number of consumers in the household, the consumer/worker ratio did not distinguish households with maize and beans in excess of consumption needs, higher mean wealth, and more coffee trees from others. Households that mobilized more labor were not those with lower dependency ratios, but those in which political patronage, education, and salaries gave access to nonhousehold labor, both hired and cooperative (Haugerud 1989: 74). As land availability becomes the major constraint on production and intensification raises yields, gross differences in household size and composition seem to become *less* important determinants of agricultural output. In Mgeta, a Tanzanian community with adequate rainfall and smallholder production of vegetables for the urban market, family size had no significant effect on acreage, and the number of adult male labor equivalents accounted for only 3 percent of the variation in farm size. Labor accounted for only an insignificant 1 percent of the variation in total production (Due and Anadajayasekeram 1984). The intensification strategy, though certainly labor-demanding, does not specify how that labor is to be mobilized, either from within or outside of the household, how much labor particular workers will apply, what proportion of members' time will be devoted to off-farm activities, and how land and capital factors of production will be managed in conjunction with labor.

Even an apparently simple smallholder, household-based system com-

TABLE IO.4
Philippine Agricultural Labor Types, 1983–84

Labor	Rice	Sweet potatoes	Vegetables (commercial)	Total
Person-hrs/mo	58.34	21.51	101.26	
Months	4–7	6–9	3	
Household labor	47%	93%	75%	63%
Cash wage labor (*poldeya*)	18%	7%	25%	17%
Payment in kind (*atang*)	24%	–	–	14%
Exchange (*oboan*)	11%	–	–	6%
Proportion of total labor	59%	22%	19%	100%

SOURCE: Wiber 1985: 430–31.

bining subsistence and cash production can exhibit remarkable and quite un-Chayanovian complexity. Kebayan, an upland Philippine community in northern Luzon, grows irrigated, terraced, double-cropped rice and hillside-swidden sweet potatoes for subsistence. Household wet-rice lands average one-quarter to one-half ha, while dry lands are one to two ha. Fully 81 percent of labor time goes into the subsistence crops of rice and sweet potatoes, while the remaining 19 percent is devoted to irrigated vegetables grown for the market, but work on vegetables is concentrated in three wet-season months, whereas the subsistence-crop labor takes place over longer periods (Table 10.4). Different crops make use of varying proportions of household labor, work paid in cash at a standard rate, in-kind payment of a percentage of the rice crop for each day worked (usually at harvest, but it can include transplanting), and exchange labor where households give each other equal amounts of work over a predetermined time period (Wiber 1985: 429–30). Wage work may be used for tasks requiring little skill but a high degree of drudgery, such as turning the soil, but it may also be used for hiring drivers and water buffalos in plowing, harrowing, and leveling tasks. Labor shortages in growing commercial vegetables may also be met with hired workers, but they rarely participate in the more specialized tasks of planting seeds, fertilizing, spraying pesticides, and harvesting.

There was a strong correlation ($R = .62$) between the net worth of the household and the amount of hired labor, while net worth and exchange labor were negatively correlated ($R = -.40$) (Wiber 1985: 434). The dependency ratio of consumer/workers was not a reliable indicator of household labor recruitment and allocation patterns. There was also no correlation between dependency ratio and household net worth. Because families do not produce enough rice to meet all their needs, they prefer to grow rice during the dry season, even though that is when market prices for vegetables peak. Avoiding the risk of commercial-crop failure and the threat of incurring debt to buy food appears to be preferable to the possibility of maximizing household cash income by specializing in vegetables. The technical and seasonal demands of labor on different crops, the emphasis on subsistence and cash strategies, and the use of a variety of non-household labor sources all make household composition a poor guide to the ways in which labor will be applied. But it is just this diversity of farming systems, market and nonmarket activities, and differences in household access to resources that characterizes most intensive agriculturalists.

It is possible, indeed necessary, to appreciate Chayanov's heuristic achievements in emphasizing the distinctive microeconomic dynamics of the farm-family household and the influences of its developmental cycle on labor without accepting the restrictive assumptions on which the model is based. His analysis "from below" neglects the immediate environment of the household and the way in which restricted resources under conditions of population density intensify labor use and increase conflicts of interest within the supposedly solidary household. But the abstraction of the peasant family from its networks of non-family labor, market systems, and politically supported property and tax structures was conceptually the more hazardous step. "Not accidentally it was his most exclusively family-centered model, the demographic one, which first fell into disuse. The only way to handle effectively contemporary social reality is through models and theories in which peasant family farms do not operate separately and where peasant economy does not merely accompany other economic forms but is inserted into and usually subsumed under a dominant political economy, different in type" (Shanin 1986: 19).

No one would deny that household production possesses great differences in the uses it can make of labor time and other resources compared to the pure capitalist firm with hired wage labor. And these differences may permit both the survival and competitive advantage of household production in economic spaces within a capitalist economy. But household production does not occur independently of the dominant capitalist mode except in the extreme (and entirely theoretical) case of pure subsistence agriculture. As soon as households buy or sell in the market place they confront prices and costs which are established in the larger capitalist economy. Moreover unless they are pure subsistence farmers they have to engage in such transactions for family survival and their economic actions can no longer be considered independent of the wider system. (Ellis 1988: 116)

It is unfortunate that Chayanov's success in illuminating a noncapitalist, but still economically rational, logic at the heart of farm-family decision making made attractive the formulation by others of a domestic mode of production that is isolated, self-sufficient, and made up of altruistic social individuals who are not economizers. Smallholder householders practicing intensive agriculture are not bound to the developmental cycle, with its changing consumer/worker ratios, and they march to a different drummer than do Kantu shifting cultivators or traditional Russian peasants. Although households remain, as they were for Chayanov, the key units for mobilizing farm labor and organizing consumption, they are neither limited to subsistence nor isolated from the market. And neither capitalism nor socialism can contradict the economic logic or deny the stubborn persistence of smallholder adaptations under conditions of high population density and scarce land.

Epilogue: Does the Smallholder Have a Future?

SOCIAL SCIENTISTS do not have a distinguished record as prognosticators, and anthropologists seldom even try to predict the future. Our forte has frequently been the marshaling of particularistic ethnographic evidence to contradict sweeping cross-cultural generalizations or the gleeful debunking of ethnocentric claims. This book in its emphasis on small-scale agriculture and household production has espoused contemporary farming systems that appear to be technologically primitive, economically undeveloped, and socially simple, and, by extension, localized, outmoded, and anachronistic. Such a negative description implies a set of evolutionary assumptions about the present and future of agriculture and rural social organization that run contrary to much of what we know about crop production, sustainable systems, labor, tenure, and incentives. I have denied the inevitability and the perfectibility of "modernization," implicitly equated with progress, in which agriculture as a whole becomes increasingly large-scale, scientific, dependent on nonrenewable energy sources, and narrowly specialized. In short, I question the exclusive application of an *industrial model* to agriculture because of its technical rigidity, its capital costs and labor savings, its energy inefficiency, its tendency to degrade natural resources, and its separation of ownership, management, and labor.

My contention is that smallholder intensive systems achieve high production, combine subsistence and market benefits, transform energy efficiently, and encourage practices of stewardship and conservation of resources. If this analysis is correct, we shall *not* everywhere witness the dispossession and demise of smallholders and their replacement by factory farms and landless wage workers. There will certainly continue to be such large-scale enterprises, and some kinds of land use, such as dry-wheat farming and ranching may never be practiced on small, independent parcels. Arable land that is abundant, cheap, geographically (and legally) accessible, and involved in a market economy will continue to be used in extensive, often destructive ways, usually by large-holding entrepreneurs or corporations. But densely settled areas of traditionally intensive production, like the great irrigated regions of Asia, will remain smallholder bastions, and zones of increasing population pressure in Africa and Latin America may move gradually in the same direction.

My argument for the existence of a distinctive smallholder socioeconomic type and for its shared ecosystemic characteristics has been based on case studies from anthropologists and geographers, more abstract analyses by economists and agronomists, historians' reconstructions of long-term processes of change, and the theoretical frameworks of Steward and Boserup. The elements that make up the smallholder system are not new, but their regular functional association in a number of societies under different political-economic regimes and in different time periods has been neglected. The characteristics of the smallholder agricultural system that I have defined often violate the conventional wisdom and contradict the industrial model of development shared by both socialists and capitalists. There are cross-culturally a set of distinctive features that smallholders share:

1. Farms that are typically small in average size, permanent in location, and with high, continuous levels of production tend to occur in areas of dense rural population, and they may serve both subsistence consumption and market demands.

2. Practices of agricultural intensification on smallholdings, including diversified crop and livestock production, manuring, terracing, irrigation, drainage, intercropping, rotation, weed control, fencing, and agricultural storage, are dependent on elaborated systems of folk knowledge, practical experience (often with specific plots of land, water sources, and local climatic variability), and an appropriate technology that is not necessarily complex.

3. The skilled, responsible, and complementary labor required for such intensive cultivation is provided largely by the farm-family household. The household is also the unit of management of the enterprise and of consumption. Implicit contracts or covenants of long-term reciprocity give an economic rationale to certain relationships among household members. Willingness to accept low marginal returns on labor and strong individual incentives for direct consumption and maintaining rights in valuable landed property allow survival of the household under circumstances when a wage-paying farm firm would not be viable.

4. Intensive agriculture demands more work regularly applied over longer periods of time than other types of farm production. The division of labor spreads this work input over household members of various ages and different genders and places a premium on their effective cooperation. Restricted land resources may mean that some household members engage in craft, commercial, or wage work, but returns per hour from

off-farm labor are often lower or less reliable than those from work on the smallholding.

5. The smallholding that depends largely on manual labor and animal traction generally shows a favorable ratio of energy input (measured in kilocalories) to output. Systems using large amounts of fossil fuels and manufactured inputs such as chemical fertilizers, pesticides, herbicides, and machines such as tractors and pumps characteristically have negative energy balances. The intensive smallholding with few unrenewable energy sources and heavy reliance on processes of regeneration and recycling of nutrients is sustainable, as evidenced by the development and maintenance of highly productive, stable multicropping by Chinese cultivators for centuries. Environmental degradation is prevented by the careful practices of intensive cultivation and by the emphasis on the long-term value and heritability of the holding.

6. Productivity per unit of land is *inversely* related to farm size. Smallholders use their fields more frequently and produce larger yields than larger landholders in the same environment and with the same technology and crops. Just as it is profitable for a big landowner to produce lower yields of specialized market crops or beef cattle by extensive, labor-saving methods, so it is to the smallholder's advantage to produce in a more diversified, continuous, skilled labor-demanding manner in order to make fullest use of more restricted resources.

7. Rights to scarce land under intensive cultivation are held by the smallholder under systems of long-term usufruct, publicly recognized exchange, and heritability that approximate private tenure. Such individualized rights may exist apart from state controls, legal tenure, and major market involvement. Even legal statuses of sharecropping and tenancy with nonworking owners give farmers a significant share in the benefits arising from improvements they make, such as orchard planting or extending irrigation. Private household property often coexists with common property institutions in resources such as pastures, forests, and waterways that must be protected, periodically reallocated, and monitored by corporate groups. Communities of smallholders have the demonstrated capacity for cooperative management of environmental resources without the untrammeled individual competition that brings on a "tragedy of the commons."

8. Inequality in the household distribution of landholdings, livestock, buildings, and money wealth is characteristic of smallholder communities. Such inequity may be owing to initial endowment, inheritance system, marriage, climatic variability, or factors of individual hard work, skills, and management abilities. Mobility over the personal career and the household developmental cycle seems to prevent classlike permanent differentiation among smallholders. Household flexibility, resilience, and

efficiency in allocating labor and consumption contribute to effective smallholder competition with larger, wage-worker farms. Societies of intensive cultivators, even in the context of capitalist market economies, do not necessarily undergo polarization into a bourgeois landlord stratum and a rural proletariat.

9. Agricultural collectives instituted for political and ideological reasons among smallholder intensive farmers, as in the case of the Chinese People's Republic, encounter severe problems of administration, incentive, and land productivity. The recent return to household responsibility for labor organization, landholding, participation in the market, and budgeting has raised total farm production, work efficiency, and rural standards of living. The industrial model of large, consolidated farms with communal ownership, mechanization, and centralized planning has been effectively scrapped in China.

If the smallholder household system of agriculture that I have advanced, the theoretical analysis on which it is built, and the ethnographic-historic cases on which it rests have some validity, it should be possible to say something substantive about the future of the smallholder's way of life. Will there be more intensively tilled little farms, is this a good thing or not, and what should the rest of us do about it? There is no gainsaying the fact that in the coming years there will be more people wanting more food and a better standard of living from a global environment whose natural resources are only in part renewable. Even for those parts of the earth that are still land-rich, an agricultural utopia based on fossil-fuel power, chemical fertilizers and bug killers, and biotechnology on factory farms is beginning to look expensive and hazardous. The most promising alternative and the one with the best historic track record of adapting safely to heavier population is the process of agricultural intensification that Ester Boserup describes and analyzes. Such a course need not be confined to densely settled, capital-poor Third World countries. Even in the United States, it is said, the number of farmers may finally increase because:

1. Historically, in all the past civilizations I am aware of, the denser the population becomes, the smaller and more numerous the farms become.

2. Financially, the economies of scale that apparently rule manufacturing do not really apply to any sustainable kind of food production—when you count all the costs, it is cheaper to raise a zucchini in your garden than on your megafarm.

3. Socially, people are beginning to understand they really are what they eat, and we are demanding quality foods that megafarms can't supply (Logsdon 1989: 65).

Even without population growth, the double bind of rapidly (and perhaps permanently) escalating energy costs and the accumulated environmental problems of massive topsoil loss, diminishing returns to fertilizers and pesticides, and the "addiction to petrofarming" (Lovins et al. 1984: 73) suggest that alternative (National Research Council 1989) and certainly more intensive agriculture is something beyond a romantic dream of the ecologists and the Greens.

Except for historical cases and a few contemporary American examples like the Amish, this book has considered smallholders in those parts of the world where they still exist and where they are likely to increase. The characteristics of industrial agriculture as enumerated by Peggy Barlett (1987a, 1989) make it unlikely that it will be widely adopted where land is scarce and nonrenewable energy is expensive. The increased use of complex technology and its energy-dependent inputs requires large amounts of capital initially, efficient distribution and maintenance of imported manufactured inputs, repair facilities and skills, and a growing appetite for oil. This is not to say that smallholders do not mechanize appropriately. In a growing economy with more off-farm jobs, and as household labor shortages increase, farming households may buy garden tractors, grain mills, and portable irrigation pumps. Demand for inputs that are not dependent on large-scale utilization like improved seeds and chemical fertilizers may far exceed supply, especially if state agencies monopolize their distribution.

The Kofyar, without benefit of any agricultural extension, had begun purchase of fertilizer for their cash-crop yams, but farmers lacked information on proper applications in interplanted fields. The experiments with new methods and the synergistic interaction with indigenous technology (Paul Richards 1985) suggest the same individualistic drive to increase production that characterizes intensive cultivators everywhere. Though the smallholder is wary of expensive heavy equipment and the debt needed to purchase it, and though big, centrally run agricultural projects have not provided adequate private returns on labor, smallholders eagerly take advantage of motorized transport for their crops and themselves.

Despite the fears of many that commercialization and a decline of self-sufficiency mean dependence on external markets and the loss of control to outside elites, smallholders want usually to participate more rather than less in the market, and they actively innovate to do just that. The ever-present risks of climatic fluctuations, disease, and price swings ensure the prudent maintenance of some subsistence production, but complete food self-sufficiency in the modern world is autarchic impoverishment. Smallholders actively seek economic comparative advantage. They are anxious as well for practical knowledge, all too often not forthcoming from agricultural experiment stations, on how to grow more crops *within their existing farming systems*. Official prejudice against smallholders, part-time farmers, and women gardeners often means that those with the most intensive system of production receive the least advice, technical assistance, and credit.

Smallholders like the Ibo and the Chinese are often avid for formal education, being well aware that the household's nonfarm income is best supplemented by children who have been to school, and agricultural earnings are often invested in such occupational diversification. My guess is that smallholders are relatively unanimous in their support for a good road network, big and honestly administered markets, and decent, accessible schools. As to what they cultivate and how they do it, when they sell their crops, and how they run their businesses, planners and bureaucrats should be cautioned against interference. A populist reliance on the good sense of peasant households in such matters and a conviction that smallholders make realistic, informed economic decisions is evident in Chayanov but was anathema to Stalin's regime, which killed him.[1]

One of the most persistent dilemmas of modern developing nations is what to do about smallholder property rights. An article of faith for both the old colonial regimes and their independent state successors has been that technologically simple indigenous farmers had only customary "communal" rights to land. Land might allegedly be worked in common, it was reallocated occasionally by lineage heads or village assemblies, and it could never be bought or sold. From this premise, it was an easy step to declaring state ownership of all land, collecting tax as rent for its usufruct, and supposedly preventing the poor from losing their land to greedy chiefs and usurious moneylenders (Shenton 1986). Clear legal indications that smallholders owned permanent rights in scarce resources that could be improved, leased, sold, and inherited, as in the Close-Settled Zone of hedged, manured farms around Kano in northern Nigeria, was conveniently ignored (Mortimore 1972; Shenton 1986). If land tenure, as I have maintained, is in fact intimately connected to land use, such legal fictions can deny the security in property and the possibility of exchange in various temporary and permanent ways that intensive land use requires. It also impedes the orderly litigation that necessarily accompanies competition for scarce and valuable means of household support.

Just as damaging as the myth of primitive communal land rights is the effort to promote economic development by mandatory land registration under a standard legal code. The contention is always that farmers will not invest in land improvement unless their property rights are protected

---

1. In 1931 Chayanov and other senior scholars were condemned for sabotaging Soviet agriculture, and they were executed in 1937 (Shanin 1990: 189, 203). Chayanov's reputation was vindicated in 1987, when his fundamental contributions to agricultural theory were recognized by the president of the Academy of Agricultural Sciences of the USSR (ibid.: 203).

by law, and that credit for seed or livestock is only available by mortgaging land held by title. I have cited considerable evidence that intensively tilled land is indeed held as private property, but this emphasizes the economic reasons for individualization, rather than suggesting that jural rights are a precondition for more intensive usage. Top-down governmental land-registry schemes, often combined with consolidation of scattered plots, are generally rigid, arbitrary, and subject to corrupt adjudication (Shipton 1988). Actual land use and such indigenous practices as lending, pawning, and exercising independent rights to trees may be disregarded. Outsiders and politically powerful residents may secure more or better-quality land than they are entitled to (Reyna 1987).

Even more destructive may be the appropriation by individuals of common property resources, in part because most modern codes have no provisions for local community holdings corporately administered by members. Smallholders and landless individuals all suffer when legal acts of enclosure expel them, and the privatized property may be exposed to environmental degradation. If more intensive and permanent resource use, with its accompanying investment and higher land values, is gathering force among smallholders, there should be institutional mechanisms whereby neighbors and co-villagers can resolve questions of boundaries, ownership, and recording on the local level with due regard to the history of occupancy, acquisition, and use over time. Such a group could also make provisions for dispute settlement under a more general state jurisdiction. Effective common property regimes seem always to require local assemblies, constitutions, and rules, allowing for elected officials and paid supervisory personnel (Ostrom 1990), like the Indian pasture guards (Wade 1988) and Swiss irrigation societies (Netting 1974a). Some devolution of authority from state bureaucracies to locally organized smallholders would seem to be necessary if both private and common property rights are to be creatively developed and fairly enforced.

Policies in developing nations that might further the interests of smallholders in crop pricing and marketing, appropriate technology, and land tenure are less likely because the models of agricultural progress remain those of industrial organization advocated by the dominant world powers.[2] One of the enduring ironies of twentieth-century history has been

2. There is, in fact, an ongoing debate in the literature of development economics between advocates of a 'unimodal' strategy of agrarian change aimed at progressive modernization of peasant smallholders (Johnston and Kilby 1975, 1982) and supporters of a 'bimodal' model that concentrates resources on a highly commercialized subsector of large-scale, mechanized farms with wage labor (Cohen 1988, 1989). Household farm units in Japan and Taiwan have been successful in adopting divisible innovations such as improved seed and fertilizer while allocating their resources so as to minimize costs. Advantageous decentralized decision making by individual producers is furthered by efficient market mechanisms (Johnston and Kilby 1982: 51), and government policies should allow struc-

the agreement between generally hostile capitalist and communist societies that agricultural progress could be based *only* on big farms; energy-intensive, labor-saving technology; scientific agronomy; and centralized management. The premise that justified such a farm enterprise and that evoked a largely unquestioning cultural allegiance was that in agriculture, as in industry, there are economies of scale. Production efficiency and lower costs of food for all were presumed to be a function of farm size, and the smallholder was therefore held to be inefficient and expensive. Studies in the United States have, on the contrary, found costs per unit of output on a fully mechanized one-man farm to be equal to or lower than those of larger operations. In California, while larger farms had greater returns on capital, medium-sized farms maximized work opportunity, total production, trade, and income (Goldschmidt 1978: xxxi). Small crop farms (averaging 142 acres) in a 1985 U.S. census tended to be more diversified and to have rates of return similar to or better than those of larger farms, while small farms also had lower costs and used fewer purchased inputs per unit of output (Strange 1988: 95–100). The increase in the number of industrialized, corporate farms and the displacement of independent smallholders in U.S. agriculture[3] is owing, not to the mythic economies of scale, but to the special advantages given by government to

tural transformation that minimizes the role of public agencies in setting producer prices, subsidizing inputs, and supporting capital-extensive production (Johnston 1991). Public investment in roads, irrigation, and drainage facilities can improve rural infrastructure. Such policies would benefit the intensive land use that is the characteristic smallholder strategy for coping with restricted resources. The alternative bimodal pattern claims that equitable, widespread state investment targeting peasant smallholders requires unimodal, government-regulated, bureaucratically implemented policies, which have led to failure in many African cases (Cohen 1988: 10). Such criticism rightly points out that in much of Africa, land is not scarce, considerable rural differentiation exists, and most food sales come from the largest one-quarter to one-third of peasant farms (ibid.: 11–23). Though such a categorization is much more appropriate for sparsely settled regions of Nigeria, such as the Kanuri area, where Ronald Cohen has worked, than for the Kofyar, it does emphasize extensive land-use systems much different from those of the smallholder type dealt with in this book. The bi- or multimodal path is seen as "adaptive, experimental, and realistic" (ibid.: 28) in allowing development at various levels, but the contention that "satisfying Africa's food needs depends on the development of a specialized, commercial agriculture" (ibid.: 23) implies the official encouragement of large-scale farming and a lack of confidence in the development potential of smallholders.

3. The total number of U.S. farms declined from 5.9 million in 1945 to slightly more than 2.2 million in 1985. Since the total number of harvested acres remained relatively constant at approximately 340 million acres, average farm size almost tripled (National Research Council 1989: 54). Though U.S. farms at the end of World War II were obviously larger than the intensively used tracts of Asian and African cultivators, they maintained elements of the smallholder pattern. Farms in the Midwest and Northeast were diversified crop-livestock operations with full-time operators. They produced forage and feed grains for their animals, using crop rotations and animal manure. Fewer inputs such as fertilizers, pesticides, and herbicides were purchased. With growing farm size, specialization, and de-

large growers. Walter Goldschmidt (1978: xxxii) summarizes these advantages as "(1) the agricultural support programs, (2) tax policies, (3) agricultural labor policies, and (4) the research-orientation of the USDA and of land grant colleges."

Another version of the same philosophy of "the larger and more mechanical the better" was enforced on the Soviet Union under Stalin and Brezhnev. The equation of size with efficiency, mechanization with effectiveness, and chemical inputs with crop outputs created a "gigantomania" reflecting "the transfer of insufficiently digested experience of heavy industry to an environment where such innovations were counterproductive" (Shanin 1990: 191). Coerced collectivization of resources, managed communal labor, and "scientific" methods were the revolutionary means to establish a modern, large-scale agriculture. "Poverty in the Russian countryside, low production within agriculture, and the underdevelopment of the country at large were . . . seen as rooted in the small and nature-dependent character of the peasant family farm" (Shanin 1990: 195). But tractors and fertilizers, plowing virgin lands, refusing rural people the right to migrate, and later depopulation of some 40 percent of existing villages achieved few increases in production while creating severe ecological damage (Shanin 1990: 190, 192, 196). Although the differences between the United States and the USSR in ideology, political economy, productivity, history, and environment are tremendous, the shared myth of agricultural economies of scale and the implicit or explicit denigration of the smallholder are evident.[4]

If land and labor are the twin poles around which our discussion of smallholder agriculture has revolved, then the concept of *alienation* may provide a means for coming to grips with the seeming contradictions of this historic and enduring way of life. Marx's claim that human beings make their own lives in their productive activity, and that they become estranged from their true natures when they are alienated from the prod-

---

pendence on nonhuman energy sources, 15 to 20 percent of all U.S. farms now produce 80 percent of all output (ibid.: 59).

4. Even in classical economic theory, Adam Smith and Karl Marx "shared the belief that commercialization would transform the peasant economy" (P. C. C. Huang 1990: 1) by means of the division of labor, specialization, and economies of scale. The hegemonic ideal of agricultural change continues to be based on a particular region and historical period. "Smith and Marx, of course, based their shared assumptions largely on the English experience. There was, after all, the empirical example of England's enclosure movement and the eighteenth-century agricultural revolution in which peasant farms gave way with commercialization to large-scale wage labor–based capitalist farms. Thus reinforced, these assumptions came in time to hold the force of a paradigm, considered almost too obviously true to require any discussion or comment" (ibid.: 2). The ecological threats of contrasting political ideologies both of which glorify economic growth at the expense of the environment are equally obvious. "The two major models at work today, capitalism and the Soviet and Chinese brands of socialism, are *industrial*" (Jackson 1987: 36; emphasis in the original).

ucts of their labor and the act of production, is as true of the farmer as the wage laborer. But the smallholder represents a bastion of resistance to the alienation of employment in capitalist industry of the nineteenth-century type. Work in the field and around the house may be arduous and pro-longed, but it is not characterized by the "routinization, specialization, and submission to external control and direction" of the factory (Preston 1986: 29). The labor of smallholders, even those in a capitalist society, is voluntary rather than coerced. The logic of intensification and the hope that energizes the farmer are dependent on the achievement of an accept-able return on the expenditure of extra time and effort and the mobiliza-tion of special skill, knowledge, and management abilities.[5] Alienation of the work product through confiscatory rents, taxes, labor services, terms of trade, or appropriation of the means of production *can* threaten and finally destroy a rural society of smallholders. But paradoxically the health and vigor of farming communities and their constituent house-holds depends on the alienability of land, the right to convey or transfer property freely. Bequest, inheritance, loan, lease, swap, mortgage, and even sale appear to be necessary for the flexible reallocation of the factors of production that mediates between scarce resources and human wants.

The Marxian notion that "the abolition of private property is the nec-essary condition of the establishment of a system of free, creative labor" (Preston 1986: 30) makes a system of alienable property responsible for the alienation of the worker. With the exception of a few ideologically or religiously distinctive communities such as some Israeli kibbutzim and Hutterite colonies, it is safe to say that smallholding intensive cultivators have clung to their individualized properties for dear life. While insisting on their rights of tenure and exchange, they have resisted land concentra-tion by estate owners, state or corporate appropriation, and forced collec-tivization. The possession of established private and common property rights in a market economy of productive smallholdings is a protection, though not a perfect one, against alienated, subservient, and degraded work.

Perhaps the most consistent and insidious threat to a recognition of the

5. An "acceptable" return is, of course, always based on a comparison of one's own life chances with those of one's peers. Children of smallholders in today's Denmark or Switz-erland are often reluctant to take over the family farm, even when government subsidies guarantee a reasonable income. Alienation may even be welcomed. "In my opinion, the point is that changes in urban technology change the way of life, which changes opinions in urban areas: and with modern means of communication urban attitudes and opinions pene-trate quickly to the countryside, so a fixed income and fixed working hours become pre-ferred to independence" (Boserup, pers. comm.). When, on the other hand, state welfare mechanisms are inadequate, urban incomes are reduced, and political breakdown interferes severely with the market, as we see in parts of the Third World and Eastern Europe, the smallholding may offer the satisfactions of a secure livelihood.

utility of a smallholder adaptation is the judgment of a modern, urbanized world that technologically simpler farmers must work too hard for miserable returns. Agricultural labor, particularly manual tasks that involve the worker with dirt, animal excrement, and exposure to the elements, must seemingly be unpleasant and are little respected. Yet smallholders persist in choosing to acquire land and to cultivate, even if only on a part-time basis, even when they have other occupational options. Are they merely ignorant and traditional? Do they have an unreasonable affection for unremitting toil? Can't they calculate their own economic advantage?[6] I have tried to demonstrate that where rural population is dense and natural resources will support intensive use—that is, where labor is relatively cheap and land dear—access to a modicum of arable soil is often the best guarantee of a reasonable livelihood and the long-term security of the household.

Smallholders are not as rich as landed aristocrats, the higher officialdom of government, the commercial elite, or urban professionals, although they may well have kinship and economic links with these classes. What they do have in their experience, their fertile land, their livestock, and their diversified production strategies is a set of defenses against the uncontrollable vagaries of weather, prices, and war. Smallholders may not always live well, but they are seasoned survivors. They may moonlight as craftsmen, petty traders, field hands, or factory workers, but they do not keep one foot on the farm out of sentiment or stupidity. And the society that dispossesses smallholders in favor of factory farms, plantations, or socialistic communes simultaneously risks a decline in agricultural production, rural unemployment, and ecological deterioration.

If smallholder farming households scrimp, scheme, and sacrifice to get another scrap of land or urge a marginally higher return from reluctant soil, can we affirm that they have a "good life"? There is, after all, nothing as bitter as peasant grudges that poison successive generations in factional conflict and family-splitting fights. But those millions who opt to remain

6. When agricultural economists broke down the costs and benefits of smallholder farming in northern Portugal, the household enterprise at first appeared to be a going concern. It required little capital and modest "tradable inputs," while achieving high returns on land. If, however, total labor is credited at the opportunity cost of market wages, small farms slip into the deficit column and appear to be inherently unprofitable (Fox and Finan 1987: 196). The stubborn desire of Portuguese rural people to invest their savings in one or two hectares of land, often paying what seem to be highly inflated prices, and then to raise a little corn, some potatoes, wine grapes, and kale, along with a few milk cows, chickens, and rabbits seems to be economically irrational. Only land consolidation into larger commercial farms and dispossession of the majority of smallholders would promote agricultural progress. But if labor costs were to reflect the much lower returns acceptable to men working part-time or as elderly, retired farmers, women combining gardening at odd hours with childcare and domestic duties, and children doing chores after school, the family farm would make economic sense (ibid.: 198–99).

as farmers or to return to a smallholding are telling us that this is the best available alternative, economically and socially, for them. The long-run returns on the cooperative labor of all the family members, coupled with a security for the household that is not finally dependent on state welfare benefits or company pensions, is still the better bet in less affluent countries. Until we have sound comparative measures of the quality of life among both rural *and* urban masses, it would be wrong to dismiss smallholder preferences for hard work and property as somehow misguided and irrational.

Perhaps the real delusion of economic growth is the hegemonic preoccupation with high food production achieved by less human labor. There is implicit agreement that people should work fewer hours and with less physical exertion, presumably saving time for such leisure activities as TV viewing, gardening, and sports. I have documented the fact that smallholders do indeed invest a great deal of year-round labor, much of it manual, but whether the actual hours exceed those of a small restaurant keeper, a street vendor, or a stockbroker is open to question. The increases in labor time are relative, not necessarily bad, and may have commensurate returns. Thus Javanese rice farmers with more than 0.5 ha of land work more than their landless neighbors but achieve higher rates of return on their time (Chapter 3). Poorer families in the Third World generally lack the opportunity to work as much as they would like and to receive consistent wages that match their highest seasonal earnings. In countries where urban unemployment and the underemployment frequent in the informal economy are rife, the labor-absorptive capacity of smallholder intensive agriculture should be welcomed as an economic stimulus. On the other hand, the heavy food imports and aid from the industrial nations that lower prices of indigenous agricultural products and encourage rural-urban migration erode this economic stimulus by reducing the potential of smallholdings to mobilize labor and invest capital productively. Growth in urban manufacturing at the expense of a supposedly stagnant farm sector is not the yellow brick road of modernization. Despite the popular prejudice, labor-saving is not the chief end of life, and farm work is not a bad thing. When labor and property rights are combined, and when the farm household organizes and schedules its own skilled activities as an independent enterprise, the relations of production are not those of alienation.

Georgia part-time farmers talked to Peggy Barlett (1987b) about loving the magic of growing things and the personal satisfaction of creating something. They appreciated the peace and contentment that comes from physical work and the high-quality food that they could raise on their own places. Supplemental income, some tax advantages, and desirable retirement work give economic support to this way of thinking, but the intangible values of raising children with responsibilities in a family-sustaining enterprise (Logsden 1989) seem to rank even higher. People *like* the bucolic tasks that Hesiod and Virgil commemorated (Montmarquet 1989). In a modern, rapidly aging society like that of Japan, the stem-family farm household may be seen as a vehicle for the filial duty and care that even the most potent industrial state cannot provide (Prindle 1984). A pattern of even temporary co-residence and cooperation weakens the classic forms of alienation, of children from parents, of youth from age, and of generation from generation (Lyon 1985).

There is no need to assume that there are not trade-offs between the narrowly conceived economic efficiency of the farm enterprise and the personal and family needs of household members. With the myriad occupational and technological options of many modern nations, individuals could often make more money with less work both on and off farm than they do. A survey of Swedish farmers found that "they strive primarily to satisfy a complex combination of social needs and environmental concerns, while obtaining personal job satisfaction. Their aspirations have an 'organic' rather than 'instrumental' orientation, i.e., for most farmers, farming means stewardship and a way of life" (Nitsch 1987: 105). The superficial homogeneity of smallholders that led Marx to compare them to potatoes in a sack is not borne out by farmers' self-images. Planning, for the Swedish farmers studied by Ulrich Nitsch, was something that went on continuously in the mind during the day's work, and management decisions originated from their own idiosyncratic farm experiences. Smallholders the world around emphasize their freedom to chart their activities and goals independently, *to be their own bosses.* Such autonomy as economic actors brings both market rewards and personal satisfactions. But success on the land seems most generally to be conceived as service to the family and its future. Lacking such motivation, Swedish "single farmers often saw little meaning in improving the farm, since they had no family to appreciate their efforts and no family member to inherit the fruits of their labor" (ibid.). The mutual support, insurance, and optimizing economic performance of the smallholder household are compatible with the altruistic, selfless love that is a universal virtue.

"Human scale" and the preservation of rural "working landscapes" may be starting to replace what E. F. Schumacher (1973: 66) called "the almost universal idolatry of giantism." But the beauties of smallness are neither merely in the comprehensibility and the aesthetics of smallholder farming nor in the agricultural productivity, efficiency, and resilience that I have been at pains to document. Perhaps one major reason for the myth that bigger is better is that the social and environmental costs of large industrial agribusiness are difficult to quantify and are therefore consigned by the economists' confraternity to the outer darkness of externalities (Strange 1988: 87–88). In his 1940s comparison of two California communities, one with medium-sized owner-operated farms, and the other with large-scale, absentee-owned operations, Walter Goldschmidt (1978 [1947]) showed convincingly that the town of smallholders (relative to U.S. standards) was a better place to live. Average family income was higher, there were more stores with more retail trade, and the town had more schools, parks, and newspapers, and better streets (Perelman 1976). Dying American rural communities reflect less a fall in the volume of production from an area than a decline in the number, and therefore an increase in the size, of the local farms (Strange 1988: 87). Neither migrant laborers nor a few foremen and equipment drivers support stable, prosperous communities. "Everyone who has done careful research on farm size, residency of agricultural landowners and social conditions in the rural community finds the same relationship: as farm size and absentee ownership increase, social conditions in the local community deteriorate" (Dean MacCannell, quoted in Strange 1988: 87).

The social ties and sentiments that bind household members to each other and to neighbors with whom they share work, tools, churches, and schools also have an explicitly temporal dimension. Smallholders cannot wittingly destroy their own resources and thereby ruin the future livelihoods of their offspring. Choices of farming practices that involve soil mining that causes run-off erosion, declining nutrient status that necessitates higher fertilizer application, increasing pesticide and herbicide use with declining effectiveness, and the pumping down of groundwater supplies contradict obvious good farming practice and leave an impaired inheritance for the next generation. Mammoth modern economies with cheap energy sources and plenty of capital, plus farm policies that subsidize surplus production and underwrite land speculation, can get rid of many smallholders, regardless of the costs. But even in the U.S. midwestern heartland of the hog and the dollar, family farms with many of the characteristics of the smallholder adaptation have endured and flourished. Sonya Salamon (1985) distinguishes two types of farming systems that have existed side by side in the same central Illinois environment with the identical basic technology for over a century. The Yankee entrepreneurial variety uses large monocropped grain fields, often rented, with careful calculation of short-term economic profits. Farmers may live in town and have little interest in passing farm property on to their children. The motivations and strategies of "yeoman" farmers of German Catholic ancestry are distinctly different. Their smaller farms have often been in the same family for generations and continue to be run as diversified dairy-hog-beef-grain operations. The yeomen use a labor-intensive strategy and practice careful stewardship, avoiding debt and husbanding the resources that will in turn support their heirs. Such farmers, though competing vigorously with each other, are proud of the churches, schools, and community services that their prosperous villages maintain. The small-scale yeoman orientation rests on beliefs that are part of a cherished ethnic heritage. Though it eschews economic maximization, a smallholder strategy that "entails close family cooperation and relatively modest financial goals may be particularly adaptive under certain economic conditions" (Salamon 1985: 338). Intensively farming smallholders don't get rich quick, but they are less likely than rootless entrepreneurs to create dust bowls or to go broke.

In the course of attempting to construct (but not deconstruct) out of ethnographic evidence and the tools of practical reason a plausible picture of smallholder households, I have made little reference to the other, psychological side of alienation. The meaning of agricultural work to the cultivator and the emotions suffusing the ideas of family and farm and village community have been given short shrift. A somewhat single-minded focus on technology, energy inputs and outputs, resource scarcity, institutional functions, and legal rules was my way of arguing that the smallholder system makes economic sense, regardless of whether it conforms to the romantic ideal or the pastoral idyll. This is not to deny that the walls and fences of Wordsworth and Frost or the tenacity and courage in the peasant novels of Knut Hamsun and Emile Zola are great affirmations of a distinctively smallholder spirit. Robert Redfield was right about the sturdy independence, reverence for land, and honoring of tradition that we admire in the smallholder. On the other hand, we are not surprised when farmers tell us with absolutely predictable pessimism about droughts and pestilence, rotten luck with cows, or the venomous spite of neighbors and the greed of moneylenders. Happy peasants (and there must be some) never talk as if they are. But in the introspective and occasionally lyrical moments when farm people respond to a beautifully tilled landscape or a fine ewe or remark on why they gave up indoor work, there is a fusion of feeling and livelihood that is genuinely moving.

Where people are plentiful and land is scarce, the distinctive adaptation of smallholder households practicing intensive agriculture will appear, just as it has for centuries in a variety of human societies. Farm families train and mobilize labor, manage their limited resources of land and livestock, claim and inherit rights of property, and produce both for their own consumption and for exchange. They have learned to understand, nurture, and renew the soil and water that sustain them, and the means that they have devised are worthy of our respect and emulation. The question of whether the practical and coherent smallholder system has a future is not in doubt. It may be more vital and necessary to *our* future than we realize.

Reference Matter

References Cited

Abélès, Marc
1981   In Search of the Monarch: Introduction of the State Among the Gamo of Ethiopia. *In* Modes of Production in Africa: The Precolonial Era. Donald Crummey and C. C. Steward, eds. Pp. 35–67. Beverly Hills, CA: Sage.

Adams, Jane H.
1988   The Decoupling of Farm and Household: Differential Consequences of Capitalist Development on Southern Illinois and Third World Family Farms. Comparative Studies in Society and History 30: 453–82.

Alexander, J., and P. Alexander
1982   Shared Poverty as an Ideology: Agrarian Relationships in Colonial Java. Man 17: 597–619.

Ali, Abu Muhammad Shajaat
1987   Intensive Paddy Agriculture in Shyampur, Bangladesh. *In* Comparative Farming Systems. B. L. Turner II and Stephen B. Brush, eds. Pp. 276–305. New York: Guilford Press.

Ali, Abu Muhammad Shajaat, and B. L. Turner II
1989   Stagnation or Growth? Agricultural Change in Six Bangladesh Villages, 1950–1985. MS.

Allan, W.
1965   The African Husbandman. Edinburgh: Oliver & Boyd.

Allen, B. J., and R. Crittenden
1987   Degradation and a Pre-Capitalist Political Economy: The Case of the New Guinea Highlands. *In* Land Degradation and Society. P. M. Blaikie and H. C. Brookfield, eds. Pp. 145–56. London: Methuen.

Allison, P. D.
1978   Measures of Inequality. American Sociological Review 42: 865–80.

Alston, Lee J., S. K. Datta, and J. B. Nugent
1984   Tenancy Choice in a Competitive Framework with Transaction Costs. Journal of Political Economy 92 (6): 1121–33.

Altieri, M. A.
1987   Agroecology: The Scientific Bases of Alternative Agriculture. Boulder, CO: Westview Press.

Altieri, M. A., and Matt Liebman
  1986   Insect, Weed, and Plant Disease Management in Multiple Cropping Systems. *In* Multiple Cropping Systems. Charles Francis, ed. Pp. 183–218. New York: Macmillan.
Amin, Samir
  1976   Unequal Development. New York: Monthly Review.
Anderson, Anthony B.
  1990   Deforestation in Amazonia: Dynamics, Causes, and Alternatives. *In* Alternatives to Deforestation: Steps Toward Sustainable Use of the Amazon Rain Forest. Anthony B. Anderson, ed. Pp. 3–23. New York: Columbia University Press.
Anderson, Edgar
  1952   Plants, Man, and Life. Boston: Little, Brown.
Anderson, E. N.
  1988   The Food of China. New Haven: Yale University Press.
  1989   The First Green Revolution: Chinese Agriculture in the Han Dynasty. *In* Food and Farm: Current Debates and Policies. Monographs in Economic Anthropology, No. 7. Christina Gladwin and Kathleen Truman, eds. Pp. 135–52. Lanham, MD: University Press of America.
Andrae, Gunilla, and Bjorn Beckman
  1986   The Wheat Trap: Bread and Underdevelopment in Nigeria. London: Zed Press.
Armillas, P.
  1971   Gardens on Swamps. Science 174: 653–61.
Attwood, Donald W.
  1992   Raising Cane: The Political Economy of Sugar in Western India. Boulder, CO: Westview Press.
Bachman, K. L., and R. P. Christensen
  1967   The Economics of Farm Size. *In* Agricultural Development and Economic Growth, H. M. Southworth and B. F. Johnston, eds. Pp. 234–57. Ithaca, NY: Cornell University Press.
Bailey, F. G., ed.
  1971   Gifts and Poison: The Politics of Reputation. Oxford: Basil Blackwell.
Baksh, Michael
  n.d.   Changes in Machiguenga Quality of Life. MS.
Baldwin, K. D. S.
  1957   The Niger Agricultural Project. Oxford: Basil Blackwell.
Barbier, Edward B.
  1987   The Concept of Sustainable Economic Development. Environmental Conservation 14 (2): 101–10.
Bardhan, P. K.
  1973   Size, Productivity, and Returns to Scale: An Analysis of Farm Level Data in Indian Agriculture. Journal of Political Economy 18: 1370–86.
Barlett, Peggy F.
  1976   Labor Efficiency and the Mechanism of Agricultural Evolution. Journal of Anthropological Research 32: 124–40.

1977 The Structure of Decision Making in Paso. American Ethnologist 4 (2): 285–307.

1980 Adaptive Strategies in Peasant Agricultural Production. Annual Review of Anthropology 9: 545–73.

1982 Agricultural Choice and Change: Economic Decisions and Agricultural Evolution in a Costa Rican Community. New Brunswick: Rutgers University Press.

1987a Industrial Agriculture in Evolutionary Perspective. Cultural Anthropology 2: 137–54.

1987b The Crisis in Family Farming. Who Will Survive? *In* Farm Work and Fieldwork: American Agriculture in Anthropological Perspective. Michael Chibnick, ed. Pp. 29–57. Ithaca, N.Y.: Cornell University Press.

1989 Industrial Agriculture. *In* Economic Anthropology. Stuart Plattner, ed. Pp. 253–91. Stanford: Stanford University Press.

Barrau, J.

1965 L'humide et le sec: An Essay on Ethnobiological Adaptation to Contrasted Environments in the Indo-Pacific Area. Journal of the Polynesian Society 74: 329–46.

Barton, Roy F.

1919 Ifugao Law. American Archaeology and Ethnology 15: 1–187.

Bates, Robert H.

1981 Markets and States in Tropical Africa: The Political Basis of Agricultural Policies. Berkeley: University of California Press.

Bayliss-Smith, T. P.

1981 Seasonality and Labour in Rural Energy Balance. *In* Seasonal Dimensions to Rural Poverty. Robert Chambers, R. Longhurst, and A. Pacey, eds. Pp. 30–38. London: Frances Pinter.

1982 The Ecology of Agricultural Systems. Cambridge: Cambridge University Press.

Becker, Gary S.

1981 A Treatise on the Family. Cambridge, MA: Harvard University Press.

Befu, H.

1968a Ecology, Residence, and Authority: The Corporate Household in Central Japan. Ethnology 7: 25–42.

1968b Origin of Large Households and Duolocal Residence in Central Japan. American Anthropologist 70 (2): 309–20.

Benigno, Francesco

1989 The Southern Italian Family in the Early Modern Period: A Discussion of Co-Residential Patterns. Continuity and Change 4: 165–94.

Bennett, John W.

1969 Northern Plainsmen: Adaptive Strategy and Agrarian Life. Chicago: Aldine.

1976 The Ecological Transition: Cultural Anthropology and Human Adaptation. Oxford: Pergamon Press.

Ben-Porath, Yoram

1980 The F-Connection: Families, Friends, and Firms and the Organization of Exchange. Population and Development Review 6: 1–30.

Bentley, Jeffery W.

1987 Economic and Ecological Approaches to Land Fragmentation: In Defense of a Much-Maligned Phenomenon. Annual Review of Anthropology 16: 31–67.

1992 Today There Is No Misery: The Ethnography of Farming in Northwest Portugal. Tucson: University of Arizona Press.

Berkes, F., D. Feeney, B. J. McCay, and J. M. Acheson

1989 The Benefits of the Commons. Nature 340: 91–93.

Berkner, L. K.

1972 The Stem Family and the Developmental Cycle of the Peasant Household: An Eighteenth Century Austrian Example. American Historical Review 77: 398–418.

Berlin, B.

1973 Folk Systematics in Relation to Biological Classification and Nomenclature. Annual Review of Ecology and Systematics 4: 259–71.

Bernstein, H.

1981 Concepts for the Analysis of Contemporary Peasantries. *In* The Political Economy of Rural Development: Peasants, International Capital and the State. R. Galli, ed. Pp. 3–24. Albany, NY: SUNY Albany Press.

Berry, R. A., and W. R. Cline

1979 Agrarian Structure and Productivity in Developing Countries. Baltimore: Johns Hopkins University Press.

Berry, Wendell

1987 Home Economics. San Francisco: North Point Press.

Biebuyck, D., ed.

1963 African Agrarian Studies. London: Oxford University Press.

Binswanger, Hans P., and John McIntire

1987 Behavioral and Material Determinants of Production Relations in Land Abundant Tropical Agriculture. Economic Development and Cultural Change 36: 73–99.

Binswanger, Hans P., and M. R. Rosenzweig

1986 Behavioural and Material Determinants of Production Relations in Agriculture. Journal of Development Studies 22: 503–39.

Binswanger, Hans P., and M. R. Rosenzweig, eds.

1984 Contractual Arrangements, Employment and Wages in Rural Labor Markets in Asia. New Haven: Yale University Press.

Blau, P. M.

1977 Inequality and Heterogeneity. New York: Free Press.

Blum, J.

1961 Landlord and Peasant in Russia from the Ninth to the Nineteenth Century. Princeton: Princeton University Press.

Bohannan, P.
1954   The Migration and Expansion of the Tiv. Africa 24: 2–16.

Boserup, Ester
1965   The Conditions of Agricultural Growth: The Economics of Agrarian Change under Population Pressure. Chicago: Aldine.

1970   Present and Potential Food Production in Developing Countries. *In* Geography and a Crowding World. W. Zelinsky, L. A. Kosinski, and R. Mansell Prothero, eds. Pp. 100–113. London: Oxford University Press.

1981   Population and Technological Change: A Study of Long-Term Trends. Chicago: University of Chicago Press.

1983   The Impact of Scarcity and Plenty on Development. Journal of Interdisciplinary History 14: 383–407.

1987   Population and Technology in Preindustrial Europe. Population and Development Review 13 (4): 691–701.

1990   Economic and Demographic Relationships in Development. Baltimore: Johns Hopkins University Press.

Boyce, J. K.
1987   Agrarian Impasse in Bengal: Institutional Constraints to Technological Change. London: Oxford University Press.

Bradley, Candice, C. C. Moore, M. L. Burton, and D. R. White
1990   A Cross-Cultural Historical Analysis of Subsistence Change. American Anthropologist 92: 447–57.

Bray, Francesca
1984   Agriculture. *In* Science and Civilization in China. Vol. 6, pt. 2. Joseph Needham, ed. Cambridge: Cambridge University Press.

1986   The Rice Economies. Cambridge: Cambridge University Press.

Brokensha, David
1989   Local Management Systems and Sustainability. *In* Food and Farm: Current Debates and Policies. Monographs in Economic Anthropology, No. 7. Christina Gladwin and Kathleen Truman, eds. Pp. 179–98. Lanham, MD: University Press of America.

Bronson, B.
1972   Farm Labor and the Evolution of Food Production. *In* Population Growth: Anthropological Implications. B. Spooner, ed. Pp. 190–218. Cambridge, MA: MIT Press.

Brookfield, H. C.
1972   Intensification and Disintensification in Pacific Agriculture: A Theoretical Approach. Pacific Viewpoint 13: 30–48.

1984   Intensification Revisited. Pacific Viewpoint 25 (1): 15–44.

Brookfield, H. C., and P. Brown
1963   Struggle for Land: Agriculture and Group Territories Among the Chimbu of the New Guinea Highlands. Melbourne: Oxford University Press.

Brookfield, H. C., and D. Hart
1971   Melanesia: A Geographical Interpretation of an Island World. London: Methuen.

Brown, Paula, and A. Podolefsky
  1976   Population Density, Agricultural Intensity, Land Tenure, and Group Size in the New Guinea Highlands. Ethnology 15: 211–38.
Brown, Paula, Harold Brookfield, and Robin Grau
  1990   Land Tenure and Transfer in Chimbu, Papua New Guinea, 1958–1984: A Study in Continuity and Change, Accommodation and Opportunism. Human Ecology 18: 21–49.
Bruce, John W.
  1988   A Perspective on Indigenous Land Tenure System and Land Concentration. *In* Land and Society in Contemporary Africa. R. E. Downs and S. P. Reyna, eds. Pp. 23–52. Hanover, NH: University Press of New England.
Brush, Stephen B.
  1976   Introduction to the Symposium "Cultural Adaptations to Mountain Ecosystems." Human Ecology 4: 125–33.
Buck, John Lossing
  1956   Land Utilization in China. New York: Council on Economic and Cultural Affairs. [First published in 1937.]
Burnham, Philip
  1980   Opportunity and Constraint in a Savanna Society. New York: Academic Press.
Burton, Michael L., and Douglas R. White
  1984   Sexual Division of Labor in Agriculture. American Anthropologist 86: 568–83.
  1987   Sexual Division of Labor in Agriculture. *In* Household Economics and Their Transformations. Morgan D. Maclachlan, ed. Monographs in Economic Anthropology, No. 3. Pp. 107–30. Lanham, MD: University Press of America.
Cain, Mead T.
  1980   The Economic Activities of Children in a Village in Bangladesh. *In* Rural Household Studies in Asia. H. P. Binswanger, R. E. Evenson, C. A. Florencia, and B. N. F. White, eds. Pp. 218–47. Singapore: Singapore University Press.
Caldwell, John C.
  1982   The Theory of Fertility Decline. New York: Academic Press.
Cancian, Frank
  1965   Economics and Prestige in a Maya Community. Stanford: Stanford University Press.
  1980   Risk and Uncertainty in Agricultural Decision Making. *In* Agricultural Decision Making. Peggy Barlett, ed. Pp. 161–76. New York: Academic Press.
  1989   Economic Behavior in Peasant Communities. *In* Economic Anthropology. Stuart Plattner, ed. Pp. 127–70. Stanford: Stanford University Press.
  1992   The Decline of Community in Zinacantan: Economy, Public Life, and Social Stratification, 1960–1987. Stanford: Stanford University Press.

Carlstein, Tommy
    1982   Time Resources, Society and Ecology. London: George Allen & Unwin.

Carter, Anthony T., and Robert S. Merrill
    1979   Household Institution and Population Dynamics. Report prepared for the Bureau for Program and Policy Coordination. Washington, DC: Agency for International Development.

Carter, Michael
    1989   Risk as Medium of Peasant Differentiation Under Unimodal Development Strategies in the Semi-Arid Tropics of West Africa. *In* Food and Farm: Current Debates and Policies. Monographs in Economic Anthropology, No. 7. Christina Gladwin and Kathleen Truman, eds. Pp. 35–58. Lanham, MD: University Press of America.

Chao, Kang
    1986   Man and Land in Chinese History: An Economic Analysis. Stanford: Stanford University Press.

Chapin, Mac
    1988   The Seduction of Models: Chinampa Agriculture in Mexico. Grassroots Development 12 (1): 8–17.
    n.d.   In Search of Ecodevelopment: An Account of Travels through Mexico. MS.

Chayanov, A. V.
    1966   The Theory of Peasant Economy. D. Thorner, B. Kerblay, and R. E. F. Smith, eds. Homewood, IL: Richard D. Irwin. [Contains translations of Peasant Farm Organization (1925) and On the Theory of Non-Capitalist Economic Systems, with commentaries by Daniel Thorner and Basile Kerblay.]
    1986   The Theory of Peasant Economy. D. Thorner, B. Kerblay, and R. E. F. Smith, eds. Madison: University of Wisconsin Press. [A new edition of Chayanov 1966.]

Chibnik, Michael
    1987   The Economic Effects of Household Demography: A Cross-Cultural Assessment of Chayanov's Theory. *In* Household Economies and Their Transformations. Morgan D. Maclachlan, ed. Monographs in Economic Anthropology, No. 3. Pp. 74–106. Lanham, MD: University Press of America.

Childe, E. Gordon
    1951   Man Makes Himself. New York: New American Library. [First published in 1936.]

Chisholm, M.
    1970   Rural Settlement and Land Use. Chicago: Aldine.

Chubb, L.
    1961   Ibo Land Tenure. Ibadan: Ibadan University Press.

Clammer, John
    1985   Anthropology and Political Economy: Theoretical and Asian Perspectives. New York: St. Martin's Press.

Clark, C., and M. R. Haswell
1967   The Economics of Subsistence Agriculture. London: Macmillan.
Clarke, W. C.
1966   From Extensive to Intensive Shifting Cultivation: A Succession for New Guinea. Ethnology 5: 347–59.
Cleave, J. H.
1974   African Farmers: Labor Use in the Development of Smallholder Agriculture. New York: Praeger.
Cleveland, David A.
1990   Development Alternatives and the African Food Crisis. *In* African Food Systems in Crisis. Rebecca Huss-Ashmore and Solomon H. Katz, eds. Pp. 181–206. New York: Gordon & Breach.
Cleveland, David A., and Daniela Soleri
1991   Food from Dryland Gardens: An Ecological, Nutritional, and Social Approach to Small-Scale Household Food Production. Tucson, AZ: Center for People, Food, and Environment.
Coe, M. D.
1964   The Chinampas of Mexico. Scientific American 211 (1): 90–98.
Cohen, Ronald
1988   Guidance and Misguidance in Africa's Food Production. *In* Satisfying Africa's Food Needs: Food Production and Commercialization in African Agriculture. Ronald Cohen, ed. Pp. 1–30. Boulder, CO: Lynne Rienner.
1989   The Unimodal Model: Solution or Cul de Sac for Rural Development? *In* Food and Farm: Current Debates and Policies. Monographs in Economic Anthropology, No. 7. Christina Gladwin and Kathleen Turner, eds. Pp. 7–23. Lanham, MD: University Press of America.
Collier, G. A.
1975   Fields of the Tzotzil: The Ecological Basis of Tradition in Highland Chiapas. Austin: University of Texas Press.
Collier, W.
1981   Agricultural Evolution in Java. *In* Agricultural and Rural Development in Indonesia. G. E. Hanson, ed. Pp. 147–73. Boulder, CO: Westview Press.
Conelly, W. Thomas
1992   Population Pressure and Changing Agropastoral Management Strategies in Western Kenya. *In* Plants, Animals, and People: Agropastoral Systems Research in the SR-CRSP Rural Sociology Program. Constance McCorkle, ed. Boulder, CO: Westview Press. [In press.]
n.d.   Livestock and Food Security Among Smallholder Farmers in Kenya. Paper presented at the annual meeting of the American Anthropological Association, Nov. 16–20, 1988, Phoenix, AZ.
Conklin, H. C.
1954   An Ethnoecological Approach to Shifting Cultivation. Transactions of the New York Academy of Sciences, 2d ser., 17 (2): 133–42.
1957   Hanunoo Agriculture. Rome: FAO.

1961 The Study of Shifting Cultivation. Current Anthropology 2: 27–61.

1967 Some Aspects of Ethnographic Research in Ifugao. Transactions of the New York Academy of Sciences, 2d ser., 30 (1): 99–121.

1980 Ethnographic Atlas of Ifugao: A Study of Environment, Culture, and Society in Northern Luzon. New Haven: Yale University Press.

Connell, John, and Michael Lipton
1977 Assessing Village Labour Situations in Developing Countries. Delhi: Oxford University Press.

Conway, Gordon R.
1985 Agroecosystem Analysis. Agricultural Administration 20: 31–55.

1987 The Properties of Agroecosystems. Agricultural Systems 24 (2): 95–118.

Conway, Gordon R., and Edward B. Barbier
1990 After the Green Revolution: Sustainable Agriculture for Development. London: Earthscan Publications.

Coquery-Vidrovitch, Catherine
1972 Research on an African Mode of Production. *In* Perspectives on the African Past. Martin Klein and G. Wesley Johnson, eds. Pp. 33–51. Boston: Little, Brown.

Cornell, Laurel L.
1987 Hajnal and the Household in Asia: A Comparativist History of the Family in Preindustrial Japan, 1600–1870. Journal of Family History 12: 143–62.

Cowgill, G. L.
1975 On Causes and Consequences of Ancient and Modern Population Changes. American Anthropologist 77: 505–25.

Croll, Elisabeth
1987 Some Implications of the Rural Economic Reforms for the Chinese Peasant Household. *In* The Re-emergence of the Chinese Peasantry: Aspects of Rural Decollectivization. Ashwani Saith, ed. Pp. 105–36. London: Croom Helm.

Crummey, D., and C. Stewart, eds.
1981 Modes of Production in Africa: The Pre-Colonial Era. Beverly Hills, CA: Sage.

Curwen, E. C., and G. Hatt
1953 Plough and Pasture. New York: Schuman.

Cutileiro, José
1971 A Portuguese Rural Society. Oxford: Clarendon Press.

Czap, P.
1982 The Perennial Multiple Family Household. Journal of Family History 7: 5–26.

1983 A Large Family: The Peasant's Greatest Wealth. Serf Households in Mishino, Russia, 1814–1858. *In* Family Forms in Historic Europe. R. Wall, J. Robin, and P. Laslett, eds. Pp. 105–51. Cambridge: Cambridge University Press.

Dalton, George
  1961   Economic Theory and Primitive Society. American Anthropologist 63:
         1–25.
Delgado, Christopher L., and C. B. Ranade
  1987   Technological Change and Agricultural Labor Use. *In* Accelerating
         Food Production in Sub-Saharan Africa, J. W. Mellor, C. L. Delgado,
         and M. J. Blackie, eds. Pp. 118–34. Baltimore: Johns Hopkins Univer-
         sity Press.
Delman, Jorgen
  1988   The Agricultural Extension System in China. Agricultural Administra-
         tion Network Paper #3. London: Overseas Development Institute.
Deqi, Jiang, Qi Leidi, and Tan Jiesheng
  1981   Soil Erosion and Conservation in the Wuding River Valley, China. *In*
         Soil Conservation: Problems and Prospects. R. P. C. Morgan, ed. Pp.
         461–79. Chichester: Wiley Press.
De Vries, J.
  1974   The Dutch Rural Economy in the Golden Age, 1500–1700. New Ha-
         ven: Yale University Press.
De Wilde, J. C.
  1967   Mali: The Office du Niger—An Experience with Irrigated Agriculture.
         In Experiences with Agricultural Development in Tropical Africa. Vol.
         2, The Case Studies. Pp. 245–300. Baltimore: Johns Hopkins Univer-
         sity Press.
Diamond, Norma
  1985   Taitou Revisited: State Policies and Social Change. *In* Chinese Rural
         Development. W. L. Parish, ed. Pp. 246–69. Armonk, NY: M. E.
         Sharpe.
Domar, Evsey
  1970   The Causes of Slavery or Serfdom: A Hypothesis. Journal of Economic
         History 30: 18–31.
Donham, Donald
  1981   Beyond the Domestic Mode of Production. Man 16 (4): 515–41.
  1990   History, Power, Ideology: Central Issues in Marxism and Anthropol-
         ogy. Cambridge: Cambridge University Press.
Donkin, R. A.
  1979   Agricultural Terracing in the Aboriginal New World. Viking Fund
         Publications in Anthropology, No. 56. Tucson: University of Arizona
         Press.
Donnelly, Fr. J.
  n.d.   Notes on the Angas People of Pankshin Division, Plateau Province.
         MS. Roman Catholic Mission, Kagoro, Nigeria.
Douglass, G. K.
  1985   When Is Agriculture Sustainable? *In* Sustainable Agriculture and Inte-
         grated Farming Systems. Eden Thomas, Cynthia Fridgen, and Susan
         Battenfield, eds. Pp. 10–21. East Lansing: Michigan University Press.

Douglass, G. K., ed.
  1984 Agricultural Sustainability in a Changing World Order. Boulder, CO: Westview Press.
Dove, M. R.
  1983 Theories of Swidden Agriculture and the Political Economy of Ignorance. Agroforestry Systems 1: 85–99.
  1984 The Chayanov Slope in a Swidden Society: Household Demography and Extensive Agriculture in West Kalimantan. *In* Chayanov, Peasants, and Economic Anthropology. E. Paul Durrenberger, ed. Pp. 97–132. New York: Academic Press.
Downing, T. E.
  1974 Irrigation and Moisture-Sensitive Periods: A Zapotec Case. *In* Irrigation's Impact on Society. T. E. Downing and McG. Gibson, eds. Pp. 113–22. Tucson: University of Arizona Press.
Downs, R. E., and S. P. Reyna, eds.
  1988 Land and Society in Contemporary Africa. Hanover, NH: University Press of New England.
Drucker, C. B.
  1977 To Inherit the Land: Descent and Decision in Northern Luzon. Ethnology 16: 1–20.
Due, J. M., and P. Anandajayasekeram
  1984 Contrasting Farming Systems in Morogoro Region, Tanzania. Canadian Journal of African Studies 18 (3): 583–91.
Dumond, D. E.
  1961 Swidden Agriculture and the Rise of Maya Civilization. Southwestern Journal of Anthropology 17: 301–16.
Durrenberger, E. P.
  1978 Agricultural Production and Household Budgets in a Shan Peasant Village in Northwestern Thailand: A Quantitative Description. Ohio University Center for International Studies, Southeast Asia Series, No. 49.
  1979 An Analysis of Shan Household Production Decisions. Journal of Anthropological Research 35 (4): 447–58.
Durrenberger, E. P., ed.
  1984 Chayanov, Peasants, and Economic Anthropology. New York: Academic Press.
Eastman, Lloyd E.
  1988 Family, Fields, and Ancestors: Constancy and Change in China's Social and Economic History. New York: Oxford University Press.
Eder, James F.
  1977 Agricultural Intensification and the Returns to Labour in the Philippine Swidden System. Pacific Viewpoint 18: 1–21.
  1982 Who Shall Succeed? Agricultural Development and Social Inequality on a Philippine Frontier. Cambridge: Cambridge University Press.
  1990 The Gardens of San Jose: The Survival of Family Farming in a Developing Philippine Community. *In* Social Change and Applied Anthro-

pology: Essays in Honor of David Brokensha. M. S. Chaiken and A. K. Fleuret, eds. Pp. 132–45. Boulder, CO: Westview Press.

Ellen, Roy
1982   Environment, Subsistence, and System: The Ecology of Small-Scale Social Formations. Cambridge: Cambridge University Press.

Ellis, Frank
1988   Peasant Economics: Farm Households and Agrarian Development. Cambridge: Cambridge University Press.

Elvin, Mark
1973   The Pattern of the Chinese Past: A Social and Economic Interpretation. Stanford: Stanford University Press.

Ember, Carol R.
1983   The Relative Decline in Women's Contribution to Agriculture with Intensification. American Anthropologist 85: 285–304.

Ember, Melvin, and Carol R. Ember
1971   The Conditions Favoring Matrilocal Versus Patrilocal Residence. American Anthropologist 73: 571–94.

Engels, F.
1972   The Origin of the Family, Private Property, and the State. New York: International Publishers. [First published in 1884.]

England, Paula, and G. Farkas
1986   Households, Employment and Gender: A Social, Economic and Demographic View. New York: Aldine.

Ensminger, Jean
1992   Making a Market: The Institutional Transformation of an African Society. Cambridge: Cambridge University Press. [Forthcoming.]

Erlanger, Steven
1989   Vietnam Still Lives Hand to Mouth, But Less Collectively. New York Times, April 9, 1989, E2.

Evenari, Michael
1988   The Problem Posed: Elements of the Agricultural Crisis. Israel Journal of Development 10 (2): 4–8.

Feder, Gershon
1985   The Relation Between Farm Size and Farm Productivity: The Role of Family Labor, Supervision, and Credit Constraints. Journal of Development Economics 18: 297–313.

Feder, Gershon, and Raymond Noronha
1987   Land Rights Stystems and Agricultural Development in Sub-Saharan Africa. Research Observer [World Bank] 2: 143–69.

Feeny, David
1988   The Demand for and Supply of Institutional Arrangements. In Rethinking Institutional Analysis and Development. Vincent Ostrom, David Feeny, and Hartmut Picht, eds. Pp. 159–209. San Francisco: ICS Press.

Fei, Hsiao-tung
1939 Peasant Life in China: A Field Study of Country Life in the Yangtze Valley. London: Routledge.
Fei, Hsiao-tung, and Chih-i Chang
1945 Earthbound China: A Study of Rural Economy in Yunnan. Chicago: University of Chicago Press.
Feil, D. K.
1987 The Evolution of Highland Papua New Guinea Societies. Cambridge: Cambridge University Press.
Fél, Edit, and Tamas Hofer
1969 Proper Peasants. Viking Fund Publications in Anthropology, No. 46. New York: Wenner-Gren Foundation for Anthropological Research.
Ferdon, E. N.
1959 Agricultural Potential and the Development of Cultures. Southwestern Journal of Anthropology 15: 1–29.
Firth, Sir Raymond
1975 The Skeptical Anthropologist? *In* Social Anthropology and Marxist Views of Society. M. Bloch, ed. Pp. 29–60. New York: Halsted.
Fisk, E. K.
1964 Planning in a Primitive Economy: From Pure Subsistence to the Production of a Market Surplus. Economic Record 40: 156–74.
Fladby, Berit
1983 Household Viability and Economic Differentiation in Gama, Sri Lanka. Occasional Papers in Social Anthropology, No. 28. Bergen: University of Bergen.
Flannery, Kent
1968 Archaeological Systems Theory and Early Mesoamerica. *In* Anthropological Archaeology in the Americas. Betty J. Meggers, ed. Pp. 67–87. Washington, DC: Anthropological Society of Washington.
Floyd, B.
1965 Terrace Agriculture in Eastern Nigeria: The Case of Maku. Journal of the Geographical Association of Nigeria 7: 91–108.
Folbre, Nancy
1984 Household Production in the Philippines: A Non-Neoclassical Approach. Economic Development and Cultural Change 32: 303–30.
1986a Cleaning House: New Perspectives on Households and Economic Development. Journal of Developmental Economics 22: 5–40.
1986b Hearts and Spades: Paradigms of Household Economy. World Development 14: 245–55.
Forde, D.
1947 The Anthropological Approach in Social Science. Advances in Science 4: 213–24.
Fortes, Meyer
1949 The Web of Kinship Among the Tallensi. London: Oxford University Press.

Foster, George M.
    1965   Peasant Society and the Image of Limited Good. American Anthropologist 67 (2): 293–315.
    1967   Introduction: What Is a Peasant? *In* Peasant Society: A Reader. Jack M. Potter, May N. Diaz, George M. Foster, eds. Pp. 2–14. Boston: Little, Brown.

Fox, Roger, and Timothy J. Finan
    1987   Patterns of Technical Change in the Northwest. *In* Portuguese Agriculture in Transition. Scott R. Pearson, ed. Pp. 187–200. Ithaca, NY: Cornell University Press.

Francis, Charles, and Garth Youngberg
    1990   Sustainable Agriculture: An Overview. *In* Sustainable Agriculture in Temperate Zones. C. Francis, C. Flora, and Larry King, eds. Pp. 1–23. New York: Wiley.

Freedman, Maurice
    1966   Chinese Lineage and Society: Fukien and Kwangtung. London: Athlone.

Freeman, J. D.
    1955   Iban Agriculture. Colonial Research Studies, No. 18. London: Colonial Office.

Friedman, Jonathan
    1974   Marxism, Structuralism, and Vulgar Materialism. Man 9: 444–69.

Gallin, Bernard, and Rita S. Gallin
    1982   Socioeconomic Life in Rural Taiwan: Twenty Years of Development and Change. Modern China 8: 205–46.

Geertz, Clifford
    1963   Agricultural Involution. Berkeley: University of California Press.
    1972   The Wet and The Dry: Traditional Irrigation in Bali and Morocco. Human Ecology 1: 23–40.
    1980   Organization of the Balinese Subak. *In* Irrigation and Agricultural Development in Asia: Perspectives from the Social Sciences. E. Walter Coward, Jr., ed. Pp. 70–90. Ithaca, NY: Cornell University Press.
    1984   Culture and Social Change: The Indonesian Case. Man 19: 511–32.

Geschiere, P.
    1985   Applications of the Lineage Mode of Production in African Studies. Canadian Journal of African Studies 19: 80–90.

Ghose, Ajit Kumar
    1987   The People's Commune: Responsibility Systems and Rural Development in China, 1965–1984. *In* The Reemergence of the Chinese Peasantry: Aspects of Rural Decollectivization. Ashwani Saith, ed. Pp. 35–80. London: Croom Helm.

Gladwin, Christina H., and Carl Zulauf
    1989   The Case for the Disappearing Mid-Sized Farm in the U.S. *In* Food and Farm: Current Debates and Policies. Monographs in Economic Anthropology, No. 7. Christina Gladwin and Kathleen Turner, eds. Pp. 259–84. Lanham, MD: University Press of America.

Gleave, M. B., and H. P. White
  1969 Population Density and Agricultural System in West Africa. *In* Environment and Land Use in Africa. M. F. Thomas and G. W. Whittington, eds. Pp. 273–300. London: Methuen.

Gliessman, S. R.
  1984 Resource Management in Traditional Tropical Agroecosystems in Southeast Mexico. *In* Agricultural Sustainability in a Changing World Order. G. Douglass, ed. Pp. 191–201. Boulder, CO: Westview Press.

Goddard, A. D., M. J. Mortimore, and D. W. Norman
  1975 Some Social and Economic Implications of Growth in Rural Hausaland. *In* Population Growth and Socio-Economic Change in West Africa. J. C. Caldwell, ed. Pp. 321–36. New York: Columbia University Press.

Goldschmidt, Walter
  1978 As You Sow: Three Studies in the Social Consequences of Agribusiness. Montclair, NJ: Allanheld, Osmun. [First published in 1947.]

Goldschmidt, Walter, and Evelyn S. Kunkel
  1971 The Structure of the Peasant Family. American Anthropologist 73: 1058–76.

Gomez-Pompa, Arturo, and Juan J. Jimenez-Osornio
  1989 Some Reflections on Intensive Traditional Agriculture. *In* Food and Farm: Current Debates and Policies. Monographs in Economic Anthropology, No. 7. Christina Gladwin and Kathleen Truman, eds. Pp. 231–53. Lanham, MD: University Press of America.

Gonzalez, Nancie L.
  1983 Household and Family in Kaixiangong: A Re-examination. China Quarterly 93: 76–89.

Goodenough, Ward H.
  1956 Residence Rules. Southwestern Journal of Anthropology 12: 22–37.

Goody, Jack
  1958 The Developmental Cycle in Domestic Groups. Cambridge: Cambridge University Press.
  1976 Production and Reproduction: A Comparative Study of the Domestic Domain. Cambridge: Cambridge University Press.

Gorecki, P.
  1979 Population Growth and Abandonment of Swamplands: A New Guinea Highland Example. Journal de la Société des océanistes 63 (35): 97–107.

Gotsch, Carl H.
  1972 Technical Change and the Distribution of Income in Rural Areas. American Journal of Agricultural Economics 34: 326–41.

Gourou, Pierre
  1966 The Tropical World: Its Social and Economic Conditions and Its Future Status. Translated by S. H. Beaver and E. D. Laborde. 4th ed. New York: Wiley. [Originally published in 1958.]

Greenberg, James B.
  1981   Santiago's Sword: Chatino Peasant Religion and Economics. Berkeley: University of California Press.
Greenhalgh, Susan
  1985   Is Inequality Demographically Induced? The Family Cycle and the Distribution of Income in Taiwan. American Anthropologist 87: 571–94.
  1988   Fertility as Mobility: Sinic Transitions. Population and Development Review 14 (4): 627–74.
Greenwood, Davydd
  1974   Political Economy and Adaptive Processes: A Framework for the Study of Peasant-States. Peasant Studies Newsletter 3: 1–10.
Griffin, Keith
  1984   Epilogue: Rural China in 1983. *In* Institutional Reform and Economic Development in the Chinese Countryside. Keith Griffin, ed. Pp. 303–29. London: Macmillan.
Grigg, D. B.
  1979   Ester Boserup's Theory of Agrarian Change: A Critical Review. Progress in Human Geography 3: 64–84.
  1980   Population Growth and Agrarian Change. Cambridge: Cambridge University Press.
Gross, Daniel R.
  1984   Time Allocation: A Tool for the Study of Cultural Behavior. Annual Review of Anthropology 13: 519–58.
Grove, A. T.
  1961   Population and Agriculture in Northern Nigeria. *In* Essays on African Population. K. M. Barbour and R. M. Prothero, eds. Pp. 115–36. London.
Guillet, David
  1981   Land Tenure, Ecological Zone, and Agricultural Regime in the Central Andes. American Ethnologist 8: 139–58.
Guo, Shutian
  1990   What's In Store for China's Grain Harvest? Beijing Review 33 (8): 18–21.
Guyer, Jane I.
  1981   Household and Community in African Studies. African Studies Review 24: 87–137.
  1986   Intra-household Processes and Farming Systems Research: Perspectives from Anthropology. *In* Understanding Africa's Rural Households and Farming Systems. Joyce L. Moock, ed. Pp. 92–104. Boulder, CO: Westview Press.
Hahn, Eduard
  1919   Von der Hacke zum Pflug. Leipzig: Quelle & Meyer.
Hahn, Natalie D.
  1986   The African Farmer and Her Husband. Farming Systems Research and Extension: Food and Feed. Farming Systems Symposium, 5–8 October 1986, Kansas State University, Manhattan, KS.

Halpern, Joel, and Barbara Halpern
1972 A Serbian Village in Historical Perspective. New York: Holt, Rinehart & Winston.

Hammel, E. A.
1972 The Zadruga as Process. *In* Household and Family in Past Time. Peter Laslett, ed. Pp. 335–74. London: Cambridge University Press.
1980 Household Structure in Fourteenth Century Macedonia. Journal of Family History 5: 242–73.
1984 On the *** of Studying Household Form and Function. *In* Households: Comparative and Historical Studies of the Domestic Group. Robert McC. Netting, Richard R. Wilk, and Eric J. Arnould, eds. Pp. 29–43. Berkeley: University of California Press.

Hammel, E. A., and Nancy Howell
1987 Research in Population and Culture: An Evolutionary Framework. Current Anthropology 28 (2): 141–60.

Hammel, E. A., and Peter Laslett
1974 Comparing Household Structure over Time and Between Cultures. Comparative Studies in Society and History 16: 73–109.

Hanks, L. M.
1972 Rice and Man: Agricultural Ecology in Southeast Asia. Chicago: Aldine.

Hardin, Garrett
1968 The Tragedy of the Commons. Science 162: 1243–48.

Harner, Michael J.
1970 Population Pressure and the Social Evolution of Agriculturalists. Southwestern Journal of Anthropology 26: 67–86.

Harris, Marvin
1969 Monistic Determinism Anti-Service. Southwestern Journal of Anthropology 25: 198–206.

Harris, Michael S.
1989 Diminishing Resources: Land Fragmentation and Inheritance in a Bangladeshi Village. Ph.D. diss., Department of Anthropology, Southern Methodist University, Dallas, TX.

Harrison, Mark
1975 Chayanov and the Economics of the Russian Peasantry. Journal of Peasant Studies 2 (4): 389–417.
1977 The Peasant Mode of Production in the Work of A. V. Chayanov. Journal of Peasant Studies 4 (4): 323–36.

Harriss, John, ed.
1982 Rural Development. London: Hutchinson.

Hart, Gillian
1980 Patterns of Household Labour Allocation in a Javanese Village. *In* Rural Household Studies in Asia. H. P. Binswanger, R. E. Evenson, A. Florencio, and B. N. F. White, eds. Pp. 188–217. Singapore: Singapore University Press.

1986   Power, Labor, and Livelihood: Processes of Change in Rural Java. Berkeley: University of California Press.

Hart, Keith
1982   The Political Economy of West African Agriculture. Cambridge: Cambridge University Press.

Hartford, Kathleen
1985   Socialist Agriculture Is Dead: Long Live Socialist Agriculture. *In* The Political Economy of Reform in Post-Mao China. Elizabeth J. Perry and Christine Wong, eds. Pp. 1–61. Cambridge, MA: Harvard University Press.

Haugerud, Angelique
1989   Land Tenure and Agrarian Change in Kenya. Africa 59 (1): 61–90.

Hawkesworth, D. P.
1981   Variations of Jointness Within the Indian Family. Folk 23: 235–50.

Hayami, Yujiro, and Masao Kikuchi
1982   Asian Village Economy at the Crossroads: An Economic Approach to Institutional Change. Tokyo: University of Tokyo Press.

Hayami, Yujiro, and Vernon Ruttan
1971   Agricultural Development: An International Perspective. Baltimore: Johns Hopkins University Press.

Helms, Mary
1976   Domestic Organization in Eastern Central America: The San Blas Cuna, Miskito and Black Carib Compared. Western Canadian Journal of Anthropology 6: 133–63.

Herring, Ronald J.
1983   Land to the Tiller: The Political Economy of Agrarian Reform in South Asia. New Haven: Yale University Press.
1984   Chayanovian Versus Neoclassical Perspectives on Land Tenure and Productivity Interactions. *In* Chayanov, Peasants, and Economic Anthropology. E. P. Durrenberger, ed. Pp. 133–49. Orlando, FL: Academic Press.

Hildyard, Nicholas
1985   Adios Amazonia? A Report from the Altimira Gathering. The Ecologist 1 (1): 2.

Hill, Polly
1982   Dry Grain Farming Families: Hausaland (Nigeria) and Karnataka (India) Compared. Cambridge: Cambridge University Press.

Hinton, William
1983   Shenfan. New York: Random House, Vintage Books.
1990   The Great Reversal: The Privatization of China, 1978–1989. New York: Monthly Review Press.

Hirschman, A. O.
1981   The Rise and Decline of Development Economics. *In* Essays in Trespassing: Economics to Politics and Beyond. Pp. 1–24. Cambridge: Cambridge University Press.

Hopkins, Nicholas S.

1987a The Agrarian Transition and the Household in Rural Egypt. *In* Household Economies and Their Transformations. Monographs in Economic Anthropology, No. 3. Morgan D. Maclachlan, ed. Pp. 155–72. Lanham, MD: University Press of America.

1987b Mechanized Irrigation in Upper Egypt: The Role of Technology and the State in Agriculture. *In* Comparative Farming Systems. B. L. Turner II and Stephen B. Brush, eds. Pp. 223–47. New York: Guilford Press.

Hostetler, John A.

1980 Amish Society. 3d ed. Baltimore: Johns Hopkins University Press. [Originally published in 1963.]

Hsu, Cho-yun

1980 Han Agriculture: The Formation of Early Chinese Agrarian Economy (206 B.C.–A.D. 220). Seattle: University of Washington Press.

Huang, Philip C. C.

1985 The Peasant Economy and Social Change in North China. Stanford: Stanford University Press.

1990 The Peasant Family and Rural Development in the Yangzi Delta, 1350–1988. Stanford: Stanford University Press.

Huang, Shu-min

1984 Market and Nonmarket Factors in Taiwanese Peasant Economy. *In* Chayanov, Peasants, and Economic Anthropology. E. Paul Durrenberger, ed. Pp. 167–81. Orlando, FL: Academic Press.

1989 The Spiral Road: Change in a Chinese Village Through the Eyes of a Communist Party Leader. Boulder, CO: Westview Press.

Hunn, E.

1982 The Utilitarian Factor of Folk Biological Classification. American Anthropologist 84 (4): 830–47.

Huntington, E.

1963 The Human Habitat. New York: Norton.

Hyden, Goran

1980 Beyond Ujamaa in Tanzania: Underdevelopment and an Uncaptured Peasantry. London: Heinemann.

Jackson, Wes

1987 Altars of Unhewn Stone: Science and the Earth. San Francisco: North Point Press.

Jansen, Eirik G.

1987 Rural Bangladesh: Competition for Scarce Resources. Dhaka: University Press.

Jewsiewicki, Bogumil

1981 Lineage Mode of Production: Social Inequalities in Equatorial Central Africa. *In* Modes of Production in Africa. D. Crummey and C. C. Stewart, eds. Pp. 92–118. Beverly Hills, CA: Sage.

Johnson, Allen W.

1971 Sharecroppers of the Sertão. Stanford: Stanford University Press.

1975   Time Allocation in a Machiguenga Community. Ethnology 14: 310–21.

1980   Comment to Does Labor Time Decrease with Industrialization? A Survey of Time Allocation Studies, Wanda Minge-Klevana. Current Anthropology 21 (3): 279–98.

Johnson, Allen W., and Timothy Earle
1987   The Evolution of Human Society. Stanford: Stanford University Press.

Johnson, Warren A., Victor Stoltzfus, and Peter Craumer
1977   Energy Conservation in Amish Agriculture. Science 198: 373–78.

Johnston, B. F.
1991   Getting Priorities Right: Structural Transformation and Strategic Notions. *In* Structural Adjustment and African Women Farmers. Christina H. Gladwin, ed. Pp. 81–99. Gainesville: University of Florida Press.

Johnston, B. F., and P. Kilby
1975   Agriculture and Structural Transformation. New York: Oxford University Press.

1982   "Unimodal" and "Bimodal" Strategies of Agrarian Change. *In* Rural Development: Theories of Peasant Economy and Agrarian Change. John Harriss, ed. Pp. 50–65. London: Hutchinson University Library.

Jones, G. I.
1949   Ibo Land Tenure. Africa 19: 309–23.

Jordan, Terry G., and Matti Kaups
1989   The American Backwoods Frontier: An Ethnic and Ecological Interpretation. Baltimore: Johns Hopkins University Press.

Kahn, Joel S.
1981   The Social Context of Technological Change in Four Malaysian Villages: A Problem for Economic Anthropology. Man 16: 542–62.

Katz, Cindi R.
n.d.   If There Weren't Seeds There Wouldn't Be Fields. MS based on author's Ph.D. diss., Department of Geography, Clark University, Worcester, MA.

Kautsky, Karl
1983   Karl Kautsky: Selected Political Writings. Edited and translated by Patrick Goode. London: Macmillan.

Kay, G.
1964   Chief Kalaba's Village. Rhodes-Livingstone Papers, No. 35. Manchester: Manchester University Press.

Keesing, Felix M.
1962   The Ethnohistory of Northern Luzon. Stanford: Stanford University Press.

Keyfitz, Nathan
1986   An East Javanese Village in 1953 and 1985. Observations on Development. Population and Development Review 12: 695–719.

Khan, A. R.
1984   The Responsibility System and Institutional Change. *In* Institutional

Reform and Economic Development in the Chinese Countryside. K. Griffin, ed. Pp. 76–120. London: Macmillan.

Khera, S.
  1972a An Austrian Peasant Village Under Rural Industrialization. Behavior Science Notes 7 (1): 29–36.
  1972b Kin Ties and Social Interaction in an Austrian Peasant Village with Divided Land Inheritance. Behavior Science Notes 7 (4): 349–65.

Khusro, A. M.
  1964 Returns to Scale in Indian Agriculture. Indian Journal of Agricultural Economics 19 (3–4): 51–80.

King, F. H.
  1911 Farmers of Forty Centuries: Permanent Agriculture in China, Korea, and Japan. Madison, WI: Mrs. F. H. King.

Kinoshita, Futoshi
  1989 Population and Household Change of a Japanese Village, 1760–1870. Unpublished Ph.D. diss., University of Arizona, Tucson.

Kula, Witold
  1976 An Economic Theory of the Feudal System: Towards a Model of the Polish Economy, 1500–1800. London: NLB. [Originally published in Polish in 1962.]

Kussmaul, Ann
  1981 Servants in Husbandry in Early Modern England. Cambridge: Cambridge University Press.

Lagemann, Johannes
  1977 Traditional African Farming Systems in Eastern Nigeria: An Analysis of Reaction to Increasing Population Pressure. Munich: Weltforum Verlag.

Lansing, J. Stephen
  1987 Balinese "Water Temples" and the Management of Irrigation. American Anthropologist 89: 326–41.

Laslett, Peter
  1972 Introduction: The History of the Family. *In* Household and Family in Past Time. Peter Laslett and Richard Wall, eds. Pp. 1–89. Cambridge: Cambridge University Press.
  1984 The Family as a Knot of Individual Interests. *In* Households: Comparative and Historical Studies of the Domestic Group. Robert McC. Netting, Richard R. Wilk, and Eric J. Arnould, eds. Pp. 353–79. Berkeley: University of California Press.

Laslett, Peter, and Richard Wall, eds.
  1972 Household and Family in Past Time. Cambridge: Cambridge University Press.

Lathrap, D. W.
  1970 The Upper Amazon. New York: Praeger.

Leach, E. R.
  1965 Political Systems of Highland Burma. Boston: Beacon Press. [First published in 1954.]

358   *References Cited*

Leach, Gerald
  1976   Energy and Food Production. Guilford, CT: IPC Science and Technology Press.
Leacock, Eleanor Burke
  1963   Introduction to Ancient Society, by Lewis Henry Morgan. Part 1 of Research in the Lines of Human Progress from Savagery through Barbarism to Civilization. Eleanor B. Leacock, ed. Pp. Ii–Ixx. Cleveland: World, Meridian Books. [Originally published in 1877.]
Leaf, Murray J.
  1973   Peasant Motivation, Ecology, and Economy in Panjab. *In* Contributions to Asian Studies, No. 3. K. Ishwaran, ed. Pp. 40–50. Leiden: E. J. Brill.
Lee, Richard B.
  1968   What Hunters Do for a Living, or How to Make Out on Scarce Resources. *In* Man the Hunter. R. B. Lee and I. DeVore, eds. Pp. 30–48. Chicago: Aldine.
  1969   !Kung Bushman Subsistence: An Input-Output Analysis. *In* Ecological Essays. National Museum of Canada [Ottawa], Bulletin No. 230. D. Damas, ed. Pp. 73–94.
  1979   The !Kung San: Men, Women, and Work in a Foraging Society. Cambridge: Cambridge University Press.
Lee, Ronald D.
  1986a Malthus and Boserup: A Dynamic Synthesis. *In* The State of Population Theory. David Coleman and Roger Schofield, eds. Pp. 96–130. Oxford: Basil Blackwell.
  1986b Was Malthus Right? Diminishing Returns, Homeostasis and Induced Technical Change. Program in Population Research Working Paper No. 21. Berkeley: University of California.
Legros, Dominique
  1977   Chance, Necessity, and Mode of Production: A Marxist Critique of Cultural Evolutionism. American Anthropologist 79: 26–41.
LeRoy Ladurie, Emmanuel
  1974   The Peasants of Languedoc. Urbana: University of Illinois Press.
Lightfoot, Clive, and Nguyen Ahn Tuan
  1990   Drawing Pictures of Farms Helps Everyone. ILEIA Newsletter for Low External Input and Sustainable Agriculture 6(3): 18–19.
Little, Daniel
  1989   Understanding Peasant China: Case Studies in the Philosophy of Social Science. New Haven: Yale University Press.
Lockeretz, William
  1984   Energy and the Sustainability of the American Agricultural System. *In* Agricultural Sustainability in a Changing World Order. G. Douglass, ed. Pp. 77–88. Boulder CO: Westview Press.
  1990   Major Issues Concerning Sustainable Agriculture. *In* Sustainable Agriculture in Temperate Zones. C. Francis, C. Flora, and L. King, eds. Pp. 423–38. New York: Wiley.

Löfgren, Orvar

1974 Family and Household Among Scandinavian Peasants: An Exploratory Essay. Ethnologia Scandinavica 74: 1–52.

1984 Family and Household: Images and Realities: Cultural Change in Swedish Societies. *In* Households. Robert McC. Netting, Richard R. Wilk, and Eric J. Arnould, eds. Pp. 446–70. Berkeley: University of California Press.

Logsdon, Gene

1989 The Future: More Farmers Not Fewer. Whole Earth Review, Spring 1989, pp. 64–70.

Loucky, James

1979 Production and Patterning of Social Relations and Values in Two Guatemalan Villages. American Ethnologist 6: 702–23.

Lovins, Amory B., L. H. Lovins, and M. Bender

1984 Energy and Agriculture. *In* Meeting the Expectations of the Land. Wes Jackson, Wendell Berry, and B. Colman, eds. Pp. 68–86. San Francisco: North Point Press.

Lowie, R. H.

1920 Primitive Society. New York: Liveright.

Lucas, John

n.d. Intensive Agriculture and Local-Central Tension in China since 1949. MS. Indiana University, Spring 1989.

Ludwig, H.-D.

1968 Permanent Farming on Ukara: The Impact of Land Shortage on Husbandry Practices. *In* Smallholder Farming and Smallholder Development in Tanzania. Afrika-Studien, No. 24. Hans Ruthenberg, ed. Pp. 87–135. Munich: Weltforum Verlag.

Lyon, John

1985 Review of Essays in Sustainable Agriculture and Stewardship, ed. Wendell Berry and Bruce Coleman. Agriculture and Human Values 2 (4): 54–57.

MacCannell, Dean

n.d. Agribusiness and the Small Community. MS. University of California, Davis.

McCay, Bonnie M., and James J. Acheson, eds.

1987 The Question of the Commons: The Culture and Ecology of Communal Resources. Tucson: University of Arizona Press.

McGough, James P.

1984 The Domestic Mode of Production and Peasant Social Organization: The Chinese Case. *In* Chayanov, Peasants, and Economic Anthropology. E. P. Durrenberger, ed. Pp. 183–201. Orlando, FL: Academic Press.

McGuire, Randall, and Robert McC. Netting

1982 Leveling Peasants? The Maintenance of Equality in a Swiss Alpine Community. American Ethnologist 9: 269–90.

McIntire, John
    1984 Policy Environments for Mechanized Agriculture in Sub-Saharan Africa with Special Reference to Animal Traction. ILCA/Systems Research Unit. Addis Ababa: International Livestock Center for Africa.
McKean, Margaret A.
    1982 The Japanese Experience with Scarcity: Management of Traditional Common Lands. Environmental Review 6: 63–88.
    1991 Defining and Dividing Property Rights in the Japanese Commons. Paper presented at the annual meeting of the International Association for the Study of Common Property, Winnipeg, Canada, Sept. 25–30, 1991.
Maclachlan, Morgan D.
    1987 From Intensification to Proletarianization. *In* Household Economies and Their Transformations. Monographs in Economic Anthropology, No. 3. M. D. Maclachlan, ed. Pp. 1–27. Lanham, MD: University Press of America.
McMillan, Della E.
    1986 Distribution of Resources and Products in Mossi Households. *In* Food in Sub-Saharan Africa. Art Hansen and Della E. McMillan, eds. Pp. 260–73. Boulder, CO: Lynne Rienner Publishers.
Malthus, Thomas R.
    1986 Essay on the Principle of Population. *In* The Works of Thomas Robert Malthus, Vol. 2. E. A. Wrigley and David Souden, eds. London: William Pickering. [First published in 1826.]
Marten, Gerald G., and Daniel M. Saltman
    1986 The Human Ecology Perspective. *In* Traditional Agriculture in Southeast Asia: A Human Ecology Perspective. Gerald G. Marten, ed. Pp. 20–53. Boulder, CO: Westview Press.
Marx, Karl
    1967 Capital. 3 vols. New York: International Publishers. [Originally published in German in 1867, 1893, 1894.]
    1971 Peasantry as a Class. Excerpts from "The Class Struggles in France, 1848–1850" and "The Eighteenth Brumaire of Louis Bonaparte." *In* Peasants and Peasant Societies. Teodor Shanin, ed. Pp. 229–37. Harmondsworth: Penguin Books.
Matlon, Peter
    1981 The Structure of Production and Rural Incomes in Northern Nigeria: Results of Three Village Case Studies. *In* The Political Economy of Income Distribution in Nigeria. Henry Bienen and V. P. Diejomaon, eds. Pp. 323–72. New York: Holmes & Meier.
Matlon, Peter J., and Dunstan S. Spencer
    1984 Increasing Food Production in Sub-Saharan Africa: Environmental Problems and Inadequate Technological Solutions. American Journal of Agricultural Economics 66: 671–76.
Mayer, Enrique
    1979 Land-Use in the Andes: Ecology and Agriculture in the Mantaro Valley

of Peru with Special Reference to Potatoes. Lima: International Potato Center, Social Science Unit.

Maynard, John
  1942   The Russian Peasant and Other Studies. New York: Collier.

Meek, R. L.
  1953   Marx and Engels on Malthus. London: Lawrence & Wishart.

Meggers, B. J.
  1971   Amazonia: Man and Culture in a Counterfeit Paradise. Chicago: Aldine.

Meggitt, Mervyn J.
  1965   The Lineage System of the Mae-Enga of New Guinea. Edinburgh: Oliver & Boyd.

Meillassoux, Claude
  1972   From Reproduction to Production: A Marxist Approach to Economic Anthropology. Economy and Society 1: 93–105.
  1978   The Economy in Agricultural Self-sustaining Societies: A Preliminary Analysis. *In* Relations of Production: Marxist Approaches to Economic Anthropology. D. Seddon, ed. Pp. 159–69. London: Frank Cass.
  1981   Maidens, Meal, and Money: Capitalism and the Domestic Community. Cambridge: Cambridge University Press.

Meilleur, B.
  1986   Alluetain Ethnoecology and Traditional Economy: The Procurement and Production of Plant Resources in the Northern French Alps. Ph.D. diss., University of Washington. Ann Arbor, MI: University Microfilms International.

Mellor, John W.
  1988   Sustainable Agriculture in Developing Countries. Environment 30 (9): 7.

Minge-Kalman, Wanda
  1977   On the Theory and Measurement of Domestic Labor Intensity. American Ethnologist 4: 273–84.

Minge-Klevana, Wanda
  1980   Does Labor Time Decrease with Industrialization? A Survey of Time-Allocation Studies. Current Anthropology 21: 279–98.

Montmarquet, James A.
  1989   The Idea of Agrarianism: From Hunter-Gatherer to Agrarian Radical in Western Culture. Moscow, ID: University of Idaho Press.

Moran, Emilio F.
  1981   Developing the Amazon. Bloomington: Indiana University Press.
  1984   Limitations and Advances in Ecosystem Research. *In* The Ecosystem Concept in Anthropology. Pp. 3–32. Boulder, CO: Westview Press.
  1990   Amazon Soils: Distribution and Agricultural Alternatives Under Indigenous and Contemporary Management. Paper for Wenner-Gren Symposium No. 109, "Amazonian Synthesis," Novo Friburgo, Brazil, June 2–10, 1989.

Morgan, Lewis Henry

1963   Ancient Society. Part 1 of Research in the Lines of Human Progress from Savagery through Barbarism to Civilization. Eleanor B. Leacock, ed. Cleveland: World, Meridian Books. [Originally published in 1877.]

Morgan, W. B.

1953   Farming Practice, Settlement Pattern, and Population Density in Southeastern Nigeria. Geographical Journal 121: 320–33.

Morren, George

1977   From Hunting to Herding: Pigs and the Control of Energy in Montane New Guinea. *In* Subsistence and Survival: Rural Ecology in the Pacific. T. Bayliss-Smith and R. Feachem, eds. Pp. 373–16. London: Academic Press.

Mortimore, Michael

1972   Land and Population Pressure in the Kano Close-Settled Zone, Northern Nigeria. *In* People and Land in Africa South of the Sahara: Readings in Social Geography. R. Mansell Prothero, ed. Pp. 60–70. New York: Oxford University Press.

Mortimore, Michael, E. A. Olofin, R. A. Cline-Cole, and A. Abdulkadir

1987   Perspectives on Land Administration and Development in Northern Nigeria. Kano: Department of Geography, Bayero University.

Murphy, Robert F.

1971   The Dialectics of Social Life: Alarms and Excursions in Anthropological Theory. New York: Basic Books.

1981   Julian Steward. *In* Totems and Teachers: Persepectives on the History of Anthropology. Sydel Silverman, ed. Pp. 171–208. New York: Columbia University Press.

Nag, Moni, B. N. F. White, and R. C. Peet

1978   An Anthropological Approach to the Study of the Economic Value of Children in Java and Nepal. Current Anthropology 19: 293–306.

Nakajima, Chihiro

1969   Subsistence and Commercial Family Farms: Some Theoretical Models of Subjective Equilibrium. *In* Subsistence Agriculture and Economic Development. C. R. Wharton, Jr., ed. Pp. 165–85. Chicago: Aldine.

National Research Council

1992   Global Environmental Change: Understanding the Human Dimensions. Washington, D.C.: National Academy Press, for the Committee on the Human Dimensions of Global Change.

National Research Council. Board of Agriculture.

1989   Alternative Agriculture. Washington, DC: National Academy Press, for the Committee on the Role of Alternative Farming Methods in Modern Production Agriculture.

Nee, Victor

1985   Peasant Household Individualism. *In* Chinese Rural Development. W. L. Parish, ed. Pp. 164–90. Armonk, NY: M. E. Sharpe.

1986   The Peasant Household Economy and Decollectivization in China. Journal of Asian and African Studies 21: 185–202.

Nell, E.
  1972 Boserup and the Intensity of Cultivation. Peasant Studies Newsletter 1: 39–44.
  1979 Population Pressure and Methods of Cultivation: A Critique of Classless Theory. *In* Toward a Marxist Anthropology. Stanley Diamond, ed. Pp. 457–68. The Hague: Mouton.
Netting, Robert McC.
  1963 Kofyar Agriculture: A Study in the Cultural Ecology of a West African People. Ph.D. diss., University of Chicago.
  1964 Beer as a Locus of Value Among the West African Kofyar. American Anthropologist 66: 375–84.
  1965 Household Organization and Intensive Agriculture: The Kofyar Case. Africa 35: 422–29.
  1968 Hill Farmers of Nigeria: Cultural Ecology of the Kofyar of the Jos Plateau. Seattle: University of Washington Press.
  1969a Ecosystems in Process: A Comparative Study of Change in Two West African Societies. *In* Ecological Essays. National Museum of Canada [Ottawa], Bulletin No. 230. D. Damas, ed. Pp. 102–12.
  1969b Women's Weapons: The Politics of Domesticity Among the Kofyar. American Anthropologist 71: 1037–46.
  1972 Sacred Power and Centralization: Some Notes on Political Adaptation in Africa. *In* Population Growth: Anthropological Implications. Brian Spooner, ed. Pp. 219–44. Cambridge, MA: MIT Press.
  1973 Fighting, Forest, and the Fly: Some Demographic Regulators Among the Kofyar. Journal of Anthropological Research 29: 164–79.
  1974a The System Nobody Knows: Village Irrigation in the Swiss Alps. *In* Irrigation's Impact on Society. T. Edmund Downing and McGuire Gibson, eds. Pp. 67–75. Tucson: University of Arizona Press.
  1974b Kofyar Armed Conflict: Social Causes and Consequences. Journal of Anthropological Research 30: 139–63.
  1976 What Alpine Peasants Have in Common: Observations on Communal Tenure in a Swiss Village. Human Ecology 4: 135–46.
  1977 Maya Subsistence: Mythologies, Analogies, Possibilities. *In* The Origins of Maya Civilization. Richard Adams, ed. Pp. 299–333. Albuquerque: University of New Mexico Press.
  1979 Household Dynamics in a Nineteenth-Century Swiss Village. Journal of Family History 4: 39–58.
  1981 Balancing on an Alp: Ecological Change and Continuity in a Swiss Mountain Community. Cambridge: Cambridge University Press.
  1982a Territory, Property and Tenure. *In* Behavioral and Social Science Research: A National Resource. Pp. 446–502. Washington, DC: National Academy Press.
  1982b Some Home Truths on Household Size and Wealth. American Behavioral Scientist 25: 641–62.
  1984 Reflections on an Alpine Village as Ecosystem. *In* The Ecosystem Con-

cept in Anthropology. Emilio F. Moran, ed. Pp. 225–35. Boulder, CO: Westview Press.

1986   Cultural Ecology. 2d ed. Prospect Heights, IL: Waveland Press. [Originally published in 1977.]

1987   Clashing Cultures, Clashing Symbols: Histories and Meanings of the Latok War. Ethnohistory 34: 352–80.

1990   Population, Permanent Agriculture and Politics: Unpacking the Evolutionary Portmonteau. *In* The Evolution of Political Systems. Steadman Upham, ed. Pp. 21–61. Cambridge: Cambridge University Press.

n.d.   Agricultural Expansion, Intensification and Market Participation Among the Kofyar, Jos Plateau, Nigeria. *In* Population Growth and Agricultural Intensification: Studies from Densely Settled Areas of Sub-Saharan Africa. B. L. Turner II, Goran Hyden, and Robert Kates, eds. Gainesville: University of Florida Press. In press.

Netting, Robert McC., M. Priscilla Stone, and Glenn D. Stone

1989   Kofyar Cash Cropping: Choice and Change in Indigenous Agricultural Development. Human Ecology 17 (3): 299–319.

Nieboer, H. H.

1971   Slavery as an Industrial System. New York: Burt Franklin.

Niñez, Vera

1987   Household Gardens: Theoretical and Policy Considerations. Agricultural Systems 23: 167–86.

Nitsch, Ulrich

1987   A Persistent Culture: Some Reflections on Swedish Family Farming. *In* Family Farming in Europe and America. Boguslaw Galeski and Eugene Wilkening, eds. Pp. 95–115. Boulder, CO: Westview Press.

Norgaard, Richard

1989   Risk and Its Management in Traditional and Modern Agroeconomic Systems. *In* Food and Farm: Current Debates and Policies. Monographs in Economic Anthropology, No. 7. Christina Gladwin and Kathleen Truman, eds. Pp. 199–216. Lanham, MD: University Press of America.

Norman, D. W.

1969   Labour Inputs of Farmers: A Case Study of the Zaria Province of the North-Central State of Nigeria. Nigerian Journal of Economic and Social Studies 11: 3–14.

Norman, D. W., E. B. Simmons, and H. M. Hays

1982   Farming Systems in the Nigerian Savanna. Boulder, CO: Westview Press.

Norman, M. J. T.

1978   Energy Inputs and Outputs of Subsistence and Cropping Systems in the Tropics. Agro-Ecosystems 4: 355–66.

North, Douglass C.

1981   Structure and Change in Economic History. New York: Norton.

Nye, D. H., and D. J. Greenland

1960   The Soil Under Shifting Cultivation. Commonwealth Bureau of Soils

Technical Communication No. 51. Farnham Royal, Bucks.: Commonwealth Agricultural Bureaux.

Oakerson, Ronald J.
1983 Reciprocity: The Political Nexus. MS. Workshop in Political Theory and Policy Analysis, Indiana University, Bloomington.
1986 A Model for the Analysis of Common Property Problems. *In* Proceedings of National Research Council Conference on Common Property Resource Management. Pp. 13–30. Washington, DC: National Academy Press.
n.d. Analyzing the Commons: A Framework. MS.

Odend'hal, Stewart
1972 Energetics of Indian Cattle in Their Environment. Human Ecology 1: 3–22.

Odum, E. P.
1969 The Strategy of Ecosystem Development. Science 164: 262–322.
1971 Fundamentals of Ecology. 3d ed. Philadelphia: W. B. Saunders. [Originally published in 1953.]

Omohundro, John T.
1987 The Folk Art of the Raised Bed. Garden 11 (3): 10–14, 32.

Orlove, Benjamin S.
1980 Ecological Anthropology. Annual Review of Anthropology 9: 235–73.

Orlove, Benjamin S., and Ricardo Godoy
1986 Sectoral Fallowing Systems in the Central Andes. Journal of Ethnobiology 6: 169–204.

Osborne, Anne
n.d. Agricultural Intensification Versus Ecological Stability: The State as Mediator in Qing China. Paper presented at the annual meeting of the American Anthropological Association, Washington, DC, Dec. 5, 1985.

Ostrom, Elinor
1990 Governing the Commons: The Evolution of Institutions for Collective Action. Cambridge: Cambridge University Press.

Padoch, Christine
1985 Labor Efficiency and Intensity of Land Use in Rice Production: An Example from Kalimantan. Human Ecology 13 (3): 271–90.

Palerm, A.
1955 The Agricultural Basis of Urban Civilizations in Mesoamerica. *In* Irrigation Civilizations: A Comparative Study. Pan Am. Union Soc. Sci. Monogr. No. 1. Julian H. Steward, ed. Pp. 28–42. Washington, DC.

Parish, William L.
1985 Introduction: Historical Background and Current Issues. *In* Chinese Rural Development: The Great Transformation. W. L. Parish, ed. Pp. 3–29. Armonk, NY: M. E. Sharpe.

Parish, William L., and Martin King Whyte
1978 Village and Family Life in Contemporary China. Chicago: University of Chicago Press.

Partsch, Gottfried
    1955   Das Mitwirkungsrecht der Familiengemeinschaft in älteren Walliser-
           recht. Geneva.
Passmore, R., and J. V. G. A. Durnin
    1955   Human Energy Expenditure. Physiological Review 35: 801–40.
Pasternak, Burton
    1969   The Role of the Frontier in Chinese Lineage Development. Journal of
           Asian Studies 28: 551–62.
Patnaik, Utsa
    1988   Three Communes and a Production Brigade: The Contract Responsi-
           bility System in China. In China: Issues in Development. Ashok Mitra,
           ed. Pp. 34–61. New Delhi: Tulika.
Pearson, Scott R., J. Dirck Stryker, and Charles P. Humphreys
    1981   Rice in West Africa. Policy and Economics. Stanford: Stanford Univer-
           sity Press.
Pelzer, K. J.
    1948   Pioneer Settlement in the Asiatic Tropics. New York: American Geo-
           graphical Society.
Perelman, Michael
    1976   Efficiency in Agriculture: The Economics of Energy. In Radical Agri-
           culture. Richard Merrill, ed. Pp. 64–86. New York: Harper Colophon
           Books.
Perkins, Dwight H.
    1969   Agricultural Development in China, 1368–1969. Chicago: Aldine.
Perkins, Dwight H., and Shahid Yusuf
    1984   Rural Development in China. Baltimore: Johns Hopkins University
           Press.
Picht, Christine, and Arun Agrawal
    1989   Corporations and Communities. Paper presented at the Mini-
           Conference, Workshop in Political Theory and Policy Analysis, Indiana
           University, May 1989.
Pimentel, David
    1984   Energy Flow in Agroecosystems. In Agricultural Ecosystems: Unify-
           ing Concepts. R. Lowrance, B. R. Stinner, and G. J. House, eds. Pp.
           121–32. New York: Wiley.
Pimentel, David, and M. Pimentel
    1979   Food, Energy and Society. London: Edward Arnold.
Pimentel, David, et al.
    1973   Food Production and the Energy Crisis. Science 182: 443–49.
Pingali, Prabhu, Yves Bigot, and Hans P. Binswanger
    1987   Agricultural Mechanization and the Evolution of Farming Systems in
           Sub-Saharan Africa. Baltimore: Johns Hopkins University Press for the
           World Bank.
Pingali, Prabhu, and Hans P. Binswanger
    1983   Population Density, Farming Intensity, Patterns of Labor Use and

Mechanization. Research Unit Report ARU 11. Washington, DC: World Bank, Agriculture and Rural Development Department.

Piot, Charles D.
1988 Fathers and Sons: Domestic Production, Conflict, and Social Forms among the Kabre. Research in Economic Anthropology 10: 269–85.

Polanyi, Karl
1944 The Great Transformation. New York: Rinehart.

Popkin, Samuel
1979 The Rational Peasant. Berkeley: University of California Press.

Pospisil, L.
1963 Kapauku Papuan Economy. Yale University Publications in Anthropology, No. 67. New Haven.

Post, K.
1972 "Peasantization" and Rural Political Movements in Western Africa. Archives européennes de sociologie 13: 223–54.

Potter, Sulamith Heins, and Jack M. Potter
1990 China's Peasants: The Anthropology of Revolution. Cambridge: Cambridge University Press.

Preston, P. W.
1986 Making Sense of Development. London: Routledge & Kegan Paul.

Prill-Brett, June
1986 The Bontok: Traditional Wet-Rice and Swidden Cultivators in the Philippines. *In* Traditional Agriculture in Southeast Asia: A Human Ecology Perspective. Gerald G. Marten, ed. Pp. 54–84. Boulder, CO: Westview Press.

Prindle, Peter H.
1984 Part-Time Farming: A Japanese Example. Journal of Anthropological Research 40: 293–305.

Pryor, Frederic L.
1977 The Origins of the Economy: A Comparative Study of Distribution in Primitive and Peasant Economies. New York: Academic Press.
1985 The Invention of the Plow. Comparative Studies in Society and History 27: 727–43.
1986 The Adoption of Agriculture: Some Theoretical and Empirical Evidence. American Anthropologist 88: 879–97.

Pryor, Frederic L., and Stephen B. Maurer
1982 On Induced Economic Change in Precapitalist Economies. Journal of Development Economics 10: 325–53.

Pudup, Mary Beth, and Michael J. Watts
1987 Growing Against the Grain: Mechanized Rice Farming in the Sacramento Valley, California. *In* Comparative Farming Systems. B. L. Turner II and Stephen B. Brush, eds. Pp. 345–84. New York: Guilford.

Putterman, Louis
1985 The Restoration of the Peasant Household as Farm Production Unit in China: Some Incentive Theoretic Analysis. *In* The Political Economy

of Reform in Post-Mao China. Elizabeth J. Perry and Christine Wong, eds. Pp. 63–82. Cambridge, MA: Council on East Asian Studies, Harvard University.

1987   The Incentive Problem and the Demise of Team Farming in China. Journal of Development Economics 26: 103–27.

1989   Entering the Post-Collective Era in North China: Dahe Township. Modern China 15: 275–320.

Rao, A. P.
1967   Size of Holdings and Productivity. Economic and Political Weekly 2: 1989–91.

Rappaport, R. A.
1968   Pigs for the Ancestors. New Haven: Yale University Press.
1971   The Flow of Energy in an Agricultural Society. Scientific American 224 (3): 116–32.

Rawski, Evelyn Sakakida
1972   Agricultural Change and the Peasant Economy of South China. Cambridge, MA: Harvard University Press.

Redfield, Robert
1941   The Folk Culture of the Yucatan. Chicago: University of Chicago Press.
1955   The Little Community: Viewpoints for the Study of a Human Whole. Chicago: University of Chicago Press.
1956   Peasant Society and Culture. Chicago: University of Chicago Press.

Revelle, R.
1976   Energy Use in Rural India. Science 192: 969–75.

Rey, Pierre-Philippe
1975   The Lineage Mode of Production. Critique of Anthropology 4: 27–79.

Reyna, S. P.
1987   The Emergence of Land Concentration in the West African Savanna. American Ethnologist 14: 523–41.

Richards, A. I.
1939   Land, Labour, and Diet in Northern Rhodesia. London: Oxford University Press.

Richards, Paul
1983   Farming Systems and Agrarian Change in West Africa. Progress in Human Geography 7: 1–39.
1985   Indigenous Agricultural Revolution: Ecology and Food Production in West Africa. London: Hutchinson.
1986   Coping with Hunger: Hazard and Experiment in an African Rice-Farming System. London: Allen & Unwin.

Richerson, P. J.
1977   Ecology and Human Ecology: A Comparison of Theories in the Biological and Social Sciences. American Ethnologist 4: 1–26.

Robinson, Gerold T.
1932   Rural Russia Under the Old Regime. London: Longmans, Green.

Robinson, W., and A. Schutjer
   1984   Agricultural Development and Demographic Change: A Generalization of the Boserup Model. Economic Development and Cultural Change 32: 355–66.
Roosevelt, Anna C.
   1984   Population, Heath, and the Evolution of Subsistence. *In* Paleopathology at the Origins of Agriculture. M. N. Cohen and G. J. Armelagos, eds. Pp. 559–83. Orlando, FL: Academic Press.
   1989   Resource Management in Amazonia Before the Conquest: Beyond Ethnographic Projection. Advances in Economic Botany 7: 30–62.
Roseberry, William
   1989   Anthropologies and Histories: Essays in Culture, History, and Political Economy. New Brunswick, NJ: Rutgers University Press.
Rosenzweig, Mark R.
   1984   From Land Abundance to Land Scarcity: The Effects of Population Growth on Production Relations in Agrarian Economies. Paper prepared for International Union for the Scientific Study of Population Seminar on Population and Rural Development, New Delhi, India, December 1984.
Rosenzweig, Mark R., and Kenneth Wolpin
   1985   Specific Experience, Household Structure and Intergenerational Transfers: Farm Family Arrangements in Developing Countries. Quarterly Journal of Economics 100: 961–87.
Rostow, W. W.
   1960   The Stages of Economic Growth: A Non-Communist Manifesto. Cambridge: Cambridge University Press.
Ruddle, K., and Gongfu Zhong
   1988   Integrated Agriculture-Aquaculture in South China: The Dike-Pond System of the Zhujiang Delta. Cambridge: Cambridge University Press.
Ruthenberg, Hans
   1968   Some Characteristics of Smallholder Farming in Tanzania. *In* Smallholder Farming and Smallholder Development in Tanzania: Ten Case Studies. Hans Ruthenberg, ed. Pp. 325–55. Munich: Weltforum Verlag.
   1976   Farming Systems in the Tropics. 2d ed. Oxford: Clarendon Press. [Originally published in 1971.]
Ruttan, V. W.
   1984   Induced Innovations and Agricultural Development. *In* Agricultural Sustainability in a Changing World Order. G. K. Douglass, ed. Pp. 107–34. Boulder, CO: Westview Press.
Sabean, David Warren
   1990   Property, Production, and Family in Neckarhausen, 1700–1870. Cambridge: Cambridge University Press.
Sahlins, Marshall D.
   1957   Land Use and the Extended Family in Moala, Fiji. American Anthropologist 59: 449–62.

1960 Political Power and the Economy in Primitive Society. *In* Essays in the Science of Culture in Honor of Leslie White. Gertrude E. Dole and R. L. Carneiro, eds. Pp. 390–415. New York: Crowell.

1961 The Segmentary Lineage: An Organization of Predatory Expansion. American Anthropologist 63: 322–43.

1968 Notes on the Original Affluent Society. *In* Man the Hunter. Richard B. Lee and Irven Devore, eds. Pp. 85–89. Chicago: Aldine.

1971 The Intensity of Domestic Production in Primitive Societies: Social Inflections of the Chayanov Slope. *In* Studies in Economic Anthropology. G. Dalton, ed. Pp. 30–51. Washington, DC: American Anthropological Society.

1972 Stone Age Economics. Chicago: Aldine-Atherton.

1976 Culture and Practical Reason. Chicago: University of Chicago Press.

Sahlins, Marshall, and E. R. Service

1960 Evolution and Culture. Ann Arbor: University of Michigan Press.

Salamon, Sonya

1985 Ethnic Communities and the Structure of Agriculture. Rural Sociology 50: 323–40.

Salehi-Isfahani, Djavad

1987 On the Generalization of the Boserup Model: Some Clarifications. Economic Development and Cultural Change 35: 875–81.

Saul, Mahir

1983 Work Parties, Wages and Accumulation in a Voltaic Village. American Ethnologist 10: 77–96.

1989 Expenditure and Intrahousehold Patterns Among the Southern Bobo of Burkina Faso. *In* The Social Economy of Consumption. B. Orlove and H. Reitz, eds. Pp. 349–78. Lanham, MD: University Press of America.

Sawhill, I.

1977 Economic Perspectives on the Family. Dædalus 106 (2): 115–26.

Schelhas, John

1991 Socio-economic and Biological Aspects of Land Use Adjacent to Braulio Carrillo National Park, Costa Rica. Ph.D. diss., School of Renewable Natural Resources, University of Arizona.

Schofield, R. S.

1970 Age-Specific Mobility in an Eighteenth Century Rural English Parish. Annales de démographie historique, 261–74.

Schumacher, E. F.

1973 Small Is Beautiful: Economics as if People Mattered. New York: Harper & Row.

Scott, James C.

1976 The Moral Economy of the Peasant: Subsistence and Rebellion in Southeast Asia. New Haven: Yale University Press.

Scott, William H.

1966 On the Cordillera: A Look at the Peoples and Cultures of the Mountain Province. Manila: MCS Enterprises.

Seavoy, Ronald E.
1986 Famine in Peasant Societies. Westport, CT.: Greenwood Press.

Segalen, Martine
1983 Love and Power in Peasant Society: Rural France in the 19th Century. Translated by Sarah Matthews. Chicago: University of Chicago Press. [Originally published in French 1980.]

Selden, Mark
1985 Income Inequality and the State. *In* Chinese Rural Development. W. L. Parish, ed. Pp. 193–218. Armonk, NY: M. E. Sharpe.

Sen, A. K.
1962 An Aspect of Indian Agriculture. Economic Weekly, Annual Number.

Service, Elman R.
1966 The Hunters. Englewood Cliffs, NJ: Prentice-Hall.

Seymore, Frances
1985 Ten Lessons Learned from Agroforestry Projects in the Philippines. Washington, DC: USAID. Mimeo.

Shah, A. M.
1974 The Household Dimension of the Family in India. Berkeley: University of California Press.

Shanin, Teodor
1971 Peasantry as a Political Factor. *In* Peasants and Peasant Societies. Teodor Shanin, ed. Pp. 238–63. Harmondsworth: Penguin Books.

1972 The Awkward Class. London: Oxford University Press.

1986 Chayanov's Message: Illuminations, Miscomprehensions and the Contemporary "Development Theory." *In* The Theory of Peasant Economy. D. Thorner, Basile Kerblay, and R. E. F. Smith, eds. Pp. 1–24. Madison: University of Wisconsin Press.

1990 Defining Peasants: Essays Concerning Rural Societies, Expolary Economies, and Learning from Them in the Contemporary World. Oxford: Basil Blackwell.

Shenton, Robert W.
1986 The Development of Capitalism in Northern Nigeria. London: James Currey.

Shepherd, John R.
1988 Rethinking Tenancy: Explaining Spatial and Temporal Variation in Late Imperial and Republican China. Comparative Studies in Society and History 30: 403–31.

Sheridan, Thomas E.
1988 Where the Dove Calls: The Political Ecology of a Peasant Corporate Community in Northwestern Mexico. Tucson: University of Arizona Press.

Shipton, Parker M.
1984 Strips and Patches: A Demographic Dimension in Some African Land-Holding and Political Systems. Man 19: 613–34.

1987 The Kenyan Land Tenure Reform. Harvard Institute for International

Development, Development Discussion Paper No. 239. Cambridge, MA: Harvard University.

1988   The Kenyan Land Tenure Reform: Misunderstandings in the Public Creation of Private Property. *In* Land and Society in Contemporary Africa. R. E. Downs and S. P. Reyna, eds. Pp. 91–135. Hanover, NH: University Press of New England.

Shue, Vivienne

1980   Peasant China in Transition: The Dynamics of Development Toward Socialism, 1949–1956. Berkeley: University of California Press.

Simon, Julian L.

1981   The Ultimate Resource. Princeton: Princeton University Press.

Skipp, V.

1978   Crisis and Development: An Ecological Case Study of the Forest of Arden, 1570–1674. Cambridge: Cambridge University Press.

Smil, Vaclav

1985   China's Food. Scientific American 253 (6): 116–24.

1987   Land Degradation in China: An Ancient Problem Getting Worse. *In* Land Degradation and Society. Piers Blackie and Harold Brookfield, eds. Pp. 214–23. London: Methuen.

Smith, Adam

1937   An Inquiry into the Nature and Causes of the Wealth of Nations. Edwin Cannon, ed. New York: Modern Library. [First published in 1776.]

Smith, C. A.

1975   Production in Western Guatemala: A Test of Von Thünen and Boserup. *In* Formal Methods in Economic Anthropology. S. Plattner, ed. Pp. 5–37. Washington, DC: American Anthropological Association.

Smith, Richard M.

1984   Some Issues Concerning Families and Their Property in Rural England, 1250–1800. *In* Land, Kinship and Life-Cycle. R. M. Smith, ed. Pp. 1–86. Cambridge: Cambridge University Press.

Smith, Thomas C.

1959   The Agrarian Origins of Modern Japan. Stanford: Stanford University Press.

1977   Nakahara: Family Farming and Population in a Japanese Village, 1717–1830. Stanford: Stanford University Press.

Sorensen, Clark W.

1988   Over the Mountains Are Mountains: Korean Peasant Households and Their Adaptations to Rapid Industrialization. Seattle: University of Washington Press.

Spencer, J. E.

1966   Shifting Cultivation in Southwestern Asia. University of California Publications in Geography, Vol. 19. Berkeley: University of California Press.

Spencer, J. E., and G. A. Hale

1961   The Origin, Nature and Distribution of Agricultural Terracing. Pacific Viewpoint 2: 1–40.

Spooner, Brian
    1974 Irrigation and Society: The Iranian Plateau. *In* Irrigation's Impact on Society. Anthropological Papers of the University of Arizona, No. 25. T. E. Downing and M. Gibson, eds. Pp. 43–57. Tucson: University of Arizona Press.

Stebler, F. G.
    1922 Die Vispertaler Sonnenberge. Jahrbuch des Schweizer Alpenclubs. Sechsundfünfzigster Jahrgang. Bern: Schweizer Alpenclub.

Steward, Julian H.
    1938 Basin-Plateau Aboriginal Sociopolitical Groups. Bureau of American Ethnology Bulletin No. 120. Washington, DC: GPO.
    1955a Theory of Culture Change. Urbana: University of Illinois Press.
    1955b Irrigation Civilizations: A Comparative Study. Pan Am. Union Soc. Sci. Monogr. No. 1. Washington, DC.

Stoler, Anne L.
    1981 Garden Use and Household Economy in Java. *In* Agricultural and Rural Development in Indonesia. Gary E. Hansen, ed. Pp. 242–54. Boulder, CO: Westview Press.

Stone, Glenn D.
    1988 Agrarian Ecology and Settlement Patterns: An Ethnoarchaeological Case Study. Ph.D. diss., University of Arizona. Ann Arbor: University Microfilms.
    1991 Settlement Ethnoarchaeology: Changing Patterns Among the Kofyar of Nigeria. Expedition 33: 16–23.

Stone, Glenn D., Robert McC. Netting, and M. Priscilla Stone
    1990 Seasonality, Labor Scheduling, and Agricultural Intensification in the Nigerian Savanna. American Anthropologist 92 (1): 7–23.

Stone, Glenn D., M. Priscilla Stone, and Robert McC. Netting
    1984 Household Variability and Inequality in Kofyar Subsistence and Cash-Cropping Economies. Journal of Anthropological Research 40: 90–108.

Stone, M. Priscilla
    1988a Women Doing Well: A Restudy of the Nigerian Kofyar. Research in Economic Anthropology 10: 287–306.
    1988b Woman, Work and Marriage: A Restudy of the Nigerian Kofyar. Ph.D. diss., Department of Anthropology, University of Arizona. Ann Arbor, MI: University Microfilms.

Stone-Ferrier, Linda A.
    1983 Dutch Prints of Daily Life. Lawrence: Spencer Museum of Art, University of Kansas.

Stover, Leon E.
    1974 The Cultural Ecology of Chinese Civilization: Peasants and Elites in the Last of the Agrarian States. New York: Mentor.

Strange, Marty
    1988 Family Farming: A New Economic Vision. Lincoln: University of Nebraska Press.

Swindell, Ken
    1985   Farm Labour. Cambridge: Cambridge University Press.
Tannenbaum, Nicola
    1984a Chayanov and Economic Anthropology. *In* Chayanov, Peasants, and
        Economic Anthropology. E. P. Durrenberger, ed. Pp. 27–38. Orlando,
        FL: Academic Press.
    1984b The Misuse of Chayanov: Chayanov's Rule and Empiricist Bias in An-
        thropology. American Anthropologist 86: 924–42.
Terra, G. J. A.
    1954   Mixed-Garden Horticulture in Java. Malayan Journal of Tropical Ge-
        ography 3: 33–43.
Terray, E.
    1972   Marxism and "Primitive" Societies. New York: Monthly Review Press.
Thorner, Daniel
    1966   Chayanov's Concept of Peasant Economy. *In* The Theory of Peasant
        Economy. D. Thorner, B. Kerblay, and R. E. F. Smith, eds. Pp. xi–
        xxiii. Homewood, IL: Richard D. Irwin.
Tilly, Charles
    1978   Peasants Against Capitalism and the State: A Review Essay. Agricul-
        tural History 52: 407–16.
Truman, Kathleen
    1989   Low-Input Mexican Agriculture: A View from the Past. *In* Food and
        Farm: Current Debates and Policies. Monographs in Economic An-
        thropology, No. 7. Christina Gladwin and Kathleen Truman, eds. Pp.
        161–76. Lanham, MD: University Press of America.
Turner, B. L., II, and S. B. Brush, eds.
    1987   Comparative Farming Systems. New York: Guilford Press.
Turner, B. L., II, and W. E. Doolittle
    1978   The Concept and Measure of Agricultural Intensity. Professional Ge-
        ographer 30: 297–301.
Turner, B. L., II, Q. Hanham, and A. V. Portararo
    1977   Population Pressure and Agricultural Intensity. Annals of the Associa-
        tion of American Geographers 67: 384–96.
Turner, B. L., II, G. Hyden, and R. W. Kates, eds.
    1992   Population Growth and Agricultural Intensification: Studies from
        Densely Settled Areas of Sub-Saharan Africa. Gainesville: University of
        Florida Press.
Udo, R. K.
    1965   Disintegration of Nucleated Settlement in Eastern Nigeria. Geograph-
        ical Review 55: 53–67.
Vayda, A. P.
    1961   Expansion and Warfare Among Swidden Agriculturalists. American
        Anthropologist 63: 346–58.
Vermeer, D. E.
    1970   Population Pressure and Crop Rotational Changes Among the Tiv of

Nigeria. Annals of the Association of American Geographers 60: 299–314.

Vermeer, E. B.
1982 Income Differentials in Rural China. China Quarterly 89: 1–33.

Viazzo, Pier Paolo
1989 Upland Communities: Environment, Population and Social Structure in the Alps Since the Sixteenth Century. Cambridge: Cambridge University Press.

Waddell, E.
1972 The Mound Builders: Agricultural Practices, Environment, and Society in the Central Highlands of New Guinea. Seattle: University of Washington Press.

Wade, Robert
1988 Village Republics: Economic Conditions for Collective Action in South India. Cambridge: Cambridge University Press.

Wallace, Ben J., R. M. Ahsan, S. H. Hussain, and E. Ahsan
1987 The Invisible Resource: Women and Work in Rural Bangladesh. Boulder, CO: Westview Press.

Wallerstein, Emmanuel
1976 The Modern World System. New York: Academic Press.

Warman, Arturo
1983 Peasant Production and Population in Mexico. *In* The Social Anthropology of Peasantry. Joan P. Mencher, ed. Pp. 121–29. Bombay: Somaiya.

Watson, Andrew M.
1983 Agricultural Innovation in the Early Islamic World: The Diffusion of Crops and Farming Techniques, 700–1100. Cambridge: Cambridge University Press.

Watson, James B.
1977 Pigs, Fodder, and the Jones Effect in Postipomoean New Guinea. Ethnology 16: 57–70.

Watts, Michael
1983 Silent Violence: Food, Famine, and Peasantry in Northern Nigeria. Berkeley: University of California Press.

1984 State, Oil, and Accumulation: From Boom to Crisis. Environment and Planning: Society and Space 2: 403–28.

Weil, Peter M.
1970 The Introduction of the Ox Plow in Central Gambia. *In* African Food Production Systems. P. F. M. McLoughlin, ed. Baltimore: Johns Hopkins University Press, 229–63.

1973 Wet Rice, Women, and Adaptation in the Gambia. Rural Africana, No. 19: 20–29.

Weiskel, Timothy
1989 The Ecological Lessons of the Past: An Anthropology of Environmental Decline. Ecologist 19 (3): 98–103.

Weisman, Steven R.
  1989   Farmers Turn Wrath on Politicians. New York Times, July 14, 1989, sec. A, 4.
Wen, Dazhong, and David Pimentel
  1986   Seventeenth Century Organic Agriculture in China: I. Cropping Systems in Jiaxing Region; II. Energy Flows Through an Agroecosystem in Jiaxing Region. Human Ecology 14 (1): 1–28.
Wheaton, Robert
  1975   Family and Kinship in Western Europe: The Problem of the Joint Family Household. Journal of Interdisciplinary History 4: 601–628.
White, Benjamin
  1976   Population, Involution, and Employment in Rural Java. Development and Change 7: 267–90.
  1979   Political Aspects of Poverty, Income Distribution, and Their Measurement. Development and Change 10: 91–114.
  1989   Problems in the Empirical Analysis of Agrarian Differentiation. *In* Agrarian Transformations: Local Processes and the State in Southeast Asia. Gillian Hart, Andrew Turton, and Benjamin White, eds. Pp. 15–30. Berkeley: University of California Press.
White, Benjamin, and Gunawan Wiradi
  1989   Agrarian and Nonagrarian Bases of Inequality in Nine Javanese Villages. *In* Agrarian Transformations: Local Processes and the State in Southeast Asia. Gillian Hart, Andrew Turton, and Benjamin White, eds. Pp. 266–302. Berkeley: University of California Press.
White, L. A.
  1943   Energy and the Evolution of Culture. American Anthropologist 45: 335–54.
  1959   The Evolution of Culture. New York: McGraw-Hill.
White, Lynn, Jr.
  1962   Medieval Technology and Social Change. London: Oxford University Press.
Wiber, Melanie G.
  1985   Dynamics of the Peasant Household Economy: Labor Recruitment and Allocation in an Upland Philippine Community. Journal of Anthropological Research 41: 427–41.
Wilk, Richard R.
  1988   Household Ecology: Tradition and Development Among the Kekchi Maya of Belize. MS.
  1989   Decision Making and Resource Flows Within the Household: Beyond the Black Box. *In* The Household Economy: Reconsidering the Domestic Mode of Production. Richard R. Wilk, ed. Pp. 23–52. Boulder, CO: Westview Press.
  1991   Household Ecology: Economic Change and Domestic Life Among the Kekchi Maya in Belize. Tucson: University of Arizona Press.
Wilk, Richard R., and Robert McC. Netting
  1984   Households: Changing Forms and Functions. *In* Households: Compar-

ative and Historical Studies of the Domestic Group. R. McC. Netting, R. R. Wilk, and E. J. Arnould, eds. Pp. 1–28. Berkeley: University of California Press.

Wilken, Gene C.

1987 Good Farmers: Traditional Agricultural and Resource Management in Mexico and Central America. Berkeley: University of California Press.

Wilkinson, R.

1973 Poverty and Progress: An Ecological Perspective on Economic Development. New York: Praeger.

Wolf, A.

1968 Adopt a Daughter, Marry a Sister: A Chinese Solution to the Problem of the Incest Taboo. American Anthropologist 70: 864–74.

Wolf, Eric R.

1955 Types of Latin American Peasantry: A Preliminary Discussion. American Anthropologist 57: 452–71.

1957 Closed Corporate Peasant Communities in Mesoamerica and Central Java. Southwestern Journal of Anthropology 13: 1–18.

1966 Peasants. Englewood Cliffs, NJ: Prentice-Hall.

1981 The Vicissitudes of the Closed Corporate Peasant Community. American Ethnologist 13: 325–29.

1982 Europe and the Peoples Without History. Berkeley: University of California Press.

Woodford-Berger, Prudence

1981 Women in Houses: The Organization of Residence and Work in Rural Ghana. Antropologiska Studier [Stockholm University] 30–31: 3–35.

Worster, D.

1990 The Ecology of Order and Chaos. Environmental History Review 14 (1–2): 1–18.

Yanagisako, Sylvia J.

1979 Family and Household: The Analysis of Domestic Groups. Annual Review of Anthropology 8: 161–205.

1984 Explicating Residence: A Cultural Analysis of Changing Households among Japanese-Americans. In Households. Robert McC. Netting, Richard R. Wilk, and Eric J. Arnould, eds. Pp. 330–52. Berkeley: University of California Press.

Yang, Martin C.

1965 A Chinese Village: Taitou, Shantung Province. New York: Columbia University Press. [First published in 1945.]

Yang, Tai-Shuenn

1989 Property Rights and Constitutional Order in Imperial China. Ph.D. diss., Department of Political Science, Indiana University, Bloomington.

Yap, Young, and Arthur Cotterell

1975 The Early Civilization of China. London: Weidenfeld & Nicolson.

Yengoyan, A. A.
   1974   Demographic and Economic Aspects of Poverty in the Rural Philippines. Comparative Studies in Society and History 16: 58–72.
Zhang, Luziang
   1956   Shen's Agricultural Book (Shenshi Nongshu). Beijing: Zhonghua Shujue. [Originally published in 1658.]
Zhong, Gongfu
   1982   The Mulberry Dike–Fish Pond Complex: A Chinese Ecosystem of Land-Water Interaction on the Pearl River Delta. Human Ecology 10: 191–202.

Index

In this index "f" after a number indicates a separate reference on the next page, and "ff" indicates separate references on the next two pages. A continuous discussion over two or more pages is indicated by a span of numbers. *Passim* is used for a cluster of references in close but not consecutive sequence.

"Household Responsibility System," 233, 250–56 *passim*. *See also* Chinese agriculture
Households: farm-family described, 58–62 *passim*; Kofyar, 58f, 198f; production economics of, 59f, 247–56 *passim*, 312–19 *passim*; ecological knowledge repository, 62f; social organization of, 63–69 *passim*, 333f; implicit contracts of, 65f; hired labor in, 72–80 *passim*, 317f; role establishment in, 80ff; stem-family organization, 81; salience of, 82–85 *passim*; size of, 85–88 *passim*, 96f; composition of, 88–100 *passim*; smallholding and, 100f; inheritance and, 167–71 *passim*; altitudinal zones and, 171f; inequality among, 189ff; Kekchi Maya management, 207; landless, 208–14 *passim*, 218–21 *passim*; social mobility of, 212ff, 229ff; equality fostered by, 221–29 *passim*; Chinese, 237–44 *passim*, 248–56 *passim*; Chayanov's model of, 297–301 *passim*, 310ff; joint or multiple-family, 307n; family labor, 321. *See also* Kinship; Smallholders; Stem-family households
Huang, Philip, 239n, 246n
Huang, Shu-min, 315n
Hungarian agriculture, 86

Ibo compound garden, 54f
Ifugao, 198
Implicit contracts: concept of, 65; smallholder, 66–69 *passim*; with daughters, 81; inheritance and, 82
Income: population density and, 111, 215f, 277; off-farm, 114–18 *passim*; farm size and, 147; seasonal employment, 202; disparity in, 205; pooling returns, 209; small vs. managerial farms, 210; family cycle and, 213f; Egyptian non-agrarian, 226; household-responsibility system and, 253n; Chinese cottage industry, 254f; differentiation, 256ff; Kofyar increase of, 293; "acceptable" returns, 329. *See also* Stratification; Wealth
Inequality: among smallholders, 2,
189ff, 197–207 *passim*, 229ff, 322f; leveling mechanisms and, 11f; in peasant societies, 20f; among unilineal descent groups, 192f; growing population and, 215f, 228; farm size impact on, 218f; intensification theory on, 227; Gamo community, 228f; in historic China, 243f. *See also* Differentiation; Polarization; Stratification
Inheritance: economic considerations of, 81f; stem-families and impartible, 95–98 *passim*; of peasant rights (*hyakusho kabu*), 98; Kofyar, 159f; of collective rights, 164ff; households and, 167–71 *passim*; Swiss alpine community, 174; ownership stability and, 186f; inequality through, 197f; mobility and, 211, 214; of tenant sharecroppers, 217, 241f; in historic China, 243f
Intensive farm practices: described, 3, 28f, 262f, 321; involution and, 3, 47n, 285–88 *passim*; population pressure and, 15, 49f, 218f, 265–74 *passim*; Asian wet-rice farming, 41–46 *passim*, 127, 129, 137–40 *passim*, 210, 267; ethnoscience of, 50ff; chinampa cultivation, 51f; garden-variety intensification, 53–56 *passim*; chronic problems in, 93ff; Amish farming, 142, 196, 324; as sustainable agriculture, 144f; farm size and, 147f; need for capital, 151–55 *passim*; property rights in, 172–78 *passim*; property administration in, 182; controlling common property, 184f; promotes social equality, 190f; "household responsibility system" and, 250–56 *passim*; African agriculture, 265f; technology and, 270–74 *passim*; environmental constraints on, 274–78 *passim*; coercion of, 282–85 *passim*; economic incentives for, 288–94 *passim*, 312–19 *passim*. *See also* Agriculture; Labor; Land; Population
Intercropping, 33f
Involution, 3, 47n, 285–88 *passim*
Involutionary commercialization, 239n

cultivation, 161f; descent-group territories, 164ff; use and tenure link, 181f, 325f; exchange values and use, 185–88 *passim*; wealth stratification and, 201f, 215ff; land-lessness, 208–14 *passim*, 218–21 *passim*, 286; Netherlands reclamation, 225; historic China distribution of, 239–44 *passim*, 243f; Communist China control over, 244–47 *passim*, 257f; Chinese private, 246f; under "Household responsibility system," 250–56 *passim*; patterns of use, 289–92 *passim*; peasant household use of, 301–6 *passim*. *See also* Farm size; Inheritance; Property; Tenure
Laslett, Peter, 82
"Law of the minimum," 84f
Leach, Gerald, 129
Lee, Ronald, 280f
Lenin, Vladimir, 214, 296n
Liebig, Justus von, 84n, 282
The Limits of Growth, 277
Lineage mode of production, 17ff, 192f
Lineage. *See* Inheritance
Livestock. *See* Domestic animals

McIntire, John, 311
McKean, Margaret, 184n
Maclachlan, Morgan, 226f
Maine, Sir Henry, 168
Malthus, Thomas R., 276–84 *passim*
Manual labor. *See* Labor
Mao Zedong, 245, 251
Market: cattle, 150f; imperfections and labor, 155f; incentives, 288–94 *passim*; zonation, 290ff; reciprocity of, 297; urban, 308; political inter-ference in, 329n. *See also* Intensive farm practices; Production
Marriages: economic considerations of, 81f; polygynous, 89f; impact of resources on, 98ff; Balinese authori-zation of, 180; corporate control of, 194; fertility, 271f, 276, 315; restric-tions on, 276. *See also* Kinship
Marx, Karl: on smallholding peas-ants, 11, 332; opposition to small-holding, 17–21 *passim*; belief in technology, 125; dogma used by,

168n; on capitalism emergence, 191; on communal equality con-cepts, 192; on population and tech-nology, 276, 281f; on alienation, 328
"Material culture," 5
Matlon, Peter, 205, 212
Matrilineage rights, 193
Meggitt, Mervyn, 164
Meillassoux, Claude, 192
Mencius's household model, 233f
Minge-Kalman, Wanda, 313n
Mobility. *See* Downward mobility; Upward mobility
Modernization: smallholders and myths of, 9–15 *passim*, 21–27 *passim*, 320, 332f; process of, 258ff, 326n
Monocropping, 32f
Morgan, Lewis Henry, 18, 125, 168
Multicropping, 150
Multiple family household, 95, 307n, 312

Nakajima, Chihiro, 92
Nee, Victor, 254
Nell, Edward, 283
Netherlands agriculture, 223ff
New Guinea Highlands, 265
New World pattern, 33n
Nieboer, H. H., 283
Nigerian Kofyar system: described, 29f; erosion protection, 30f; fertile soil restoration, 31f; diversification of, 32ff; weeding practices, 34; eth-noscience of, 50f; land use, 56, 291–94 *passim*; households described, 58f, 88–95 *passim*, 269f; reciprocal beer work parties, 73, 194ff, 229; labor schedules, 119–22 *passim*; property rights in, 158–61 *passim*; inequality among, 198f; population adaptability of, 267
Night soil. *See* Soil fertility
Nitsch, Ulrich, 332
"Noodle land," 258. *See also* Chinese agriculture
North, Douglass C., 73n

Oakerson, Ronald, 66n
Occupations, 117, 325

Library of Congress Cataloging-in-Publication Data

Netting, Robert McC.
    Smallholders, householders: farm families and the ecology of intensive, sustainable
agriculture / Robert McC. Netting.
        p. cm.
Includes bibliographical references and index.
ISBN 0–8047–2061–4 (alk. paper):
ISBN 0–8047–2102–5 (pbk.: alk. paper):
    1. Traditional farming. 2. Agricultural ecology. 3. Agriculture—Economic aspects.
4. Sustainable agriculture. 5. Family farms.
I. Title.
GN407.4.N48 1993
630—dc20
92-20376
CIP

CL

630
NET